Methods in Molecular Biology

For further volumes:
http://www.springer.com/series/7651

For over 35 years, biological scientists have come to rely on the research protocols and methodologies in the critically acclaimed *Methods in Molecular Biology* series. The series was the first to introduce the step-by-step protocols approach that has become the standard in all biomedical protocol publishing. Each protocol is provided in readily-reproducible step-by-step fashion, opening with an introductory overview, a list of the materials and reagents needed to complete the experiment, and followed by a detailed procedure that is supported with a helpful notes section offering tips and tricks of the trade as well as troubleshooting advice. These hallmark features were introduced by series editor Dr. John Walker and constitute the key ingredient in each and every volume of the *Methods in Molecular Biology* series. Tested and trusted, comprehensive and reliable, all protocols from the series are indexed in PubMed.

Mechanobiology

Methods and Protocols

Edited by

Ronen Zaidel-Bar

Department of Cell and Developmental Biology, Faculty of Medicine, Tel Aviv University, Tel Aviv, Israel

Humana Press

Editor
Ronen Zaidel-Bar
Department of Cell and Developmental
Biology, Faculty of Medicine
Tel Aviv University
Tel Aviv, Israel

ISSN 1064-3745 ISSN 1940-6029 (electronic)
Methods in Molecular Biology
ISBN 978-1-0716-2850-8 ISBN 978-1-0716-2851-5 (eBook)
https://doi.org/10.1007/978-1-0716-2851-5

Cover Illustration Caption: Image created by Hanna Grobe.

This Humana imprint is published by the registered company Springer Science+Business Media, LLC, part of Springer Nature.
The registered company address is: 1 New York Plaza, New York, NY 10004, U.S.A.

Preface

Mechanics is an essential part of multicellular life. The mechanical properties of cells and tissues determine their shape and their response to external mechanical forces. Cells also generate forces, pulling and pushing on their surroundings, be it neighboring cells or the extracellular matrix. And cells can sense and respond to the forces that are applied to them, through molecular processes called mechanosensing and mechanotransduction. In all these activities, the cytoskeleton and cell adhesion take center stage. Research in the field of mechanobiology has revealed that, similar to biochemical cues, mechanical signaling contributes to cell-cell communication and cellular proprioception. As such, mechanosensing is involved in cell fate specification and in controlling single and collective cell behaviors such as directed migration, morphogenesis, wound healing, and the immune response. The field of mechanobiology aims to quantify the mechanical properties of cells, to determine the forces cells are exposed to, to measure the forces they generate, and to elucidate the molecular mechanisms of mechanotransduction, from mechanosensing to gene expression.

Insights into the role mechanics play at the cell, tissue, and organism level have gone hand in hand with advances in methodologies to apply and measure forces at the cellular and molecular scale outside and inside cells and tissues. In this volume of *Methods in Molecular Biology*, various methods to quantify the mechanical properties of cells and proteins are described, such as microscope-based force spectroscopy, as well as single-molecule atomic force microscopy and microindentation. A variety of approaches to expose cells to external mechanical forces are also described. These include shear stress and cyclic stretching, which simulate the forces endothelial cells are exposed to in our vascular system, and pulling of cell-cell junctions, as occurs in epithelial tissues during morphogenesis and body movement.

This volume contains tried and tested methods for quantitatively characterizing mechano-responses at various scales, from the opening of ion channels, through stress fiber repair, to cell durotaxis and plant root rigidity sensing. One of the most challenging areas in mechanobiology is the measurement of forces at the piconewton range within and around cells. As a result, different approaches have been developed to measure forces generated by cells. Here, protocols are provided to facilitate measurement of traction forces, at the cellular scale as well as the single integrin level. Moreover, tools for measuring forces within cells using molecular tension sensors are provided both for cultured cells and in a living model organism.

Besides providing an overview of the most cutting-edge techniques to measure cellular forces, this book includes methods to study signaling downstream of mechanical cues, using RNA sequencing and supported lipid bilayers. After collecting quantitative data on cell mechanics, it is often informative to integrate and put our understanding of cell mechanics to the test by building biophysical models. We therefore provide protocols for simulating 3D cell shape with the cellular potts model, as well as a mathematical model for the interpretation of long-term fluorescence recovery after photobleaching experiments that reflect protein dynamics.

This volume of *Methods in Molecular Biology* aims to provide meaningful tools for cell and developmental biologists approaching the study of cell and tissue dynamics from a mechanobiological perspective, molecular biologists interested in the effects of force on proteins, as well as cancer biologists and biophysicists.

Tel Aviv, Israel ***Ronen Zaidel-Bar***

Contents

Contributors

THOMAS ALLINE • *Université Paris Cité, CNRS, Laboratoire Matière et Systèmes Complexes, UMR 7057, Paris, France*

MALAK AMER • *Department of Genetics and Developmental Biology, Rappaport Faculty of Medicine, Technion – Israel Institute of Technology, Haifa, Israel*

ATEF ASNACIOS • *Université Paris Cité, CNRS, Laboratoire Matière et Systèmes Complexes, UMR 7057, Paris, France*

MARTIN BERGERT • *Cell Biology and Biophysics Unit, European Molecular Biology Laboratory, Heidelberg, Germany*

MARIE-EDITH CHABOUTÉ • *Institut de biologie moléculaire des plantes, UPR2357 CNRS, Université de Strasbourg, Strasbourg, France*

CLAIRE CHARDÈS • *CNRS, IBDM, UMR7288, Turing Center for Living Systems, Aix-Marseille Université, Marseille, France*

ZHONGWEN CHEN • *Interdisciplinary Research Center on Biology and Chemistry, Shanghai Institute of Organic Chemistry, Chinese Academy of Sciences, Shanghai, China*

RAPHAËL CLÉMENT • *CNRS, IBDM, UMR7288, Turing Center for Living Systems, Aix-Marseille Université, Marseille, France*

ALISA CLYNE • *University of Maryland, College Park, MD, USA*

SHINJI DEGUCHI • *Division of Bioengineering, Graduate School of Engineering Science, Osaka University, Osaka, Japan*

MARK D. DISTEFANO • *Department of Chemistry, University of Minnesota, Minneapolis, MN, USA*

ALBA DIZ-MUÑOZ • *Cell Biology and Biophysics Unit, European Molecular Biology Laboratory, Heidelberg, Germany*

TAL DUANIS-ASSAF • *The Institute of Chemistry, The Hebrew University of Jerusalem, Jerusalem, Israel; The Center for Nanoscience and Nanotechnology, The Hebrew University of Jerusalem, Jerusalem, Israel*

ALEXANDER R. DUNN • *Department of Chemical Engineering, Stanford University, Stanford, CA, USA*

CHAOYU FU • *Department of Physics, National University of Singapore, Singapore, Singapore; Mechanobiology Institute, National University of Singapore, Singapore, Singapore*

MARGARET L. GARDEL • *Department of Physics, Institute for Biophysical Dynamics, and James Franck Institute, University of Chicago, Chicago, IL, USA*

BENJAMIN GEIGER • *Department of Immunology and Regenerative Biology, Weizmann Institute of Science, Rehovot, Israel*

MIRIAM B. GOODMAN • *Molecular and Cellular Physiology, Stanford University, Stanford, CA, USA*

CARSTEN GRASHOFF • *Department of Quantitative Cell Biology, Institute of Integrative Cell Biology and Physiology, University of Münster, Münster, Germany*

BERND HOFFMANN • *Forschungszentrum Jülich, Institute of Biological Information Processing, IBI-2: Mechanobiology, Jülich, Germany*

WENMAO HUANG • *Department of Physics, National University of Singapore, Singapore, Singapore; Mechanobiology Institute, National University of Singapore, Singapore, Singapore*

JULIEN HUSSON • *LadHyX, CNRS, Ecole polytechnique, Institut Polytechnique de Paris, Palaiseau, France*

JENS KONRAD • *Forschungszentrum Jülich, Institute of Biological Information Processing, IBI-2: Mechanobiology, Jülich, Germany*

MICHAEL KRIEG • *Neurophotonics and Mechanical Systems Biology, ICFO, Institut de Ciències Fotòniques, ICFO, Castelldefels, Spain*

PIERRE-FRANÇOIS LENNE • *CNRS, IBDM, UMR7288, Turing Center for Living Systems, Aix-Marseille Université, Marseille, France*

LI-CHUN LIN • *Neurophotonics and Mechanical Systems Biology, ICFO, Institut de Ciències Fotòniques, ICFO, Castelldefels, Spain*

RABEA LINK • *Institute for Theoretical Physics, Heidelberg University, Heidelberg, Germany; BioQuant, Heidelberg University, Heidelberg, Germany*

DAIKI MATSUNAGA • *Division of Bioengineering, Graduate School of Engineering Science, Osaka University, Osaka, Japan*

RUDOLF MERKEL • *Forschungszentrum Jülich, Institute of Biological Information Processing, IBI-2: Mechanobiology, Jülich, Germany*

YEKATERINA A. MIROSHNIKOVA • *National Institute of Diabetes and Digestive and Kidney Diseases, National Institute of Health, Bethesda, MD, USA*

KENJI NISHIZAWA • *CNRS, IBDM, UMR7288, Turing Center for Living Systems, Aix-Marseille Université, Marseille, France*

PATRICK W. OAKES • *Department of Cell & Molecular Physiology, Stritch School of Medicine, Loyola University Chicago, Maywood, IL, USA*

DAVID J. ODDE • *Department of Biomedical Engineering, University of Minnesota, Minneapolis, MN, USA*

KEUN-YOUNG PARK • *McNeill Baur PLLC, Cambridge, MA, USA*

DAVID PEREIRA • *Université Paris Cité, CNRS, Laboratoire Matière et Systèmes Complexes, UMR 7057, Paris, France*

KATE POOLE • *EMBL Australia Node in Single Molecule Science and Cellular and Systems Physiology, School of Biomedical Sciences, Faculty of Medicine & Health, University of New South Wales, Sydney, NSW, Australia*

ANDREW VAE PRIEST • *Department of Biomedical Engineering, University of California, Davis, CA, USA*

ROBIN PÜLLEN • *Forschungszentrum Jülich, Institute of Biological Information Processing, IBI-2: Mechanobiology, Jülich, Germany*

MEITAL RECHES • *The Institute of Chemistry, The Hebrew University of Jerusalem, Jerusalem, Israel; The Center for Nanoscience and Nanotechnology, The Hebrew University of Jerusalem, Jerusalem, Israel*

JESSICA RICHARDSON • *EMBL Australia Node in Single Molecule Science and Cellular and Systems Physiology, School of Biomedical Sciences, Faculty of Medicine & Health, University of New South Wales, Sydney, NSW, Australia*

TAKUMI SAITO • *Division of Bioengineering, Graduate School of Engineering Science, Osaka University, Osaka, Japan; Graduate School of Biomedical Engineering, Tohoku University, Miyagi, Japan; JSPS Research Fellowship for Young Scientists, Osaka, Japan*

STEFANO SALA • *Department of Cell & Molecular Physiology, Stritch School of Medicine, Loyola University Chicago, Maywood, IL, USA*

NEUS SANFELIU-CERDÁN • *Neurophotonics and Mechanical Systems Biology, ICFO, Institut de Ciències Fotòniques, ICFO, Castelldefels, Spain*
ULRICH S. SCHWARZ • *Institute for Theoretical Physics, Heidelberg University, Heidelberg, Germany; BioQuant, Heidelberg University, Heidelberg, Germany*
JASON SEDLAK • *Drexel University, Philadelphia, PA, USA*
SHAILAJA SEETHARAMAN • *Department of Physics, Institute for Biophysical Dynamics, and James Franck Institute, University of Chicago, Chicago, IL, USA*
SURABHI SHRESTHA • *EMBL Australia Node in Single Molecule Science and Cellular and Systems Physiology, School of Biomedical Sciences, Faculty of Medicine & Health, University of New South Wales, Sydney, NSW, Australia*
GAURAV SINGH • *Institut de biologie moléculaire des plantes, UPR2357 CNRS, Université de Strasbourg, Strasbourg, France*
SANJEEVI SIVASANKAR • *Department of Biomedical Engineering, University of California, Davis, CA, USA*
YING TU • *Department of Physics and Astronomy, Iowa State University, Ames, IA, USA*
JUBINA BALAN VENGHATERI • *Department of Immunology and Regenerative Biology, Weizmann Institute of Science, Rehovot, Israel*
XUEFENG WANG • *Department of Physics and Astronomy, Iowa State University, Ames, IA, USA*
LUKAS WINDGASSE • *Department of Quantitative Cell Biology, Institute of Integrative Cell Biology and Physiology, University of Münster, Münster, Germany*
HAGUY WOLFENSON • *Department of Genetics and Developmental Biology, Rappaport Faculty of Medicine, Technion – Israel Institute of Technology, Haifa, Israel*
JIE YAN • *Department of Physics, National University of Singapore, Singapore, Singapore; Mechanobiology Institute, National University of Singapore, Singapore, Singapore; Centre for Bioimaging Sciences, National University of Singapore, Singapore, Singapore*
YUAN YAO • *Regenerative Nanomedicine Lab, Department of Chemical Engineering, University of Waterloo, Waterloo, ON, Canada*
EVELYN K. F. YIM • *Regenerative Nanomedicine Lab, Department of Chemical Engineering, University of Waterloo, Waterloo, ON, Canada*

Part I

Quantifying the Mechanical Properties of Cells and Adhesion Proteins

Chapter 1

Measuring Cell Mechanical Properties Using Microindentation

Julien Husson

Abstract

Quantifying cell mechanical properties is of interest to better understand both physiological and pathological cellular processes. Cell mechanical properties are quantified by a finite set of parameters such as the effective Young's modulus or the effective viscosity. These parameters can be extracted by applying controlled forces to a cell and by quantifying the resulting deformation of the cell.

Microindentation consists in pressing a cell with a calibrated spring terminated by a rigid tip and by measuring the resulting indentation of the cell. We have developed a microindentation technique that uses a flexible micropipette as a spring. The micropipette has a microbead at its tip, and this spherical geometry allows using analytical models to extract cell mechanical properties from microindentation experiments. We use another micropipette to hold the cell to be indented, which makes this technique well suited to study nonadherent cells, but we also describe how to use this technique on adherent cells.

Key words Microindentation, Cell mechanics, Micromanipulations, Micropipettes, Contact mechanics

1 Introduction

Mechanical properties of cells regulate their shape and dynamics so they are important in many biological processes, including cell migration, division, immune reaction, and development. Furthermore, many pathologies, including cancer, are associated with altered cell mechanical properties [1]. As a result, quantifying cell mechanical properties is of interest to better understand both physiological and pathological cellular processes and might help opening routes for new diagnostic strategies [2].

Cell mechanical properties are quantified by a finite set of parameters such as the effective Young's modulus of the cell (its elasticity – the higher the Young's modulus, the stiffer the cell), its tension, viscosity, or viscoelastic modulus. These parameters relate an applied force to a resulting deformation, where both force and deformations can be time-dependent. The principle to extract these

Ronen Zaidel-Bar (ed.), *Mechanobiology: Methods and Protocols*,
Methods in Molecular Biology, vol. 2600, https://doi.org/10.1007/978-1-0716-2851-5_1,

parameters is quite universal and consists in applying a controlled constant force and measuring the resulting deformation or reversely in prescribing a deformation and in quantifying the force needed to keep this deformation constant.

Microindentation techniques usually use a spring terminated by a rigid tip. This spring is used to apply a known force when pressed against a cell. The applied force is measured by measuring the deformation (deflection) of the spring and multiplying it by the spring stiffness (in N/m). The cell deformation is quantified through the measurements of how much the probe's tip indents the cell. A gold standard for microindentation is the atomic force microscopy (AFM), a technique that is highly resolved in space, force, and time. However, this technique is relatively expensive and presents a surprisingly major difficulty in its design because the AFM probe is in the optical path. As a result, if one wants to quantify the morphology of the cell being indented using light microscopy, the AFM probe hides precisely what one would like to observe.

We have developed a relatively less expensive alternative to perform microindentation of cells. It uses micropipettes as flexible cantilevers, in an approach that we called profile microindentation, or microindentation for short (Fig. 1) [3, 4]. Our technique was built on elegant works by others in other biological contexts [5–7] including the micropipette force sensor by Backholm et al. [8], developed decades after the idea of using a hollow micropipette as a flexible cantilever was introduced [9]. Flexible micropipettes behave like a linear spring, i.e., with a linear relationship between the amplitude of their bending and the force applied perpendicular to their tip. Micropipettes can have bending stiffness as low as 0.2 nN/μm, 50 times lower than the most flexible AFM tips. This extreme flexibility compensates for a more limited spatial resolution when detecting micropipette bending directly from a camera (as opposed to the great accuracy of laser diodes used in AFM). This leads to force resolution comparable to AFM (few piconewtons in our case). The microindentation technique presented here is that it can readily be coupled with another micropipette that is used to hold a cell, which makes this technique well suited to study nonadherent cells. However, although this requires to grow cells on a support (dextran beads, see below), it can also be used on adherent cells [3].

To make use of the few analytical models that exist to extract cell mechanical properties from microindentation experiments, we developed ways to place microbeads at the tip of flexible micropipettes, with two alternative techniques. The first technique consists in using an open micropipette to aspirate a microbead at its tip. The interest of this technique is that once a bead has been used against a cell, both beads and cells can be discarded, and a new cell can be indented by a new bead. The second technique consists in melting

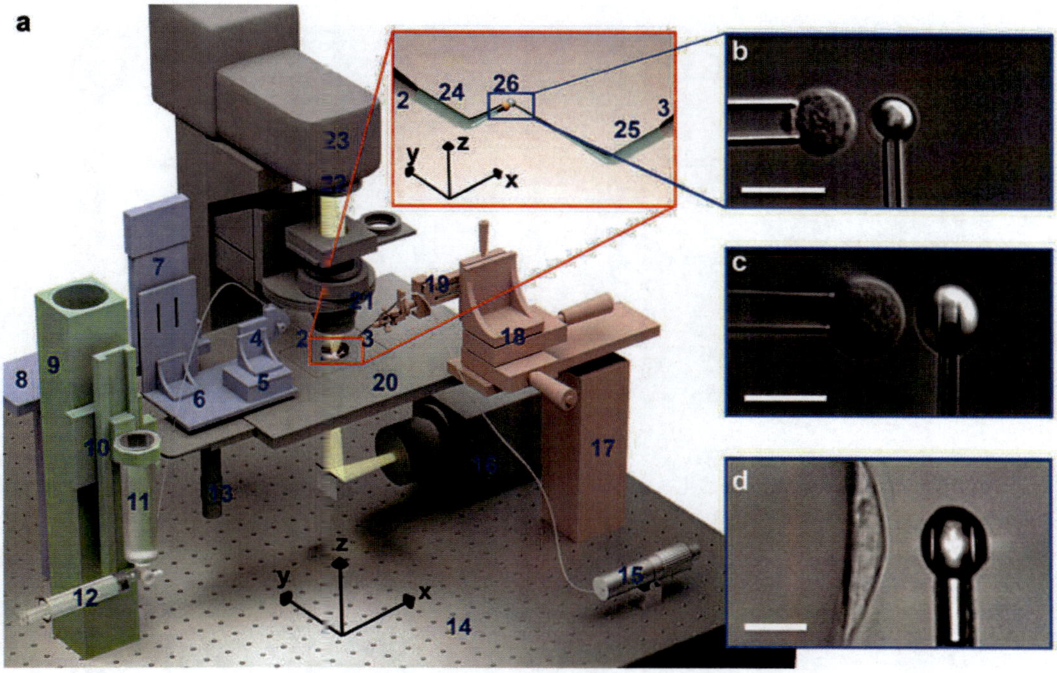

Fig. 1 Experimental setup for profile microindentation. (**a**) Schematic of the main components of the setup: **1** Glass-bottom Petri dish with buffer, cells, and eventually microbeads; **2** micropipette holder (for the cell-holding micropipette); **3** micropipette holder (for the microindenter/flexible micropipette); **4** right-angle bracket and rod-holding clamp; **5** uniaxial piezoelectric stage; **6** right-angle bracket and plates linking the 3-axis motorized micromanipulator and piezoelectric stage; **7** motorized micromanipulator; **8** rigid stand with platform supporting the head of the motorized micromanipulator; **9** pillar and mounting plates; **10** vertical manual stage supporting the water reservoir, with coarse and fine translation; **11** 50 mL plastic syringe used as the body of a water reservoir **12** auxiliary syringe used to adjust water level and exert strong pressure connected via a four-way stopcock; **13** x-y stage translation knobs; **14** antivibration optical table; **15** microinjecter; **16** camera; **17** rigid stand with platform; **18** three-axis manual translation stage; **19** rod-holding joint with tilt adjustment mechanism, and attachment pieces to the three-axis manual translation stage; **20** microscope stage; **21** condenser aperture diaphragm; **22** field diaphragm; and **23** microscope illumination. The inset shows a schematic view of the micropipette tips (not to scale); **24** cell-holding micropipette; **25** flexible micropipette or microindenter; **26** cell and microindenter/bead aspirated by the flexible micropipette. Part of the microscope body is not shown, including the oculars, the objective, and an eventual fluorescent light source. (**b**–**d**): Microscopy image in DIC (**b**–**c**) and brightfield illumination (**c**) of different cells being studied. (**b**) Microscopy image of a leukocyte (human CD4 T lymphocyte) held aspirated by the stiff micropipette (left) and flexible micropipette holding an antibody-coated microbead aspirated (right). Scale bar: 10 μm. See [4, 11] for similar configuration. (**c**) Microscopy image of a leukocyte (neutrophil-like PLB cell) held aspirated by the stiff micropipette (left) and microindenter on the right. Scale bar: 10 μm. See Ref. [10] for similar configuration. (**d**) Endothelial cell grown on a 200 μm dextran bead held by a stiff micropipette on its left extremity and microindenter (right). Scale bar: 10 μm. See Ref. [3] for similar configuration

the tip of a flexible micropipette into a glass microbead (Fig. 2). For simplicity, herein, we will call both types of modified micropipettes microindenters, except when distinction between them is needed.

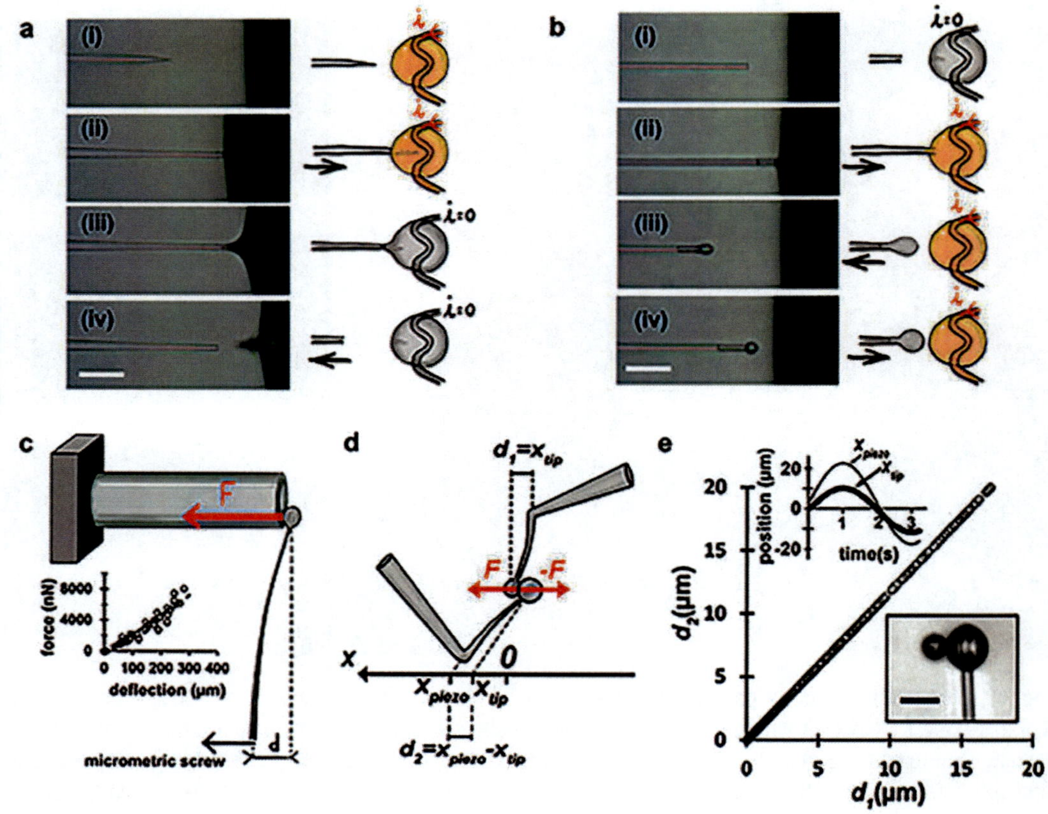

Fig. 2 Micropipette fabrication and calibration. (**a**) Once they are pulled, micropipettes are cut using a microforge. The tip of the micropipette is dipped into a drop of molten glass present on the heating filament of the microforge. A sequence of steps (i–iv) is followed to cut the tip at a desired diameter. Left column: microscopy image taken on the microforge (bar: 10 μm). Right column: schematic view (not to scale). Steps (i–iv) are repeated a few times until the desired diameter (typically 3 μm inner diameter) is reached. A demonstration video can be found online (https:/cellmechanics.jimdofree.com/videos/). (**b**) The microforge is used to form a glass microbead at the tip of a flexible micropipette. A micropipette is dipped into the drop of molten glass and quickly retracted to carry some molten glass with it. Once a droplet has been deposited at the tip of the micropipette (i–iii), it can be made more spherical by heating it at distance (iv). Steps (i–iv) can be repeated to form a larger bead. Left column: microscopy image taken on the microforge (bar: 10 μm). Right column: schematic view (not to scale). A demonstration video can be found online (https:/cellmechanics.jimdofree.com/videos/). (**c**) The measurement of the bending stiffness of a reference microindenter is performed by pressing it against a commercial force probe. The probe and microindenter are schematically represented (not to scale). Inset: the compressive force F exerted by the microindenter on the probe is plotted versus the deflection d of the microindenter. A linear fit is superimposed. The bending stiffness is the slope of this force-deflection curve. Given the force resolution of the commercial probe, we aim for reference microindenters with a stiffness of about 20 nN/μm. **(d)** Microindenters used routinely are calibrated against a reference microindenter by measuring the ratio of their respective deflection d_1 and d_2 when pressed one against the other. We use a dedicated setup that is similar to the one used during experiments on cells, except that it is equipped with a 40× air objective. The reference microindenter of bending stiffness k_1 is fixed, so the microindenter deflection d_1 equals x_{tip}, where x_{tip} is the position of its tip. The base of the microindenter of unknown bending stiffness k_2 is translated by a piezoelectric stage whose position x_{piezo} is recorded. The deflection of this microindenter is $d_2 = x_{piezo} - x_{tip}$. **(e)** The translation of the microindenter is slow enough so that inertia can be neglected and forces exerted by the tip of each microindenter equilibrate. As a result,

We have applied this technique to quantify the viscoelastic properties of endothelial cells [3] and of several types of leukocytes: T lymphocytes (and Jurkat T cells) [4, 10–12], B lymphocytes [10, 13], neutrophils, and neutrophil-like PLB cells [10, 14].

2 Materials

2.1 Consumable

1. Cells, cultured in an incubator at 37 °C under 5% CO_2.
2. P1000 and P200 pipettes with cones.
3. Glass-bottom Petri dishes (35 mm diameter, 23 mm well).
4. For adherent cells: dextran microcarrier beads (average bead size 175 μm).
5. Borosilicate glass capillaries (1 mm OD, 0.78 mm ID). *See* **Note 1**.
6. Deionized water for the reservoir and tubing.
7. 70° Ethanol to clean surfaces, tubing, and reservoirs.
8. 50 mL Luer lock plastic syringe for reservoir (Fig. 1a).
9. 30 mL Luer lock auxiliary plastic syringe for reservoir (#12 in Fig. 1a).
10. 10 mL Luer lock auxiliary plastic syringe to fill micropipettes.
11. Blunt fill needle to mount on the 10 mL auxiliary syringe (18 G).
12. Four-way stopcock for plugging an auxiliary pipette on the reservoir.
13. Cell buffers. We usually perform experiments in RPMI 1640-GlutaMax-I supplemented with 10% fetal calf serum (FCS), 1% penicillin-streptomycin. The presence of FCS is important to avoid cell-pipette and cell-bottom adhesion. FCS can be replaced with bovine serum albumin for its anti-adhesive effect (*see* **Note 2**). We keep buffer stocks at 4 °C.
14. Dulbecco's Phosphate Buffer Saline (DPBS).
15. Large Petri dish with thick adhesive rubber used as a micropipette and microindenter storage box to keep them safe from dust and collisions.
16. 15 mL plastic tubes to store cell buffer and DPBS close to the experimental setup.

Fig. 2 (continued) $k_1 d_1 = k_2 d_2$, so by plotting d_2 against d_1, one gets a line of slope k_1/k_2. As k_1 is known, k_2 can be calculated. Lower right inset: image of indenter tips under the microscope; the scale bar represents 10 μm. Upper left inset: x_{piezo} and x_{tip} are plotted against time

17. Capillary to fill the back of the micropipettes with water.
18. Kimwipes.

2.2 Microscope

1. Inverted microscope equipped with differential interference contrast (DIC). DIC is not necessary, but it is a great addition to observe cell morphology (intracellular deformation, fine membrane structures, etc.).
2. Air suspension anti-vibration table. *See* **Note 3** on vibrations.
3. Microscope objectives: 100× (Plan Fluor, 1.3 numerical aperture, DIC, oil) for experiments and eventually lower magnification air objectives (40×, 20×, 10×, and 4×) for micropipette positioning. Experiments could be performed with a lower magnification at the expense of space and force resolution.
4. Matlab software and the free open-source Micro-Manager software (https://micro-manager.org/) [15].
5. CMOS Camera is connected to the microscope controlled using a Matlab code calling the Micro-Manager software.
6. If fluorescence is used, LED fluorescence source and filter cube are added to the microscope. We control an LED fluorescent light source through transistor-transistor logic (TTL) via the Micro-Manager that is in itself controlled by a Matlab code. TTL control is ensured by an Arduino Uno card.
7. Computer.

2.3 Micro-manipulation-Related Equipment

1. Motorized micromanipulator (#7 in Fig. 1a) is mounted on a rigid stand with platform (#8 in Fig. 1a). A second motorized micromanipulator might be needed when using a third pipette, e.g., to flow a molecule of interest (such as an inhibitor) on the cell being indented [3] or when indenting a neutrophil in the back, while it phagocytes a bead at its other extremity [10, 14].
2. Custom metal plates and right-angle bracket (#6 in Fig. 1a) to support the piezoelectric stage.
3. Closed-loop piezoelectric stage is connected by USB cables to the computer. Matlab drivers for this piezoelectric stages are available on the manufacturer website.
4. Second right-angle bracket to support a rod-holding clamp (#4 in Fig. 1a).
5. Rod-holding joint with tilt adjustment mechanism and attachment pieces to the manual 3-axis micropositioner (#19 in Fig. 1a).
6. Manual 3-axis micropositioner (#18 in Fig. 1a) is placed on top of a rigid stand with platform (#17 in Fig. 1a).
7. Two micropipette holders. One holder (#2 in Fig. 1a) has its tubing connected to the water reservoir and held at a 45° angle

on the motorized micromanipulator (#7 in Fig. 1a). The other holder is either coming with the microinjecter (#15 in Fig. 1a) and is connected with the flexible micropipette to apply aspiration pressure on the air inside (no need for a water reservoir in that case) or does not need tubing in case a closed microindenter is used. More sophisticated reservoir systems are used by others to apply hydrostatic pressure to the cell [16], but they require more tubing hence more potential troubleshooting. As we only need to apply one value of pressure throughout an experiment, we go for a simpler and robust system with only one reservoir and a micropipette holder connected to it.

8. Straight metal rod or alternatively a piece of piano cord to remove glass debris when a pipette is broken inside the body of the micropipette holder.
9. Micropipette puller.

2.4 Water Reservoir

1. Vertical manual stage supporting the water reservoir, with coarse and fine translation.
2. Pillar made of an optical rail is equipped with a mounting plate and is mounted on the optical table with a vertical mounting plate.

2.5 Micropipette Manufacturing

1. Microforge to cut pipettes open and to melt their tip. A glass bead of about 0.5 mm is placed on the microforge filament and melts when heated by the current passing through the platinum filament of the microforge.
2. Microforge is used to bend pipettes.
3. Microinjector to apply negative pressure when using an open flexible micropipette.
4. Short micropipette holder.
5. Commercial force probe to calibrate the bending stiffness of reference microindenters. To facilitate the placement of the head of the probe in front of a low (4×, 10×, or 20×) magnification objective, you can place the head of the probe on a miniature manual stage which can then be placed on the stage of the inverted microscope. It might be helpful to connect the output voltage of the probe to an oscilloscope to visualize the signal and make sure it is stable over time (*see* **Note 4**).

3 Methods

Glass micropipettes can break so be mindful when manipulating them, and take all required cautions when manipulating cells.

3.1 Fabrication of Cell-Holding Micropipettes

1. Pull a borosilicate glass capillary with the micropipette puller with the appropriate program for stiff pipettes (on our micropipette puller: P = 500, heat = 474(= ramp + 4), pull = 50, vel. = 140, time = 100).
2. Cut the extremity of the pulled capillary to the desired diameter using the microforge (Fig. 2a). Measure the inner diameter using calibrated graduations in the microforge ocular. Demonstration videos can be found on our web page (https://cellmechanics.jimdofree.com/videos/) and on Narishige's website (http://products.narishige-group.com/group1/MF-900/injection/english.html).
3. Bend the micropipette at a 45° angle so that when held by the 45°-angled micropipette holder, the tip lies parallel to the bottom of the dish. To do so, place the micropipette at 20–100 μm away from the microforge filament, where the pipette is 50 μm thick. Heat the filament enough (~80% level) for a short period of time (about one second) so the filament starts bending. Repeat this heating step until the desired 45° angle is reached. During the process, adjust the pipette location if it becomes too close to the heating wire. Demonstration videos can be found on our web page (https://cellmechanics.jimdofree.com/videos/) and on Narishige's website (http://products.narishige-group.com/group1/MF-900/injection/english.html). Beware, if you bend the capillary where it is only 20 μm thick or less, instead of bending downward, it will bend upward. Measure the bending angle using the angular grid in the microforge ocular.

3.2 Fabrication of Flexible Micropipettes and Microindenters

1. Pull a borosilicate glass capillary with the micropipette puller; use the appropriate program for flexible micropipettes (on our micropipette puller: P = 400, heat = 517 (= ramp + 47), pull = 50, vel. = 140, time = 89).
2. Cut the taper of the pulled capillary at a tip diameter of typically 1–2 μm. Heat the tip for a few seconds while it is a few microns from the microforge glass bead to make the borders of the micropipette's tip smooth. This will help forming a tight seal when aspirating a microbead.
3. To know if you need to further shorten the micropipette, roughly estimate the bending stiffness of this micropipette by pressing its tip against the tip of a microindenter of known stiffness (*see* Subheading 3.4). If the stiffness is still too low (<0.1 nN/μm), bring back the freshly cut micropipette, and shorten it by typically 50 μm. Repeat these steps until the stiffness is of the desired value, typically 0.2–0.5 nN/μm (*see* **Note 5**).

4. Once this is achieved, if you aim for an open flexible micropipette, bend it at a 45° angle where it is 50 μm thick on the microforge. If you aim for a microindenter, first plunge the pipette tip in the liquid droplet inside the microforge and remove it quickly; the viscous liquid will be carried out on the pipette. Then heat this droplet at distance (~10 μm) to make the droplet of molten glass round up. Stop the current in the filament as soon as you reach the desired shape (Fig. 2b). Adapt the velocity when withdrawing the pipette from the molten glass to get different sized droplet (the faster the retraction, the larger the droplet). For droplets of only ~5 μm in diameter, you may just melt the tip at distance and wait for the droplet to become round enough. Measure the diameter of the indenter using the graduations in the microforge ocular. Usually we aim for indenter tips that are 5–10 μm in diameter.
5. Determine more accurately the microindenter tip diameter using a 40× or 100× objective.

3.3 Calibrate a Reference Microindenter Using a Commercial Probe

1. Place the commercial probe on the stage of a microscope. One can also perform this calibration under the monitoring of a simple USB camera, as the deflection of the microindenter is done by measuring the position of its base through a manual translation stage (with few microns accuracy), not based on the camera image.
2. Place the microindenter on the manual micromanipulator so that it can be pressed against the tip of the probe (Fig. 2c).
3. Move the base of the microindenter by a known distance imposed by the manual translation stage, check that the analog output signal of the probe is stable by visualizing it on an oscilloscope. Check visually through the camera image that there is no slippage of the microindenter tip. The probe will not deflect, so the microindenter deflection equals the position of its base minus the position when contact with the probe is established (as detected by a deflection of the signal from its baseline value).
4. Establish a voltage-deflection curve, which should be linear. Verify that there is no hysteresis. Perform a linear regression to determine its slope. Divide the V/μm slope by the volt-to-newton conversion factor provided by the probe manufacturer to obtain the bending stiffness of the microindenter, in N/m. Aim for a microindenter soft enough so that it will not be too stiff when compared to the stiffness of microindenters used to indent cells. However, it should be stiff enough to lead to an accurate calibration with the commercial probe. Aim for typically 20 nN/μm.

5. This 20 nN/μm stiffness is too high to compare directly with a 0.5 nN/μm microindenter, so using this reference indenter of so-called generation 1 calibrates another indenter, of generation 2, with a stiffness of typically 2 nN/μm. To do so, use the technique described in Subheading 3.4. Although this will propagate an experimental error, this propagated error is typically less than 5%. This generation 2 microindenter will be the one to calibrate your microindenters used for cell indentation.
6. Keep carefully your reference indenters in a large closed Petri dish to avoid dust and shocks.

3.4 Calibration of the Bending Stiffness

To measure the bending stiffness of a newly pulled microindenter or flexible micropipette, we press its tip against the tip of a reference microindenter of known bending stiffness. It is important that this reference microindenter has a glass bead at its tip because this will ensure that is it not broken (a broken micropipette might look a lot like an intact micropipette). The base of the microindenter of unknown stiffness is translated to press its tip against the tip of the reference microindenter (Fig. 2d). The forces exerted by each cantilever in quasi-static conditions balance each other, so that $k_1 d_1 = k_2 d_2$, where k_1 and k_2 are the bending stiffness of the reference and new microindenter, respectively, and d_1 and d_2 their respective deflection. By plotting d_2 vs. d_1, one gets a linear curve (Fig. 2d) of slope k_1/k_2, from which k_2 is deduced based on the already known value of k_1. It is immediate that any error on k_1 is propagated to k_2. In practice we dedicate a small setup for calibration. It has a piezoelectric stage which performs an oscillatory translation of the base of the microindenter of unknown stiffness. For calibration we use a 40× air objective. No Petri dish is needed at this step. Beware that in air, micropipette will stick quite well due to electrostatic interactions. One can use the same setup as for experiments, using a dedicated calibration is only a matter of user-friendliness. Other groups have developed interesting alternatives to calibrate pipette stiffness such as using a hanging droplet [8]. It is also possible to use small weights such as pieces of metallic wires [17, 18] or pieces of paper [3] to a calibrated reference fiber, but this procedure might be more difficult in practice.

3.5 Automated Detection of the Force and Indentation, and Feedback Loop

The position of the tip of the microindenter is measured each time the camera acquires a new image (at a rate of typically 400 Hz, *see* **Note 6**). In the configuration described in Fig. 1, the cell is pushed against the indenter, so that indenter deflection is the difference between the tip's updated position and its position under no force measured at the start of the experiment (Fig. 3; *see* **Note 7**). The force applied to the cell is the product of the microindenter deflection and the microindenter bending stiffness k. An algorithm implemented in Matlab runs a feedback loop (*see* **Note 8**) that

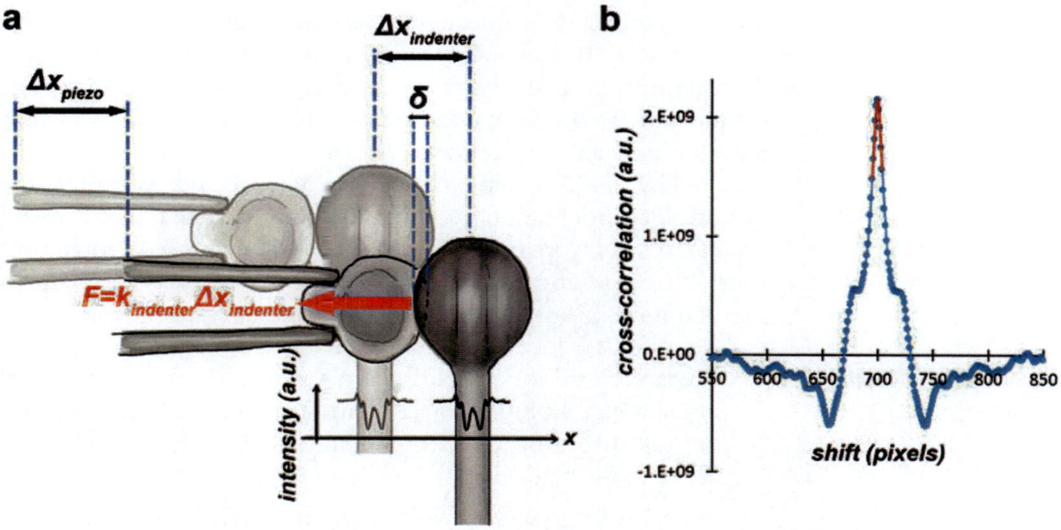

Fig. 3 Measuring force and cell indentation. (**a**) Schematic view of a stiff micropipette holding a cell (left) and a microindenter (right). The setup is shown before contact (light gray) and upon cell indentation (darker gray) following a translation Δx_{piezo} of the piezoelectric stage along the horizontal *x*-axis. When the cell is pressed against the tip of the microindenter, the latter bends. Its deflection equals the translation of the tip of the microindenter, $\Delta x_{\text{indenter}}$. The force *F* exerted by the microindenter on the cell is obtained by multiplying the microindenter deflection with the microindenter bending stiffness, k_{indenter}: $F = k_{\text{indenter}}\Delta x_{\text{indenter}}$. The indentation of the cell, δ, is the difference between the piezoelectric stage translation and the indenter deflection: $\delta = \Delta x_{\text{piezo}} - \Delta x_{\text{indenter}}$. To measure the position of the tip of the microindenter, the intensity profile along a line crossing the shaft of the microindenter (*x*-axis shown on the figure) is analyzed at an acquisition frequency of roughly 400 Hz (*see* **Note 6**). An intensity profile acquired at the beginning of an experiment, $I_{\text{template}}(x)$ is kept in memory. It is then used as a template that is compared at each time step with the actualized intensity profile, $I_{\text{live}}(x)$. The comparison is done by calculating a cross-correlation shown in **b**. (**b**) The cross-correlation $C(s)$ of $I_{\text{template}}(x)$ and $I_{\text{live}}(x)$ is $C(s) = \int (I_{\text{live}}(x) - \langle I_{\text{live}}\rangle)(I_{\text{template}}(x - s) - \langle I_{\text{template}}\rangle)dx$, where *s* is the shift of $I_{\text{template}}(x)$ relative to the live profile $I_{\text{live}}(x)$ and brackets represent an average over *x*. Before cross-correlating them, the average value of I_{live} and I_{template} is subtracted. A typical correlation function is shown on the graph. It presents a maximum at a shift that corresponds to the location of the microindenter tip relative to its position when the template was acquired. The shift takes integer values corresponding to an integer number of pixels. However, by performing a parabolic fit over typically 10 pixels (in red on the graph) around the maximum of the cross-correlation function, the position of the microindenter tip is measured with subpixel accuracy. We obtain an accuracy of about 20 nm for a pixel representing 59 nm in the specimen plane using a 100× magnification objective

controls the piezoelectric stage position based on the analysis of each new image. The experiment is decomposed in several phases, each one defining a different behavior of the feedback loop. We start an experiment with the cell facing the microindenter's tip, a few microns away from it (Fig. 1b–d). During the first phase, the cell is moved at constant velocity, typically 0.25–0.5 μm/s, toward the microindenter. Once -cell-indenter contact is made (which is detected upon reaching a microindenter deflection threshold of typically 0.1 μm), the compressive force applied to the cell increases, until a force threshold is reached (typically 0.1–0.5 nN on leukocytes, up to few nN on stiffer cells such as endothelial cells). The algorithm then automatically switches to a second phase, during which various options are possible: applying an oscillatory force around an average value at a constant frequency (typically 1 Hz) [**10, 14**], keeping the force constant and recording the evolution of indentation over time (creep experiment), or keeping the indentation constant and measuring the decrease in force over time (force relaxation experiment). Once this phase is over (after a preset duration is over), the cell is retracted at constant velocity.

3.6 Microindentation of Nonadherent Cells

1. Prepare 10 mL of cell buffer and a Petri dish. Put a droplet of immersion oil on the objective and place the dish on the 100× oil objective. Add 3 mL of medium at room temperature (*see* **Note 9**) in the dish.
2. Fill the water reservoir with ultrapure water. Position the four-way stopcock to fill the auxiliary micropipette connected to the reservoir. Make sure no bubble is trapped in the four-way stopcock.
3. Switch the four-way stopcock so the auxiliary pipette filled with water is directly connected to the tubing of the micropipette holder, and fill the tubing with water.
4. Connect a syringe to a blunt fill needle with a piece of rubber joint at its tip. Connect this auxiliary syringe to the back of a pre-forged stiff micropipette, and dip the tip of this micropipette in a tube of DPBS while making sure not to touch the walls of this tube. Use a 15 or 50 mL tube that you might be able to use for several days, as contamination in pure DPBS at room temperature is limited. Exert aspiration pressure with the syringe (*see* **Note 10**). Complete the filling of the micropipette through its back using a syringe filled with water and connected to a long capillary. You should be able to avoid any bubble, but if you trap one (which will be obvious by eye inspection), gently flick the micropipette with the nail several times. You will see the bubble deform and eventually detach from the inner wall of the micropipette. Use again the capillary to make sure the bubble doesn't stay trapped at the meniscus at the back of the capillary.
5. Connect this liquid-filled pipette with the micropipette holder. Hold the latter below the water level in the reservoir so drops start flowing. Make the meniscus at the back of the pipette coalesce with a droplet dripping from the capillary tip in order to avoid trapping any bubble. Bubbles will forbid any efficient aspiration and will create hysteresis when changing the height of the water reservoir (*see* **Note 11**).

6. Place the micropipette tip on the microscope and center it in the field of view. We usually use the camera with stretched histogram to do so. *See* **Note 12** for finding your micropipette on the microscope. Place the tip typically 30 μm above the bottom (on our microscope, a full revolution of the fine z knob corresponds to 100 μm).
7. Set Köhler illumination, which is important to get the best out of the optical setting.
8. The difference in height Δh between the water level in the reservoir and the water level in the dish sets an hydrostatic pressure difference, $\Delta P = \rho g \Delta h$, where ρ is the volume mass of the water in the reservoir and tubbing and of the buffer in the dish (both being approximated as equal to 1000 kg/m^3), g is the acceleration of the earth gravity field on the earth surface ($g \approx 9.8$ m/s^2). As a result, 1 mm of water difference represents 9.8 Pa of pressure difference. Put an aspiration corresponding to $\Delta h \approx 5$ mm. This will let the micropipette aspirate some medium and let the serum present in the medium coat the pipette walls. This will minimize adhesion of cells to the surface of the pipette (*see* **Note 2**).
9. Place your microindenter (*see* **Note 13**). Place the tip of the indenter at about 30 μm above the bottom of the dish (*see* **Note 14**).
10. Bring the cells; in our case no need to concentrate them (they are at 0.5–1.0 × 10^6/mL), inject 50–100 μL of cell solution (*see* **Note 15**). Inject at the bottom of the chamber, in the "north" half of the dish (higher values on the y-axis following axis convention shown in Fig. 1a). The "south" part is less accessible given the angle of the indenter (Fig. 1a) that will hit the border of the Petri dish if you are too far toward the south. Gently press the tip of the cone on the bottom of the dish, then gently expulse cell solution, so that the cells are not too diluted in the whole dish (*see* **Note 16**). Wait a couple of minutes for cells to sediment, then find the zone where cells are more concentrated.
11. Set the zero of aspiration pressure by changing the height of the reservoir. It is very helpful to know already quite well (within a millimeter) where the zero should be (put a mark on you reservoir). Adjust the reservoir height using the fine translation (#10 in Fig. 1a) so that the micropipette tip neither aspirates nor blows. To probe if the pipette is aspirating or blowing, you can use a bead that is not stuck to the bottom (or unstick it by scratching the pipette tip against it), or a cell, but being careful not to aspirate it too much so that its membrane stays intact. You can also use a cell debris. Once zero is

known, set the aspiration level for the experiment, Δh is typically 3–5 mm.

12. Aspirate a cell, and place it in front of the tip of the indenter (Fig. 1b–d).
13. Prepare in advance the parameters of the experiment (*see* **Note 17**).
14. Add all necessary remarks in a note book: unusual aspect of cell, cell-bead misalignment, etc.
15. You may need to slightly adjust the relative cell-bead position in-between experiments.
16. Once a cell has been probed, to avoid retesting it by mistake, move the microscope stage (*see* **Note 18**) to go to extreme north position in the dish (by moving microscope stage toward the increasing values of *y*). Gently nail-flick the cell holding pipette to expulse the cell, and move back to a zone where there are many cells to run another experiment (*see* **Note 19**).
17. Scrutinize the aspects of cells. With no control of pH, do not expect to run experiments typically more than 2 h. After this time lapse, take out the pipette and the indenter (fix them in a safe place next to the microscope); then change the dish. When taking the dish out, clean the oil on the objective (Kimwipes or equivalent is fine), and put a new oil droplet; experience shows that this minimizes the appearance of bubbles in the oil.

3.7 Microindentation of Adherent Cells

We have developed a strategy to use profile microindentation on adherent cells by culturing them on large (200 μm) dextran beads, which allows observing cell deformation upon indentation (Fig. 1d). During cell indentation, the dextran is held in place using a micropipette with a tip of 10–50 μm [3](*see* **Note 20**).

3.8 Microindentation Using an Activating Microbead

The techniques described above use an indenting surface whose role is only to press on the cell to deform it. However, one might want to follow mechanical changes induced by the engagement of receptors at the surface of the cell. To do so, instead of a micro-indenter, one can use a micropipette aspirating a microbead that is coated with molecules of interest, e.g., antibodies targeting some specific receptors on the cells. We have used this strategy to follow mechanical changes in leukocytes during their activation following the engagement of their receptors. In that case, the indenter becomes an adhesive surface, and the bead cannot be detached after the first contact. One can apply force modulation once this first contact was made. This allows extracting the initial Young's modulus of the cell and the evolution of the complex stiffness of the cell-bead contact over time [10, 14].

3.9 End of Experiments

1. Remove the micropipette and the microindenter (*see* **Note 21**).
2. Throw out the Petri dish in a trash suited for the biological material you are working with.
3. Remove the oil on the oil objective.
4. Empty the water reservoir using the auxiliary syringe; then rinse the reservoir with ethanol and flow it in the tubing and in the micropipette holder.
5. Clean the microscope stage with ethanol.
6. Save all data if not done already, and turn everything off (camera, computer, microscope lamp, etc.).

3.10 Data and Statistical Analysis

After acquisition of raw data (Fig. 4), we analyze them using a custom Python code. One task is to determine the contact point, when force starts increasing, via post-processing of raw data (several strategies exist, see, e.g., ref. [19]). Finite thickness correction can be used when calculating the Young's modulus [20].

4 Notes

1. We use thin-walled glass capillaries to make both stiff and flexible micropipettes, but thick wall could also be used after adapting the puller settings. Capillaries do not need to have a filament. A filament might actually become a problem for the tight seal between the flexible micropipette and a bead.
2. Although some groups performing micropipette aspiration use micropipette passivation [22], we always work with serum present in the medium (usually 10% fetal calf serum) and rely on the albumin present in the serum to passivate glass surfaces and make them nonadhesive. Yet, cells might become slightly adherent through discrete spots/tubes, and it might not be immediate to expulse a cell from the cell-holding pipette. In that case nail-flicking the pipette holder will usually create a vibration strong enough to detach the cell. The same issue might happen because the bead held by the flexible micropipette can be a bit sticky. In that case a stronger nail-flicking might be needed, but beware that this might propagate to the surrounding of the flexible micropipette and expulse the cell held aspirated by the other micropipette.
3. Avoiding vibrations is important for signal quality, even more so when using soft microindenters and high-magnification objective. Avoid any tension in wires, e.g., in the possibly heavy wires linking the camera to the computer. Avoid as much as possible placing objects with fans on the insulating table. It might be useful to enclose the whole setup inside an acoustic insulation case, although this will make the access to

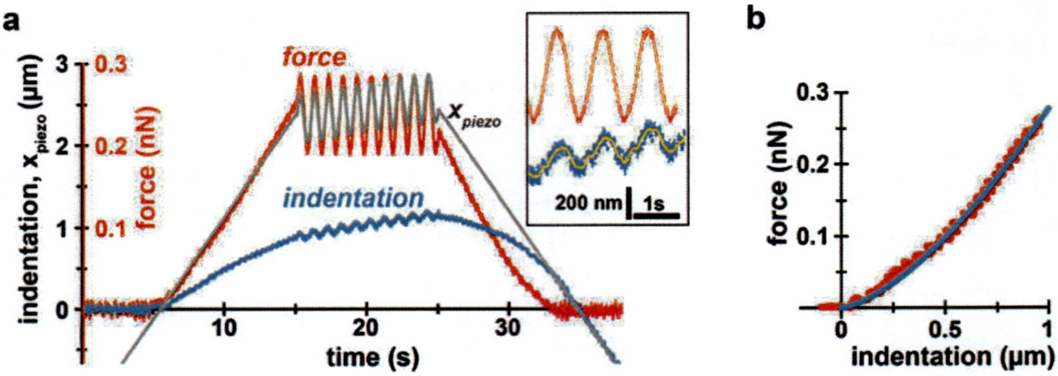

Fig. 4 Time traces and force-indentation curve. (**a**) Force (red), indentation (blue), and piezoelectric stage translation (x_{piezo}, gray) during the indentation of a T lymphocyte (see an example in Fig. 1b). A first compression phase is performed by translating the piezoelectric stage at constant velocity ($dx_{piezo}/dt = 0.25$ µm/s) until a preset force threshold is reached, after which an oscillatory force is applied through the use of a live force feedback loop. After a preset duration is over under oscillatory force, the piezoelectric stage is withdrawn at constant velocity (here again 0.25 µm/s). The inset shows a closer view on the force and indentation signals, with sliding-window-averaged values superimposed to raw data acquired at roughly 400 Hz. (**b**) Force-indentation curve (red) during the initial compressive phase. Data are fitted with the Hertz model $F = \frac{4}{3}\frac{E}{1-\nu^2}\sqrt{R_{eff}}\delta^{3/2}$ [3, 21] (blue curve), where E is the effective stiffness, ν is the Poisson's ratio taken as 0.5 (see suppl. Mat. in [3]), δ is the indentation of the cell, and R_{eff} an effective radius defined by $1/R_{eff} = 1/R_{indenter} + 1/R_{cell}$, where $R_{indenter}$ is the radius of the microindenter tip or micropipette bead and R_{cell} is the cell radius

the setup less user-friendly – we do not use such case. Avoid the vicinity of air conditioning and open window in the room.

4. What we describe here is a setup that has benefited from about 10 years of accumulation of hardware. One can start with a less comfortable but less expensive version, for instance, by using only manual micromanipulators, no commercial force probe, a simpler water reservoir with only a coarse translation, and a less expensive microscope stage (even a custom simple microscope stage can be applicable).

5. Adjust the stiffness of the cantilever by adjusting the puller settings, cutting the micropipette longer or shorter. See the "pipette cookbook" on Sutter Instrument website for many useful tips (https://www.sutter.com/MICROPIPETTE/cookbook.html).

6. Fast camera-to-computer transfer rate ensures an optimized acquisition rate. Provided the exposure time is not the limiting step (we expose 1–2 milliseconds only), this rate depends on the camera and connectivity (USB3 or Camera Link interface card, for instance) and also importantly on the size of the region of interest (ROI) selected in the Matlab code via Micro-Manager. For an ROI of 600 × 400 pixels, we typically get a rate of acquisition of 400 Hz. Note that not every image is

stored, which would dramatically decrease the acquisition rate: only a line is analyzed to determine the position of the indenter on the fly through a cross-correlation operation (Fig. 3). This raw data can then be smoothed during post-processing using, for instance, a sliding-window average. We keep a full image every 1 or 2 s to monitor the morphology of the cell.

7. When designing over which micropipette holder you should place the piezoelectric stage, two options are possible: either pushing the cell against the indenter or pushing the indenter against the cell. Although this might seem equivalent, there will be differences, e.g., the hydrodynamic drag will be higher on the flexible micropipette if it is the one being translated by the piezoelectric stage. In our later evolution of the setup presented here, we push the cell against the microindenter. If the cell moves, the deflection of the indenter will simply be the displacement of its tip relative to its initial resting position. The cell indentation will be obtained by measuring the difference between the displacement of the piezo stage and the position of the indenter's tip. On the contrary, if the indenter is the one being pushed against the cells, the deflection of the indenter will be obtained by measuring the difference between the position of the piezoelectric stage and the position of the indenter's tip. In the latter case, the indentation of the cell will simply be the position of the indenter's tip relative to its position when contacting the cell.

8. It is possible to implement a live feedback loop to apply an arbitrary controlled force to the cell. The same principles apply to design a feedback loop on cell indentation. We introduce an error signal that is the difference between the actual microindenter deflection and the desired microindenter deflection (the latter being the ratio between the target force and the microindenter stiffness). At each time step, the new piezoelectric position is set to its actual position plus a quantity that is the error signal multiplied with a constant proportional coefficient. This is a very simple implementation that has been proven to work satisfactorily in our setup. However, much more sophisticated feedback algorithms can be implemented (including a PID controller where a term proportional to the integral and the derivative of the error signal are added) [23], at the expense of more complexity of the algorithms and potential difficulties to tune the additional feedback parameters.

9. Working at 37 °C is a challenge using the microindentation technique described here. The dish can be heated using a heating plate, eventually complemented with an objective heater. However, because we work in a relative large volume of 3 mL in an open dish, this type of heating is quite inefficient and will lead to strong temperature gradients inside the dish.

To work at 37 °C, we would advise to adapt the setup to work in a smaller volume and eventually to work on a way to circulate medium at 37 °C (but stopping circulation during experiments). Another option is to enclose the whole setup in a temperature-controlled chamber, but beware that the electronics (piezo stage, micromanipulator) might get damaged over time, especially if you also control humidity. Such an enclosure might not prevent the presence of strong temperature gradients, and air flow within the chamber might disturb measurements. Using large metallic pillars as we do is problematic, because of the large dilation and thermal that can occur in the pillar. If the tubing of the micropipette is also at 37 °C, bubbles will form much more often than at room temperature, which might block cell aspiration and require reinstalling pipettes regularly. Consider sonicating the medium to expulse gas before experiment in that case. We have shown in the past that leukocyte dynamics are faster when close to 37 °C although qualitatively they behave the same [10]. Given these considerations, we generally work at room temperature.

10. Hydraulic resistance varies with a power 4 of the pipette tip diameter, so filling the tip of micropipette is much harder when it is 2 μm than 3 μm in diameter. When using tips of diameter smaller than 2 μm, be patient. Tips of 1 μm and less are in practice not possible to fill within a reasonable time and in that case using capillaries with filaments will be required.

11. If you see that passing to a positive pressure by raising the water reservoir does not stop aspiration, this is mostly due to a bubble present inside the micropipette holder or its tubing. If you can live with inaccurate aspiration (some experiments allow that), you can make sure the pipette aspirates using a very large height difference of the water reservoir. However, the best is usually to take out pipette, unplug it, eventually cut its back few millimeters with a diamond (a cheap synthetic diamond will be enough), and replace it, or even use a new micropipette.

12. Micropipette placement can be done using only a 100× oil objective, without the need to switch to an air objective of lower magnification. However, this asks for a bit of training. The key to find and center micropipettes is to close the condenser aperture diaphragm (#21 in Fig.1a), eventually to increase exposure time to about 50 ms to optimize contrast and to use the image on screen with histogram stretching so as to see the “shadow” of the pipette passing well above the focus plane (up to hundreds of microns) when moving it along the *x*-axis past the center of the objective (axis convention shown in Fig. 1a) with the pipette being well passed the center toward positive y values. This will ensure that the part of the

micropipette shaft that crosses the field of view when moving it along the *x*-axis is thick, hence more easily seen. Once the tip was found and centered in the field of view, the micropipette has to be lowered in order to find where the bottom of the Petri dish exactly is. When lowering the pipette, if it hits the bottom, it will start moving in the x-y plane because it is getting bent. One advantage of using pipettes whose tip is parallel to the bottom and that are relatively flexible is that pipettes will tolerate bending in contact with the bottom before breaking. Alternatively, if one prefers switching objectives, beware of bubbles that can form when switching back to the oil objective. Do not forget to reopen the condenser aperture diaphragm when ready for running the experiment, with proper Köhler illumination setting.

13. The microindenter is more valuable than the pipette so place the micropipette first, it will minimize the risk of breaking the microindenter while searching for the bottom of the dish.

14. Working about 30 μm above the dish is a trade-off between a good image quality (the better the closer from the bottom of the dish) and not having the image disturbed by cells present at the bottom of the dish. Make sure the microindenter shaft does not touch another cell outside of the field of view.

15. Some groups cut 1 mL pipette cones to minimize shear, but in our hands, injecting with unmodified 200 μL cone is not a problem.

16. An alternative to inject cells is the falling drop technique. Inertia will mix the droplet, but it will disperse and dilute cells. Cells might take a longer time to sediment, and you might need to add more cells using this technique.

17. The most important parameters to be set in a code are: data log (date, cell#, file path, list of parameters used); frequency for force modulation experiments; piezo velocity for approach/retraction phase; number of approach-retraction cycles; duration at contact once the maximal force threshold is reached (can be 0 to several minutes); indenter bending stiffness; detection threshold to roughly determine if cell-microindenter contact was made (~0.1 μm); maximal force (0.1–0.5 nN); average force and amplitude when using force modulation (typically $f = 0.2 \pm 0.03$ nN); exposure time; proportional coefficient for feedback loop; pixel size corresponding to the objective (in μm); dimension of the region of interest (w*h in pixels).

18. We avoid motorized x-y stages on the microscope. In our hands a manual x-y translation is more reactive (and cheaper) than a motorize one. An experienced user can have a surprising micron-size accuracy with his/her bare hands on the x-y

manual translation knobs when centering the stage on a chosen cell at the bottom of the dish.

19. The expected data throughput with the technique will be relatively low but will depend on the type of experiment. For a single indentation of a cell to measure its effective Young's modulus, one can go down to 2 min per cell. But experiments can extend up to 30 min per cell. Beyond this duration, drift of the micropipettes might become too large (especially if working above room temperature).

20. Several extensions can be added to the setup, for instance, a third micropipette used to flow a solution of interest on the cell being indented [3]. Fluorescence imaging can be added, and snapshots under fluorescence excitation can be acquired automatically in-between indentations by modifying the Matlab code.

21. You can try reusing flexible pipettes, but beware that upon drying they might get clogged and contaminated. We usually use one flexible pipette for the whole day. We usually reuse microindenters several times provided they do not get dirty due to adhered cells or cell debris.

References

1. Guck J (2019) Some thoughts on the future of cell mechanics. Biophys Rev 11:667–670
2. Guillou L, Sheybani R, Jensen AE, Di Carlo D, Caffery TS, Thomas CB, Shah AM, Tse HTK, O'Nea HR (2021) Development and validation of a cellular host response test as an early diagnostic for sepsis. PLoS One 16:1–17
3. Guillou L, Babataheri A, Puech P-H, Barakat AI, Husson J (2016) Dynamic monitoring of cell mechanical properties using profile microindentation. Sci Rep 6:21529
4. Sawicka A, Babataheri A, Dogniaux S, Barakat AI, Gonzalez-Rodriguez D, Hivroz C, Husson J (2017) Micropipette Force Probe to quantify single-cell force generation: application to T cell activation. Mol Biol Cell 28:mbc.E17-06-0385
5. Colbert M-JJ, Raegen AN, Fradin C, Dalnoki-Veress K (2009) Adhesion and membrane tension of single vesicles and living cells using a micropipette-based technique. Eur Phys J E Soft Matter 30:117–121
6. Colbert M-JJ, Brochard-Wyart F, Fradin C, Dalnoki-Veress K (2010) Squeezing and detachment of living cells. Biophys J 99: 3555–3562
7. Backholm M, Ryu WS, Dalnoki-Veress K (2013) Viscoelastic properties of the nematode Caenorhabditis elegans, a self-similar, shear-thinning worm. Proc Natl Acad Sci 110: 4528–4533
8. Backholm M, Bäumchen O (2019) Micropipette force sensors for in vivo force measurements on single cells and multicellular microorganisms. Nat Protoc 14:594–615
9. Francis GW, Fisher LR, Gamble RA, Gingell D (1987) Direct measurement of cell detachment force on single cells using a new electromechanical method. J Cell Sci 87(Pt 4):519–523
10. Zak A, Merino-Cortés SV, Sadoun A, Mustapha F, Babataheri A, Dogniaux S, Dupré-Crochet S, Hudik E, He H-T, Barakat AI, Carrasco YR, Hamon Y, Puech P-H, Hivroz C, Nüsse O, Husson J (2021) Rapid viscoelastic changes are a hallmark of early leukocyte activation. Biophys J 120:1692–1704
11. Guillou L, Babataheri A, Saitakis M, Bohineust A, Dogniaux S, Hivroz C, Barakat AI, Husson J (2016) T-lymphocyte passive deformation is controlled by unfolding of membrane surface reservoirs. Mol Biol Cell 27:3574–3582
12. Samassa F, Ferrari ML, Husson J, Mikhailova A, Porat Z, Sidaner F, Brunner K, Teo T, Frigimelica E, Tinevez J, Sansonetti PJ, Thoulouze M, Phalipon A (2020) Shigella

impairs human T lymphocyte responsiveness by hijacking actin cytoskeleton dynamics and T cell receptor vesicular trafficking. Cell Microbiol 22:1–15

13. Merino-Cortés SV, Gardeta SR, Roman-Garcia S, Martínez-Riaño A, Pineau J, Liebana R, Merida I, Dumenil A-ML, Pierobon P, Husson J, Alarcon B, Carrasco YR (2020) Diacylglycerol kinase ζ promotes actin cytoskeleton remodeling and mechanical forces at the B cell immune synapse. Sci Signal 13:eaaw8214
14. Zak A, Dupré-Crochet S, Hudik E, Babataheri A, Barakat AIAI, Nüsse O, Husson J (2022) Distinct timing of neutrophil spreading and stiffening during phagocytosis. Biophys J:1–14
15. Edelstein AD, Tsuchida NA, Pinkard H, Vale RD, Stuurman N (2014) Advanced methods of microscope control using μManager software. J Biol Methods 1:10
16. Lam J, Herant M, Dembo M, Heinrich V (2009) Baseline mechanical characterization of J774 macrophages. Biophys J 96:248–254
17. Shimamoto Y, Kapoor TM (2012) Microneedle-based analysis of the micromechanics of the metaphase spindle assembled in Xenopus laevis egg extracts. Nat Protoc 7:959–969
18. Desprat N, Guiroy A, Asnacios A (2006) Microplates-based rheometer for a single living cell. Rev Sci Instrum 77:055111
19. Lin DC, Dimitriadis EK, Horkay F (2007) Robust strategies for automated AFM force curve analysis – I. Non-adhesive indentation of soft, inhomogeneous materials. J Biomech Eng 129:430–440
20. Dimitriadis EK, Horkay F, Maresca J, Kachar B, Chadwick RS (2002) Determination of elastic moduli of thin layers of soft material using the atomic force microscope. Biophys J 82:2798–2810
21. Hertz H (1882) Ueber die Berührung fester elastischer Körper. J für die reine und Angew Math 92:156–171
22. Guevorkian K, Maître J-L (2017) Micropipette aspiration. In: Methods in cell biology. Elsevier, New York, pp 187–201
23. Bechhoefer J (2005) Feedback for physicists: a tutorial essay on control. Rev Mod Phys 77: 783–836

Chapter 2

In Situ Measurements of Cell Mechanical Properties Using Force Spectroscopy

Meital Reches and Tal Duanis-Assaf

Abstract

Mechanobiology focuses on how physical forces and the mechanical properties of cells and whole tissues affect their function. The mechanical properties of cells are of particular interest to developmental biology and stem cell differentiation, lymphocyte activation and phagocytic action in phagocytes, and development of malignant tumors and metastases. These properties can be measured on whole tissue and cell culture. Advances in instrument sensitivity and design, as well as improved techniques and scientific know-how achieved over the past few decades, allow researchers to study the mechanical properties of single cells and even at the subcellular level. Particularly, nanoindentation measurements using atomic force microscopy (AFM) mechanically probes single cells and even allows mapping of these traits. This chapter discusses these measurements from the experimental design to the analysis.

Key words Cell mechanics, Cell elasticity, Atomic force microscopy, Force spectroscopy, Nanoindentation, Single cell

1 Introduction

The mechanical properties of cells affect many biological processes, such as stem cell differentiation [1–3], immune response [4, 5] and tumor malignancy [6–9]. As such, studying cell contact mechanics is important for cell biology and understanding various biological processes. Moreover, the mechanical traits of living cells have implications in medicinal science, for the development of new therapeutics as well as diagnostic procedures [6, 10].

The arsenal of biomechanical research offers different methods. Among these methods are atomic force microscopy (AFM) [11], optical tweezers [12], optical stretcher [13], magnetic tweezers [14], micropipette force probe [15], biomembrane force probe [16], deformability cytometry [17], and cell monolayer rheology [18]. Some methods target single cells, while others target whole communities. Most of the methods mentioned above target

Ronen Zaidel-Bar (ed.), *Mechanobiology: Methods and Protocols*,
Methods in Molecular Biology, vol. 2600, https://doi.org/10.1007/978-1-0716-2851-5_2,

suspended cells. While this is convenient for planktonic cells, it complicates the measurement of adherent cells. Extra steps are required for suspending the cells, which changes the cell environment during the measurement, thus, may affect the results [19].

AFM allows for mechanical measurements to be performed in situ with adherent cell cultures, without changing the culture conditions significantly. As a consequence, along with its high versatility, AFM is one of the most common tools used for mechanobiological studies. This chapter offers entry point information for designing and performing nanoindentation experiments using AFM to study the mechanical properties of cells and discusses some of the considerations for such experiments.

2 Materials

Whenever performing AFM measurements, choosing a probe that fits the purpose of the measurement is always crucial. This is especially true for indentation measurements. When performing indentation on soft cells, soft cantilevers with spring constants typically below 1 $N{\cdot}m^{-1}$ are usually suited for this measurement. In our previous report by Brill-Karniely et al. [6], we used cantilevers with a spring constant of 0.2 ± 0.12 $N{\cdot}m^{-1}$ to determine the elastic modulus of human bladder cancer cells.

Choosing the correct spring constant is not sufficient for this type of measurement. The tip geometry also affects the resulting Young's modulus. Tips with smaller radii are more susceptible to local heterogeneity of the cell surface and often result in higher elastic moduli [20] Therefore, it is common to use colloidal probes, where a spherical bead is attached at the edge of the cantilever [21]. Nevertheless, when the local heterogeneity or a correlation to the topography is of interest, sharp conical or pyramidal probes with a tip radius of 1–10 nm are better suited [22].

3 Methods

3.1 Cell Culture Preparation

One of the most appealing traits of AFM is that the measurement is performed on the cell culture in situ, and no particular method of preparation is necessary. However, some properties of the sample are pivotal.

1. AFM is a scanning probe apparatus, which means it operates on surfaces; therefore, adherent cells are required. For measurements with nonadherent cells, it is required to first immobilize the cells on the substrate. One approach is to fabricate a porous surface with pore size similar to that of the cells. The pores then trap the cells, preventing them from sliding horizontally while

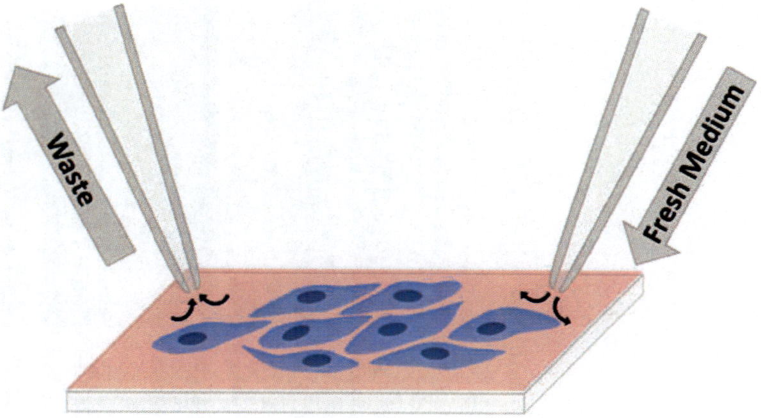

Fig. 1 Illustration of a simple technique for gently washing cell culture before measurement, using two pipettes

force is being applied [23]. Another approach is to functionalize the substrate with antigens that promote specific adhesion of the cells [5].

2. A transparent sample/substrate is required for optical probe alignment.
3. The substrate on which the cell culture is prepared should be considered.
   - Glass (glass slides or coverslips) is a good choice of rigid support for cell culture preparation.
   - Thick soft polymers and gel-like supports, such as polydopamine and agar may interfere with the measurement.
4. It is advised to wash the sample with fresh medium before the measurement.
   - Gently add fresh medium with a pipette from one side of the sample while removing excess fluids from the other side with a second pipette (Fig. 1).
   - Take caution when washing; avoid irritating the cells as much as possible.

3.2 Probe Calibration

Probe calibration is built into the software of most AFM manufacturers. While some procedures allow the calibration of the deflection sensitivity with no surface contact, in most cases, the procedure follows these general steps.

1. Place the probe above a clean hard substrate (glass slide or cover slip, *see* **Note 1**)
2. Align the LASER.
3. Approach the surface.
4. Acquire a single force curve.

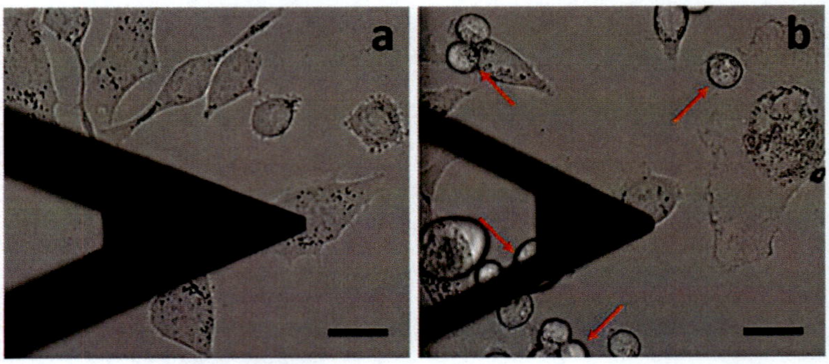

Fig. 2 Optical alignment of an AFM colloidal probe above human melanoma cells (**a**) and metastatic prostate cancer cells (**b**). (**a**) The probe is aligned above a cell in preparation for measurements. (**b**) Impossible alignment due to detached cells below the cantilever. The red arrows highlight cells that are detaching from the substrate and cannot be probed. The scale bar represents 25 μm. Unpublished results

5. The deflection sensitivity can be calculated from the slope of the force curve at contact with the surface.
6. Retract the cantilever from the surface to allow measurement of thermal noise (100 μm is sufficient).
7. Apply the thermal fluctuation method to acquire the cantilever spring constant [24, 25].

3.3 Optical Probe Alignment

1. Find a cell that is well adhered to the surface (Fig. 2).
2. Move the probe above the chosen cell.
3. Ensure no floating or detaching cells or debris interfering with cantilever movement (Fig. 2b).
4. Place the probe above the area of interest.
5. Make sure the tip or bead (depending on the type of probe used) is in the desired position (*see* Fig. 2a).

3.4 Topographical Probe Alignment

When accurate submicron-scale localization of spectra is necessary, such as studying local features of the cell surface or cytoplasm, it is advised to first acquire an image of the cell or an area of interest. Then, force spectra can be taken at exact positions on the cell surface. This approach is also useful when using one AFM tip for surface characterization and a second probe for mechanical measurements, thus preventing probe contamination before the measurement [26].

1. Optically align the probe above the cell or area of interest.
2. Take a quick (low resolution) AFM image to outline the cell surface (*see* **Note 2**). The high lateral forces, applied during contact-mode AFM, may result in artifacts and disturbances to the cell membrane. Therefore, tapping mode (also called AC or

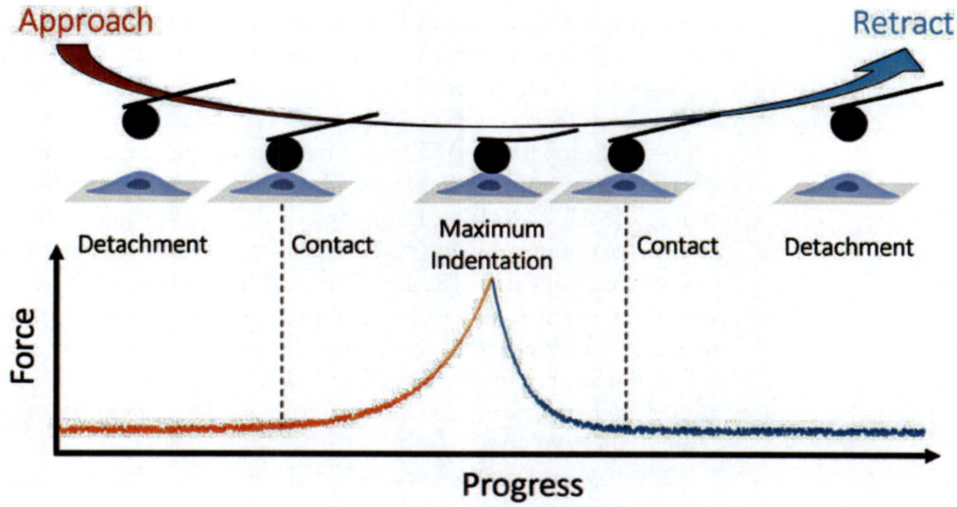

Fig. 3 Scheme of an indentation experiment cycle and corresponding schematic force curve

intermittent contact mode) or force-spectroscopy-based imaging (*see* **Note 3**) is commonly used.

3. Position the probe on the exact area of interest for further measurement.

3.5 Nanoindentation Measurements

Mechanobiological force spectroscopy experiments comprise ramp cycles: First, the probe is approached to the sample until indentation of the cell surface can occur. Indentation stops when a predefined repulsive force setpoint is achieved. Next, the probe is retracted, until separation with a predefined ramp size (distance along the z-axis). Each of these cycles produces a force curve (Fig. 3). This measurement is typically repeated multiple times to obtain sufficient data for statistical analysis.

3.5.1 Experimental Parameters Optimization

The force setpoint, ramp size, and scanner speed can affect the quality of the data, as well as the results. Therefore, it is essential to optimize the experimental parameters at the beginning of the study.

The force setpoint defines the amount of indentation achieved by the probe. This is particularly important as different indentation depths affect the measured features of the cell. Shallower indentation probes the membrane and cytoplasm, while higher deformation can be affected by organelles and the underlying substrate. Typical force setpoints vary between 100s pN [5, 20] and approximately 10 nN [20, 27].

The ramp size and scanner speed determine the rate at which force curves are obtained. There is a correlation between the indentation frequency and the derived elastic modulus [28, 29]. While maximizing the rate of data acquisition is desirable, a similar probing rate should be used to compare the results between experiments, as well as to other reports. However, it is often necessary to increase the ramp size, to ensure complete detachment between subsequent force curves. This is especially true when the cells adhere strongly to the AFM tip. The scanner speed can be increased to compensate, although at very high velocities, this could result in artifacts [30]. A probing frequency of 0.5–2 Hz (indentations per second) is common [6, 20, 27, 31], but 100-fold higher frequencies are also applicable [5, 22] with short ramp sizes. The ramp size should be as small as possible but long enough to ensure complete separation between the probe and the cell (*see* **Note 4**); values of 0.5–5 μm are common.[6, 9, 20] Lastly, the probe can be paused, after detachment from the cell, between subsequent ramp cycles to reduce cell perturbation.

These values can be used as initial parameters for optimization:

- Force setpoint – 1 nN
- Ramp size (Z-length) – 1 μm
- Scanner speed – 5 $\mu m \cdot s^{-1}$
- Data sampling rate (depending on the scanner speed) – 2048 Hz
- $$\text{points per curve segment} = \frac{\text{sampling rate} \cdot \text{ramp size}}{\text{scanner speed}}$$
- Pause after retract – no pause

3.5.2 Acquisition of Force Spectra

1. Approach the substrate.
2. Retract 100 μm.
3. Calibrate the deflection sensitivity and cantilever spring constant.
4. Align the probe above the cell or area of interest using optical or topographic alignment (*see* **Notes 5** and **6**).
5. Set the experimental parameters.
   - Force setpoint
   - Ramp size (Z-length)
   - Scanner speed (*see* **Note 7**)
   - Data sampling rate
   - Pause after retract
   - Pause on contact (may be useful for adhesion measurements)
6. Set up a measurement area (0.5×0.5–$5 \times 5\ \mu m^2$) with 25–50 individual positions for indentation.
7. Approach the cantilever on top of the cell.

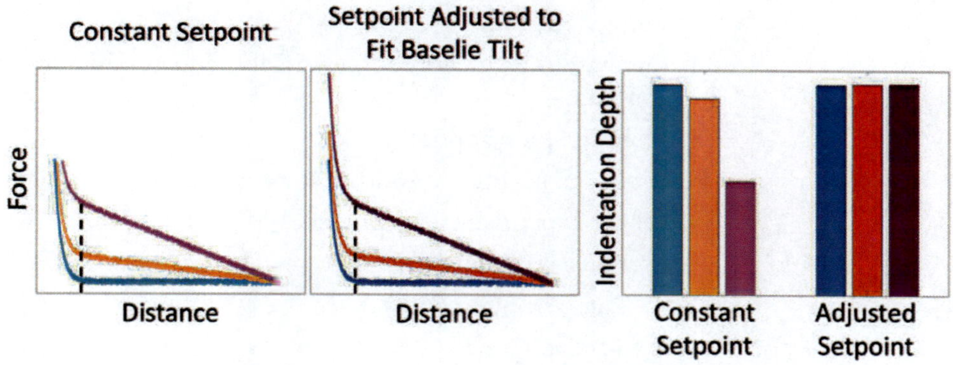

Fig. 4 Effect of tilted baseline on setpoint selection. The left and middle panels show schematic spectra with increasing baseline tilt, with a constant setpoint (left) and setpoint adjusted to fit the baseline tilt, such that the amount of indentation remains similar (middle). The dashed black lines represent the contact point. The right panel shows the indentation depth calculated at the point of maximal force load

8. Run measurement.
9. Monitor the spectra for artifacts.
10. Follow the cell optically to ensure that no cell movement occurs during the course of the measurement and that its viability is not compromised.
11. Repeat stages **3–10** with several different cells.

3.5.3 Handling Tilted Baseline

One of the most common artifacts in force spectroscopy is a tilted baseline. This artifact is easily overcome during analysis, by subtracting a linear function fitted to the tail of the force curve. However, a highly tilted baseline can affect the force exerted on the cell, and as a result the amount of indentation. This is particularly problematic in nanoindentation assays and single-cell force spectroscopy, where the ramp cycle is typically long, and therefore, higher tilt is achieved. Figure 4 shows mock spectra simulated with varying tilt angles. With a large tilt, the indentation depth is notably reduced, which in turn may affect the results. When the baseline is significantly tilted, it is recommended to adjust the setpoint such that the force gap between the contact point and setpoint remains similar (Fig. 4 middle panel).

3.6 Force-Spectra-Based Imaging

Force-spectroscopy-based imaging (*see* **Note 3**) is an excellent method for correlating mechanical properties with position on the cell surface as well as cell topography and adhesive tendencies [26, 32].

3.6.1 Mechanical Mapping of Cell

1. Mount either a sharp probe (conical or pyramidal) or a colloidal probe with a small bead (diameter ~1 μm).
2. Approach the substrate.

3. Retract 100 μm.
4. Calibrate the deflection sensitivity and cantilever spring constant.
5. Align the probe optically.
6. Set image height and width to capture the entire cell.
7. Set the experimental parameters:
   - Force setpoint
   - Ramp size (z-length)
   - Scanner speed
   - Data sampling rate
   - Scanner pause times during the ramp cycle, at contact with the cell, and/or after retracting the probe. Caution, this may cause image acquisition to be extremely slow!
8. Set image resolution (number of pixels) to fit features of interest (*see* **Note 8**).
9. Acquire an image.
10. Monitor the image and spectra for artifacts.
11. Follow the cell optically to ensure no cell movement occurs during the course of the measurement, and that its viability is not compromised.

3.6.2 High-Resolution Nanoindentation Imaging of Specific Features

1. Mount a sharp conical or pyramidal AFM probe.
2. Approach the substrate.
3. Retract 100 μm.
4. Calibrate the deflection sensitivity and cantilever spring constant.
5. Align the probe using a topographical image.
6. Select image location and size according to located features from the topographical image (*see* **Notes 5** and **6**).
7. Set the experimental parameters:
   - Force setpoint
   - Ramp size (z-length)
   - Scanner speed
   - Data sampling rate
   - Scanner pause times during the ramp cycle, at contact with the cell, and/or after retracting the probe. Caution: this may cause image acquisition to be extremely slow!
   - Image resolution (*see* **Note 8**)
8. Acquire an image.
9. Monitor the image and spectra for artifacts.

10. Follow the cell optically to ensure that no cell movement occurs during the course of the measurement and that its viability is not compromised.

3.7 Duration of Experimental Sessions

The duration of any experiment performed on live cells depends on cell viability. The cells should be monitored throughout the experiment, and signs of stress or compromised viability should be considered. Furthermore, the mechanical properties of the cells may change due to internal or environmental cues. This is especially true if the experimental conditions are not favorable for the culture (e.g., measurements in minimal media or buffer). Altogether, the experimental conditions alongside long measurements may further increase the variability of the results, on top of the high variability inherent to biological assays and samples. Therefore, it is advised to use fresh cell cultures and keep experimental sessions as short as possible, keeping in mind the need for good statistics (i.e., enough force spectra).

3.8 Data Analysis

Most commercial AFM manufacturers offer data analysis software capable of analyzing nanoindentation data. Moreover, there are several open-source projects, developed by the scientific community [33–36] devoted to this task. Generally, the analysis comprises several steps, baseline subtraction, contact point alignment, conversion of the force vs. distance curve to force vs. indentation curve, and modeling.

3.8.1 Curve Smoothing

Smoothing the data is often useful in eliminating noise; however, it may also disrupt the analyzed signal. Nevertheless, smoothing noisy data is common practice and can benefit the analysis. Moreover, some methods rely on smoothed data. Different smoothing algorithms can be applied, such as the Savitzky–Golay filter [37] and local regression (LOWESS) [38], as well as simpler methods such as the moving average.

This method relies on averaging out noise by calculating the mean of neighboring data. The number of neighboring data used to average determines the smoothing "strength." The more points used to average, the smoother the signal will be, but also the more destructive to the signal. A simplified method of this calculation can be performed as follows:

1. Define the averaging window span (S) as number of neighboring points on each side of each point to average.
2. Apply the following calculation for each point (in position i) in a force-curve force array of N elements:
   2.1. Obtain the unsmoothed force values of S points on each side of the point in position i. Including the point i itself, obtain an array of $2S + 1$ points.

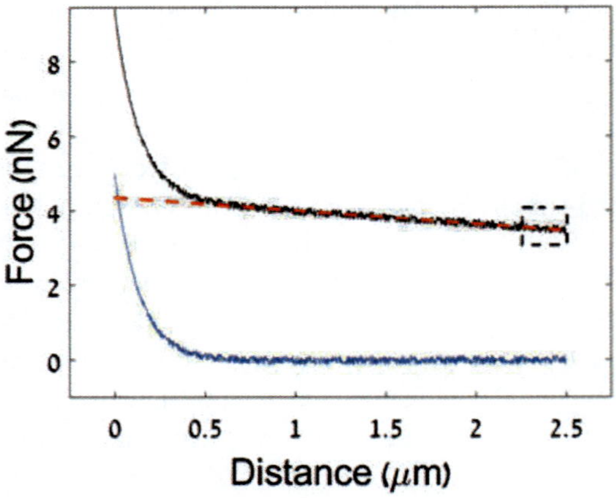

Fig. 5 Baseline subtraction via force curve tail analysis. The black plot represents raw data simulated with a moderately tilted baseline. The red dashed line is the baseline predicted by linear fit performed on the tail of the curve (last 10% of curve, highlighted by the window marked as dashed black line). The blue line is the curve after subtracting predicted baseline

2.2. Calculate the arithmetic mean of the obtained force array:

$$F_{i\text{smooth}} = \frac{\sum_{k=i-S}^{i+S} F_k}{2S+1}$$

2.3. Set the smoothed force value ($F_{i\text{smooth}}$) into the ith position of the smoothed force array.

3.8.2 Baseline Detection by Force-Curve Tail Analysis

At long distances, there should be no interactions between the AFM tip and the sample. We, therefore, expect the tail of the curve to be affected only by noise, or in other words be straight. In this case, a simple linear fit is the most common method to evaluate the baseline whether it is flat or tilted (Fig. 5).

For each curve, perform the following steps separately:

1. Obtain the force and distance data from the farthest part of the force curve (the last 10–20% of the curve tail is usually sufficient).
2. Perform a linear fit to this data array.
3. The amplitude of noise can be evaluated from the standard deviation or mean of the residuals.
4. Subtract the fitted linear function from the entire curve.

3.8.3 Baseline Detection by Force Distribution Analysis

In some cases, the tail of the curve features nonlinear baseline and is difficult to overcome. This is more common with long ramp cycles often used for nanoindentation assays. When this happens, a simple window for detecting a linear region is sometimes difficult to

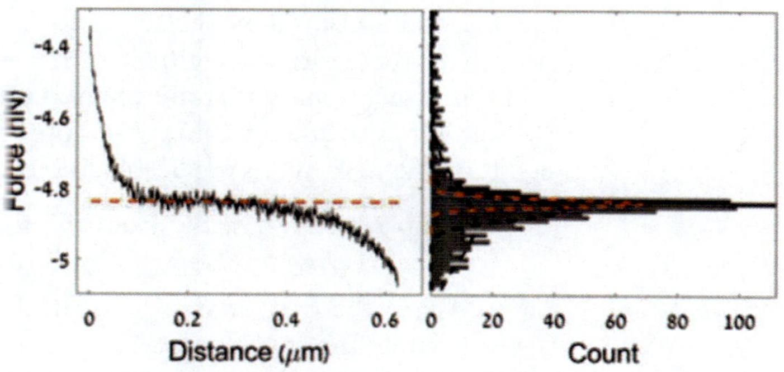

Fig. 6 Baseline analysis via force distribution analysis. The left plot shows a force vs distance curve featuring a nonlinear baseline. The black line represents raw data, and the dashed red line represents baseline prediction by force-distribution analysis. The right plot shows the force histogram and corresponding fitted Gaussian series as dashed lines. The Gaussian corresponding to predicted baseline is highlighted in red

define. A more robust method of evaluating the baseline relies on the distribution of the force signal (Fig. 6) [39–41]. Note that this method works best when the baseline features minimal linear tilt. Baseline detection using this method is performed as described below.

1. Generate a histogram of the force signal from the curve segment being analyzed (*see* **Note 9**).
2. If adhesive interactions are present, go to **step 4**; otherwise, go to **step 3**.
3. Locate the highest peak in the histogram. The baseline shift is determined by the peak position, and the mean amplitude of the noise is evaluated from the FWHM. Skip to **step 10**.
4. If adhesive interactions are present, fit a Gaussian series to the force histogram.
5. Each Gaussian typically represents a plateau in the force curve.
6. Ignore Gaussians with amplitudes lower than 20% of the highest Gaussian amplitude.
7. Choose the Gaussian with the highest mean force value.
8. The baseline shift is determined by the Gaussian mean value.
9. The mean amplitude of noise is evaluated from the Gaussian standard deviation.
10. Subtract the baseline shift from the curve.

3.8.4 Contact Point Alignment by Point of First Repulsive Force

Correct alignment of the contact point is crucial because it determines the amount of cell surface indentation for further analysis.

The simplest method of determining the contact point is the point where repulsive forces (positive force) first appear. This method is also often well suited for nanoindentation experiments, where the approach segment of the force curve is analyzed. It is worth mentioning that this method is not well suited if the cantilever jumps to contact during the approach. Still, this method is easy to implement and is applicable in most situations.

1. Subtract the baseline from the force data of the entire curve (see baseline detection and subtraction).
2. Find the first point where the force is higher than half the amplitude of the noise.
   - The first point I.E when following the curve from long distances towards short distances.
   - If the signal-to-noise ratio is poor, consider using a lower threshold than half noise amplitude.
3. Align the curve by subtracting the distance at contact from the curve distance data.

3.8.5 Contact Point Alignment by Smoothed Slope Method

Another approach to finding the contact point utilizes the slope of the curve to more accurately evaluate the change in trend toward repulsive forces, which is expected to occur during contact (Fig. 7). This method is relatively simple and is more robust as it monitors the slope rather than the absolute force values [34]. This method is capable of correctly positioning the contact point when "jump to contact" is present; however, it is more complicated to implement on the retract segment where adhesive interactions may be present.

The following steps should be taken:

1. Subtract the baseline from the force data of the entire curve (see baseline detection and subtraction).
2. Smooth the curve (see curve smoothing).

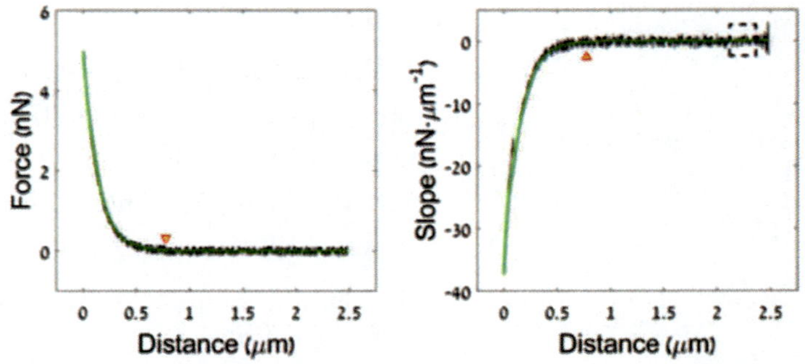

Fig. 7 Contact point identification using the smoothed slope method. The left panel shows the force vs. distance curve (black line), overlayed with the smoothed curve (green line). The right panel shows the slope of the force vs. distance curve (black line) and the smoothed slope (green line). The dashed black line displays the window used for evaluating slope noise. The identified contact point is marked by a red triangle in both plots

3. Calculate the slope of the curve at each point using simple two-point numeric differentiation:

$$\frac{dF_i}{dZ_i} = \frac{F_{i+1} - F_i}{Z_{i+1} - Z_i}$$

4. Evaluate the noise level of the slope by calculating the standard deviation of the slope in the flat region. Omit the end of the tail due to greater oscillations (Fig. 7).
5. Smooth the slope curve (see curve smoothing).
6. The point of contact is the position where the smoothed slope is lower than the noise amplitude of the slope ($\text{slope}_{\text{smoothed}} \geq -0.5 \cdot \text{noise}_{\text{slope}}$).
7. Align the curve by subtracting the distance at contact from the curve distance data.

3.8.6 Calculate Indentation Depth

The data acquired by the AFM in a force spectroscopy experiment comprises two main parameters, the amount of cantilever deflection (force) and the piezo height (distance). Importantly, due to the cantilever deflection, the measured piezo height does not correspond to the exact indentation depth (Fig. 8). Therefore, it is essential to calculate the exact depth of indentation before any modeling can be performed. The indentation depth can be calculated by subtracting the cantilever deflection (Fig. 8a) from the aligned piezo height as follows [22, 42]:

$$\delta = -(Z - Z_0) - d = -(Z - Z_0) - \frac{F}{k_c}$$

where δ is the cell indentation depth, Z is the piezo height, Z_0 is the piezo height at contact, and d is the vertical cantilever deflection. If only the force data is available, the cantilever deflection can be calculated using the spring constant according to the above relation, where k_c is the cantilever spring constant and F is the force corresponding to the vertical cantilever deflection, after correction of baseline shift. To obtain the actual depth of indentation into the cell, the deflection is subtracted from the inverse values of the piezo height. This ensures that the calculated indentation depth is a positive value.

3.8.7 Fitting Contact Model

The last stage of contact mechanics analysis is curve modeling. The Hertz model [43], solved for a spherical indenter, and deriving models for different indenter shapes [44, 45] are among the most common contact mechanics models used to describe cell elasticity. These models do not address adhesion and are mostly applicable when interactions between the tip and the cell are much smaller than the maximal applied load [42]. Hence, these models mostly best fit the approach segment of the curve. The Hertz model assumes that the sample is homogeneous, isotropic, and purely

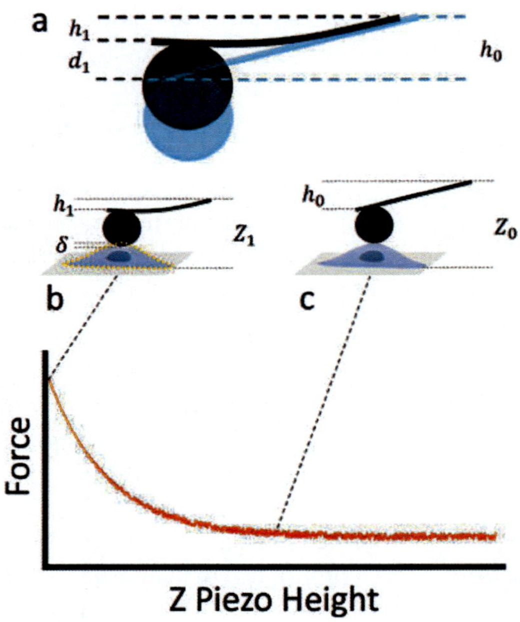

Fig. 8 Schematic representation of how the cantilever deflection affects the calculated indentation depth. (**a**) The cantilever at maximum load (black) and at resting point (light blue). The cantilever height resulting from its angle at maximal load and resting point is represented by h_1 and h_0, respectively, and the cantilever deflection is represented by d. (**b**-**c**) Scheme of cantilever position at maximal load (**b**) and contact point (**c**) corresponding to positions in the schematic force curve. The piezo height is represented by Z. (**b**) The outline of the undisturbed cell is highlighted by the yellow dashed line, and the indentation depth is represented by δ

elastic. These properties are seldom fully satisfied by living cells [46]; however, these models are still routinely used with satisfactory results.

Adhesive samples that have strong interactions with the tip should not be fitted with Hertzian models, which assume no adhesion. Instead, the DMT [47] or JKR [48] models which consider adhesive contact can be used. With these models, the analysis is performed on the retract segment of the force curve, which further complicated the analysis if strong interactions are present.

Other models were developed to account for different features, such as the presence of a polymer brush on the cell surface [27, 31], and sensing of the rigid supporting substrate [49–51] (i.e., for thin regions of the cell or very high indentation depths). Unfortunately, no model combines all of these features. Moreover, none of the existing models fully capture the complexity of biological samples. Therefore, choosing a mechanical model often involves compromises and assumptions.

Fitting the Hertz-Sneddon model can be procedurally performed as follows:

1. Subtract the baseline from the force data of the entire curve (see baseline detection and subtraction).
2. Align contact point (see contact point alignment).
3. Convert measured piezo height to indentation depth.
4. Fit the contact domain of the force vs. indentation curve to the physical model force equation:

   Hertz model (spherical indenter) [52]: $F = \frac{4E}{3(1-\nu^2)}\sqrt{r\delta^3}$

   Sneddon model (conical indenter) [53]: $F = \frac{2E}{\pi(1-\nu^2)}\tan(\alpha)\delta^2$

   where E is the elastic (Young's) modulus, ν is the Poisson's ratio (typically assumed to be 0.5, corresponding to incompressible liquid), δ is the indentation depth, r is the colloidal probe bead radius, and α is the conical probes half cone angle (often 15°).
5. Calculate the mean elastic modulus using a simple average, or if enough data is available, plot a histogram and extract the most prevalent value by fitting a distribution probability function.

4 Notes

1. Contact-based calibration of the deflection sensitivity should be performed on a rigid substrate, such as a glass slide or coverslip. On soft substrates, the deformation/indentation of the surface results in incorrect calculation of deflection sensitivity.
2. Topographical probe alignment requires AFM images of each cell to be taken prior to measurement, which may be time-consuming. The mechanical properties of the cells may change over the course of the experiment due to internal or environmental cues. Thus, time is of the essence when performing these experiments. It is therefore best to reduce acquisition times as much as possible. One method for faster imaging is to reduce the resolution. When coarse alignment is sufficient, image resolution as low as 64 × 64 can be used for successful probe positioning.
3. Force-spectroscopy-based imaging is an imaging method where each pixel corresponds to a single force curve. The topography is then determined by the scanner height at a given force amplitude or at the zero-force point derived from the force curve, which is equivalent to the point of contact between the tip and sample. This method is available in most commercial AFMs under different names, for instance,

Quantitative Imaging (QI™) in JPK NanoWizard, Fast Force Mapping (FFM) in Asylum Research Cypher, FastForce Volume, and PeakForce Tapping in Bruker AFMs.

4. The ramp size of an indentation experiment should be as small as possible. Beside maximizing the experiment throughput, it may result in better spectra. Longer z-lengths may increase thermal drift, as well as noise and the risk of other artifacts, complicating the data analysis. That said, it is also crucial to take long enough spectra to ensure complete separation between the probe and the cell. Otherwise, lateral forces and continuous contact can result in unreliable spectra and even damage to the sample or AFM tip.

5. Special caution should be taken when performing measurements close to the cell edge, where it is thinner. Measurements performed on thin areas of the cell are affected by the mechanical properties of the substrate. This should be considered when analyzing the data.

6. The cell nucleus often has different mechanical properties compared to the surrounding cytoplasm [54, 55]. Therefore, the measurement area should be carefully chosen to answer the research question. When the nucleus is difficult to make out (Fig. 2a), specific nucleus stains, such as Hoechst, 4′,6-diamidino-2-phenylindole (DAPI) [56, 57] and SiR–Hoechst [58], can be used to improve the contrast.

7. Force spectroscopy scanner speed can be set high while maintaining good spectra quality. However, higher velocities increase the hydrodynamic drag on the cantilever, which may cause artifacts in the measured force. The hydrodynamic drag can be estimated and subtracted from the spectra during analysis [59].

8. The resolution of cell mechanical property images should be determined with caution. When using sharp tips (1–10 nm tip radii), the resolution can be set very high. However, when using blunted cones or colloidal probes, the area of contact between the tip and cell can be very large. Therefore, when using these probes, increasing the resolution may be misleading, as measurements are simply repeated multiple times in almost identical positions. Another factor that should be considered is the indentation depth, which also increases the contact area.

9. The force distribution method for baseline analysis relies on the correct representation of the force distribution by the generated histogram. If the generated histogram is overly smoothed, the baseline shift prediction may be inaccurate, as well as the noise amplitude. Common histogram generation rules such as Sturges and Freedman–Diaconis often result in histograms that

are too smooth for this analysis. It is usually better to generate a histogram according to the square root rule, where the number of bins in the histogram is equal to the square root of the number of data samples in the analyzed segment of the force curve. $N_{\text{Bins}} = \sqrt[2]{N_{\text{Data Points}}}$

References

1. Gossett DR, Tse HTK, Lee SA et al (2012) Hydrodynamic stretching of single cells for large population mechanical phenotyping. Proc Natl Acad Sci U S A 109(20): 7630–7635. https://doi.org/10.1073/pnas.1200107109
2. Pajerowski JD, Dahl KN, Zhong FL et al (2007) Physical plasticity of the nucleus in stem cell differentiation. Proc Natl Acad Sci U S A 104(40):15619–15624. https://doi.org/10.1073/pnas.0702576104
3. Titushkin I, Cho M (2007) Modulation of cellular mechanics during osteogenic differentiation of human mesenchymal stem cells. Biophys J 93(10):3693–3702. https://doi.org/10.1529/biophysj.107.107797
4. Patel NR, Bole M, Chen C et al (2012) Cell elasticity determines macrophage function. PLoS One 9(7):e41024. https://doi.org/10.1371/journal.pone.0041024
5. Razvag Y, Neve-Oz Y, Sajman J et al (2018) Nanoscale kinetic segregation of TCR and CD45 in engaged microvilli facilitates early T cell activation. Nat Commun 9(1):1–17
6. Brill-Karniely Y, Dror D, Duanis-Assaf T et al (2020) Triangular correlation (TrC) between cancer aggressiveness, cell uptake capability, and cell deformability. Sci Adv 6(3):eaax2861
7. Antmen E, Demirci U, Hasirci V (2021) Micropatterned surfaces expose the coupling between actin cytoskeleton-lamin/nesprin and nuclear deformability of breast cancer cells with different malignancies. Adv Biol 5(1):2000048
8. Chen J, Zhou W, Jia Q et al (2016) Efficient extravasation of tumor-repopulating cells depends on cell deformability. Sci Rep 6(1): 19304. https://doi.org/10.1038/srep19304
9. Mandal K, Pogoda K, Nandi S et al (2019) Role of a kinesin motor in cancer cell mechanics. Nano Lett 19(11):7691–7702
10. Nematbakhsh Y, Lim CT (2015) Cell biomechanics and its applications in human disease diagnosis. Acta Mech Sinica 31(2):268–273. https://doi.org/10.1007/s10409-015-0412-y
11. Skamrahl M, Colin-York H, Barbieri L, Fritzsche M (2019) Simultaneous quantification of the interplay between molecular turnover and cell mechanics by AFM–FRAP. Small 15(40):1902202
12. Titushkin I, Cho M (2006) Distinct membrane mechanical properties of human mesenchymal stem cells determined using laser optical tweezers. Biophys J 90(7):2582–2591. https://doi.org/10.1529/biophysj.105.073775
13. Guck J, Ananthakrishnan R, Mahmood H et al (2001) The optical stretcher: a novel laser tool to micromanipulate cells. Biophys J 81(2): 767–784. https://doi.org/10.1016/S0006-3495(01)75740-2
14. Marjoram RJ, Guilluy C, Burridge K (2016) Using magnets and magnetic beads to dissect signaling pathways activated by mechanical tension applied to cells. Methods 94:19–26. https://doi.org/10.1016/j.ymeth.2015.09.025
15. Sawicka A, Babataheri A, Dogniaux S et al (2017) Micropipette force probe to quantify single-cell force generation: application to T-cell activation. Mol Biol Cell 28(23): 3229–3239
16. Husson J, Chemin K, Bohineust A et al (2011) Force generation upon T cell receptor engagement. PLoS ONE 6(5):e19680. https://doi.org/10.1371/journal.pone.0019680
17. Urbanska M, Muñoz HE, Shaw Bagnall J et al (2020) A comparison of microfluidic methods for high-throughput cell deformability measurements. Nat Methods 17(6):587–593. https://doi.org/10.1038/s41592-020-0818-8
18. Fernández P, Heymann L, Ott A et al (2007) Shear rheology of a cell monolayer. New J Phys 9(11):419–419. https://doi.org/10.1088/1367-2630/9/11/419
19. Saha P, Duanis-Assaf T, Reches M (2020) Fundamentals and applications of FluidFM technology in single-cell studies. Adv Mater Interfaces 7(23):2001115
20. Wu P-H, Aroush DR-B, Asnacios A et al (2018) A comparison of methods to assess cell mechanical properties. Nat Methods 15:491–498

21. Septiadi D, Crippa F, Moore TL et al (2018) Nanoparticle–cell interaction: a cell mechanics perspective. Adv Mater 30(19):1704463. https://doi.org/10.1002/adma.201704463
22. Razvag Y, Neve-Oz Y, Sherman E, Reches M (2018) Nanoscale topography–rigidity correlation at the surface of T cells. ACS Nano 13(1): 346–356
23. Guillaume-Gentil O, Gäbelein CG, Schmieder S (2022) Injection into and extraction from single fungal cells. Commun Biol 5(1):180. https://doi.org/10.1038/s42003-022-03127-z
24. Hutter JL, Bechhoefer J (1993) Calibration of atomic-force microscope tips. Rev Sci Instrum 64(7):1868–1873
25. Butt HJ, Jaschke M (1995) Calculation of thermal noise in atomic force microscopy. Nanotechnology 6(1):1–7. https://doi.org/10.1088/0957-4484/6/1/001
26. Dufrêne YF, Martínez-Martín D, Medalsy I et al (2013) Multiparametric imaging of biological systems by force-distance curve–based AFM. Nat Methods 10(9):847–854. https://doi.org/10.1038/nmeth.2602
27. Iyer S, Gaikwad R, Subba-Rao V et al (2009) Atomic force microscopy detects differences in the surface brush of normal and cancerous cells. Nat Nanotechnol 4(6):389–393
28. Alcaraz J, Buscemi L, Grabulosa M et al (2003) Microrheology of human lung epithelial cells measured by atomic force microscopy. Biophys J 84(3):2071–2079
29. Rigato A, Miyagi A, Scheuring S, Rico F (2017) High-frequency microrheology reveals cytoskeleton dynamics in living cells. Nat Phys 13(8):771–775. https://doi.org/10.1038/nphys4104
30. Gavara N (2017) A beginner's guide to atomic force microscopy probing for cell mechanics. Microsc Res Tech 80(1):75–84
31. Guz N, Dokukin M, Kalaparthi V, Sokolov I (2014) If cell mechanics can be described by elastic modulus: study of different models and probes used in indentation experiments. Biophys J 107(3):564–575
32. Dufrêne YF, Ando T, Garcia R et al (2017) Imaging modes of atomic force microscopy for application in molecular and cell biology. Nat Nanotechnol 12(4):295–307. https://doi.org/10.1038/nnano.2017.45
33. Roduit C, Saha B, Alonso-Sarduy L et al (2012) OpenFovea: open-source AFM data processing software. Nat Methods 9(8):774
34. Dinarelli S, Girasole M, Longo G (2018) FC_analysis: a tool for investigating atomic force microscopy maps of force curves. BMC Bioinf 19(1):258. https://doi.org/10.1186/s12859-018-2265-4
35. Partola KR, Lykotrafitis G (2016) FRAME (force review automation environment): MATLAB-based AFM data processor. J Biomech 49(7):1221–1224
36. Hermanowicz P, Sarna M, Burda K, Gabryś H (2014) AtomicJ: an open source software for analysis of force curves. Rev Sci Instrum 85(6): 063703
37. Savitzky A, Golay MJE (1964) Smoothing and differentiation of data by simplified least squares procedures. Anal Chem 36(8): 1627–1639. https://doi.org/10.1021/ac60214a047
38. Cleveland WS (1979) Robust locally weighted regression and smoothing scatterplots. J Am Stat Assoc 74(368):829–836. https://doi.org/10.1080/01621459.1979.10481038
39. Duanis-Assaf T, Razvag Y, Reches M (2019) ForSDAT: an automated platform for analyzing force spectroscopy measurements. Anal Methods. https://doi.org/10.1039/C9AY01150A
40. Hodgkinson T, Tsimbouri PM, Llopis-Hernandez V et al (2021) The use of nanovibration to discover specific and potent bioactive metabolites that stimulate osteogenic differentiation in mesenchymal stem cells. Sci Adv 7(9):eabb7921. https://doi.org/10.1126/sciadv.abb7921
41. Duanis-Assaf T, Hu T, Lavie M et al (2022) Understanding the adhesion mechanism of hydroxyapatite-binding peptide. Langmuir 38(3):968–978. https://doi.org/10.1021/acs.langmuir.1c02293
42. Li M, Dang D, Liu L et al (2017) Atomic force microscopy in characterizing cell mechanics for biomedical applications: a review. IEEE Trans Nanobiosci 16(6):523–540. https://doi.org/10.1109/TNB.2017.2714462
43. Hertz H (1881) Ueber die Berührung fester elastischer Körper (On the contact of elastic solids). J fur Reine Angew Math 92:156–171
44. Sneddon IN (1965) The relation between load and penetration in the axisymmetric boussinesq problem for a punch of arbitrary profile. Int J Eng Sci 3(1):47–57. https://doi.org/10.1016/0020-7225(65)90019-4
45. Bilodeau G (1992) Regular pyramid punch problem. J Appl Mech 59(3):519–524. https://doi.org/10.1115/1.2893754
46. Guimarães CF, Gasperini L, Marques AP, Reis RL (2020) The stiffness of living tissues and its implications for tissue engineering. Nat Rev Mater 5(5):351–370. https://doi.org/10.1038/s41578-019-0169-1

47. Derjaguin BV, Muller VM, Toporov YP (1975) Effect of contact deformations on the adhesion of particles. J Colloid Interface Sci 53(2): 314–326
48. Johnson KL, Kendall K, Roberts A (1971) Surface energy and the contact of elastic solids. Proc Math Phys Eng Sci 324(1558):301–313
49. Dimitriadis EK, Horkay F, Maresca J et al (2002) Determination of elastic moduli of thin layers of soft material using the atomic force microscope. Biophys J 82(5): 2798–2810. https://doi.org/10.1016/S0006-3495(02)75620-8
50. Gavara N, Chadwick RS (2012) Determination of the elastic moduli of thin samples and adherent cells using conical atomic force microscope tips. Nat Nanotechnol 7(11):733–736. https://doi.org/10.1038/nnano.2012.163
51. Garcia PD, Garcia R (2018) Determination of the elastic moduli of a single cell cultured on a rigid support by force microscopy. Biophys J 114(12):2923–2932. https://doi.org/10.1016/j.bpj.2018.05.012
52. Kirmizis D, Logothetidis S (2010) Atomic force microscopy probing in the measurement of cell mechanics. Int J Nanomedicine 5:137
53. Chang Y-R, Raghunathan VK, Garland SP et al (2014) Automated AFM force curve analysis for determining elastic modulus of biomaterials and biological samples. J Mech Behav Biomed Mater 37:209–218. https://doi.org/10.1016/j.jmbbm.2014.05.027
54. Haga H, Sasaki S, Kawabata K et al (2000) Elasticity mapping of living fibroblasts by AFM and immunofluorescence observation of the cytoskeleton. Ultramicroscopy 82(1-4): 253–258
55. Krieg M, Fläschner G, Alsteens D et al (2019) Atomic force microscopy-based mechanobiology. Nat Rev Phys 1(1):41–57. https://doi.org/10.1038/s42254-018-0001-7
56. Zink D, Sadoni N, Stelzer E (2003) Visualizing chromatin and chromosomes in living cells. Methods 29(1):42–50. https://doi.org/10.1016/S1046-2023(02)00289-X
57. Jež M, Bas T, Veber M et al (2013) The hazards of DAPI photoconversion: effects of dye, mounting media and fixative, and how to minimize the problem. Histochem Cell Biol 139(1):195–204. https://doi.org/10.1007/s00418-012-1039-8
58. Lukinavičius G, Blaukopf C, Pershagen E et al (2015) SiR–Hoechst is a far-red DNA stain for live-cell nanoscopy. Nat Commun 6(1):8497. https://doi.org/10.1038/ncomms9497
59. Alcaraz J, Buscemi L, Puig-de-Morales M et al (2002) Correction of microrheological measurements of soft samples with atomic force microscopy for the hydrodynamic drag on the cantilever. Langmuir 18(3):716–721

Chapter 3

Quantification of Apparent Membrane Tension and Membrane-to-Cortex Attachment in Animal Cells Using Atomic Force Microscopy-Based Force Spectroscopy

Martin Bergert and Alba Diz-Muñoz

Abstract

Reliable quantification of a cell's biophysical properties is key for understanding the role of mechanics in cell biology. Plasma membrane tension, the energetic cost of increasing the surface area of the plasma membrane, has been shown to regulate a plethora of cellular processes, ranging from leading edge formation to phagocytosis and membrane trafficking. Here, we describe the measurement of this key mechanical property of the cell surface using atomic force microscopy (AFM)-based force spectroscopy. Depending on the nature of the force curve acquisition, AFM measurements can quantify various membrane tension components, such as apparent membrane tension and membrane-to-cortex attachment (MCA). We discuss the biophysical background (1), required materials (2), sample preparation (3.1), AFM-probe calibration and functionalization (3.2), force curve acquisition (3.3) and data analysis and representation (3.4).

Key words Membrane tension, Tether, Membrane-to-cortex attachment, MCA, Atomic force microscopy, Atomic force spectroscopy, AFM, Cantilever

1 Introduction

The shape of cells is governed by their surface mechanics. Structurally the animal cell surface is a composite interface, comprising the plasma membrane, the underlying actomyosin cortex and the 'glue' that binds them together – the membrane-to-cortex attachment (MCA). The molecular composition and architecture of these diverse layers drive cell-scale mechanics (reviewed in [1–3]). Mechanically, both the plasma membrane and the actomyosin cortex can be characterized by their tension, the energetic cost of increasing the surface area (Fig. 1). For details on the nature of cortical tension and how to measure it, see [4]. The plasma membrane also has an apparent tension, resulting from the sum of the in-plane membrane tension of the lipid bilayer with associated

Ronen Zaidel-Bar (ed.), *Mechanobiology: Methods and Protocols*,
Methods in Molecular Biology, vol. 2600, https://doi.org/10.1007/978-1-0716-2851-5_3,

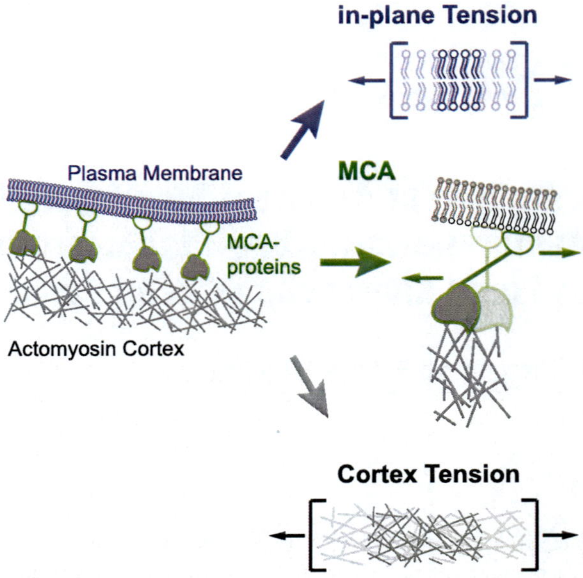

Fig. 1 Mechanics at the cell surface

proteins (Tm) and MCA, which provides further resistance to plasma membrane area changes (reviewed in [2]).

This protocol focuses on quantifying apparent membrane tension and MCA in animal cells using atomic force microscopy (AFM)-based force spectroscopy. In brief, an AFM is used to measure the force required to pull and hold nanotubes of plasma membrane called tethers (Fig. 2a). To that end, an adhesive functionalized AFM-probe (or cantilever) contacts the cell with a well-defined force for a short period of time and then retracts with a defined velocity. The retraction of the cantilever causes the formation of a thin membrane nanotube (~50–100 nm diameter [5]), which is then elongated as retraction progresses. The force required for the formation and extension of such a tether is then measured as a step/jump in the force curves at the moment the tether breaks (Fig. 2b). Depending on whether the Z-height of the cantilever (and thus the length of the tether) is held constant at the moment of tether rupture, or increasing, static or dynamic tethers can be pulled, respectively (Fig. 2c). While the first option allows quantification of apparent membrane tension [6], the second option enables the specific measurement of MCA [7].

1.1 Rationale of Static Tether Pulling

Static tether pulling is termed static as tether rupture occurs while the Z-height of the cantilever is constant (velocity 0 μm/s). Therefore, the tether length is fixed, and the flow of lipids has equilibrated. Static tether force (F_s) is proportional to apparent membrane tension and bending rigidity [8].

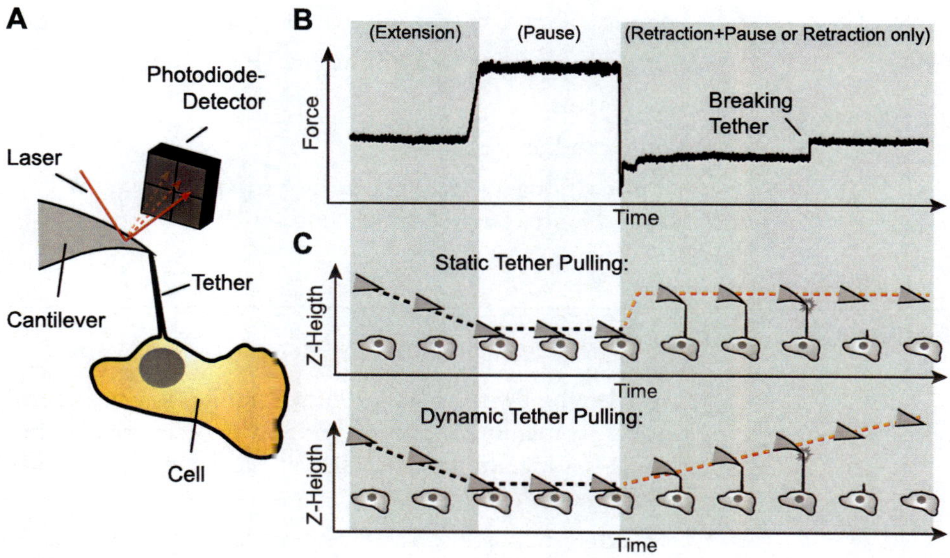

Fig. 2 AFM-based membrane tension measurements. (**a**) Schematic of tether pulling. (**b**) Exemplary force curve with key events highlighted. (**c**) System operation for static and dynamic tether pulling

$$T = T_m + \gamma = F_s^2/8B\pi^2 \tag{1}$$

T – apparent membrane tension

T_m – in-plane membrane tension

γ – MCA

F_s – static tether force

B – bending rigidity

1.2 Rationale of Dynamic Tether Pulling

Dynamic tether pulling allows the measurement of MCA. In this setup, the tethers are breaking while they are growing in length, as the Z-height of the cantilever is continuously increasing. The higher the retraction velocity, the higher is the resistance to lipid flow into the tethers, applied by the MCA linkers (viscosity as a function of velocity). Pulling tethers at different velocities and fitting the data with an appropriate model (e.g. Brochard-Wyart model [7], Eqs. 2 and 3) allows the measurement of MCA.

$$F_d^3 - F_d F_0^2 = L\,\alpha \tag{2}$$

$$\alpha = (2\pi)^3\, 2B^2\, \nu\eta_e \ln\left(R/r_t\right) \tag{3}$$

F_d – dynamic tether force

F_0 – force at velocity 0 μm/s

L – retraction velocity

B – bending rigidity

ν – density of binders (contributes to MCA)

η_e – effective viscosity (contributes to MCA)

R – cell radius

r_t – tether radius

α – proportionality factor that is proportional to MCA if B, R and r_t are assumed to be constant

2 Materials

1. AFM system with at least 20 μm Z-range (for measurement of apparent membrane tension) or 100 μm Z-range (for measurement of membrane-to-cortex attachment, e.g. Bruker JPK CellHesion® 200).
2. Inverted microscope setup with at least basic brightfield imaging capabilities, preferentially with a camera that can operate without fan cooling or is water cooled.
3. Active damping table with acoustically isolated hood (that hosts the microscope and the AFM) to minimize vibrations (= noise in the AFM force curves).
4. Temperature control of the isolation hood.
5. CO_2-controlled sample chamber.
6. Cantilever (AFM-probe): Bruker OBL-10 (Fig. 3) or similar (e.g. NANOSENSORS qp-SCONT).
7. 1x PBS without Ca^{2+} and Mg^{2+}.
8. Concanavalin A (ConA, Sigma #C5275), dissolved in 1x PBS at 2.5–4 mg/ml to functionalize the cantilevers. (This solution can be stored for several months at −20 °C to −80 °C.). The lectin Concanavalin A is a carbohydrate-binding protein that binds to glycosylated integral membrane proteins.

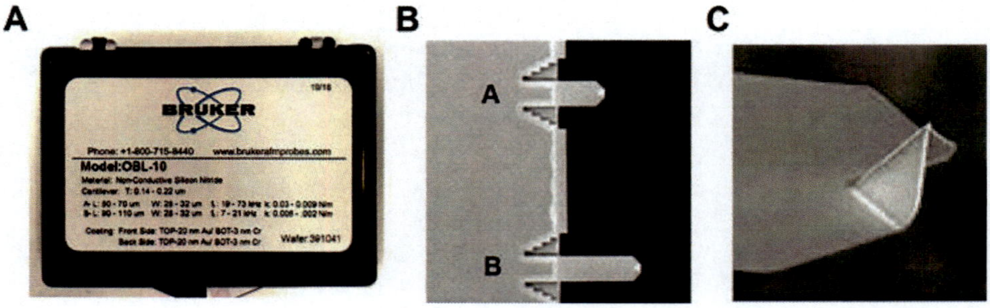

Fig. 3 Details OBL-10 cantilever. (**a**) Backside of cantilever box with cantilever details. (**b**) Close-up of cantilevers on the cantilever chip. (**c**) Close-up of cantilever tip. (Images in **b** and **c** from www.brukerafmprobes.com)

9. 35 mm glass bottom dishes (or other cell culture containers, *see* **Note 1** for required dimensions).
10. Standard glass microscopy slides.
11. Low-lint tissue (e.g. KIMTECH Science Precision Wipes).
12. Animal cells (e.g. NIH 3 T3 fibroblasts, HL60 leukaemia cells, MDCK cells, mouse embryonic stem cells).
13. Appropriate cell medium (*see* **Note 2** for required FBS concentration).
14. Cell adhesion coating (optional, *see* **Note 3**).
15. Detergent for cleaning optical instrumentation (e.g. Hellmanex® III).
16. Software Screen Markers (RollingData) or simply scotch tape.
17. Data processing software such as JPK Data Processing software (Bruker) or Matlab (Mathworks).

3 Methods

3.1 Sample Preparation: Animal Cells

1. Seed cells of interest in 35 mm glass bottom dishes (or other AFM and microscope-stage compatible cell culture container, *see* **Note 1**), and incubate until proper adhesion to the substrate is established (e.g. apply coating with fibronectin or similar if required, *see* **Note 3**). Cell density should be such that individual cells are clearly separated. If your cells of interest are growing in clusters/colonies, please note that due to the contact-based nature of an AFM, you will only be able to access the top-most layer of cell clusters/colonies and will not be able to assess the mechanical properties of cells below.
2. 30–60 min before measurement, exchange medium to remove as much floating debris as possible, as debris leads to noise in the force curves. Use this step also to lower FBS concentration to 2% (v/v) if your standard cell culture medium contains more than 2% (*see* **Note 2** for details).

3.2 AFM Preparation: Calibration and Functionalization

1. Heat the AFM + microscope setup to 37 °C (or temperature required for cell/specimen of interest). Please note that this is not required for calibration and functionalization, but as the system needs time to warm up and equilibrate, it is advised to heat several hours in advance to ensure maximum stability of the setup (*see* also **Note 4**).
2. If required, clean the AFM cantilever holder (e.g. Bruker JPK glass block) by sonication in a detergent (e.g. 1% Hellmanex® III) for 5–10 min. Rinse thoroughly with Milli-Q water and dry in a stream of dust-free gas (e.g. nitrogen gas).

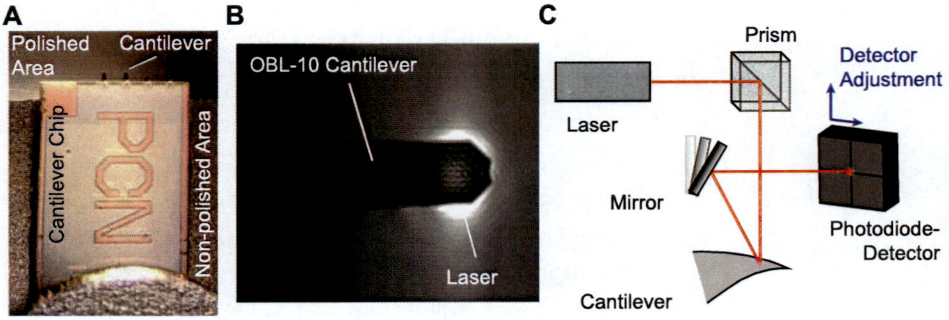

Fig. 4 AFM – Cantilever mounting and laser alignment. (**a**) Mounting a chip with OBL-10 cantilevers on a Bruker JPK Cantilever holder (glass block). (**b**) Laser alignment principle. (**c**) Photodiode window in the Bruker JPK software with successfully aligned laser

3. Mount your cantilever of choice (see Materials) on your AFM cantilever holder according to the manufacturer's instructions (Fig. 4a).
4. Mount AFM cantilever holder into AFM head, add evaporation shielding, position AFM laser on cantilever (Fig. 4b) and align laser to photodiode (Fig. 4c).
5. Mount a standard glass microscopy slide on your microscope.
6. For all actions performed with the AFM, operate it with enabled Z-closed loop (*see* **Note 5** for details). Record all force curves with 2000 Hz sampling rate.
7. Calibrate your cantilever in air. Here we provide our protocol of choice:
   (a) Use the approach function to lower the cantilever safely in Z until it touches the glass surface.
   (b) Run a calibration force curve on a glass slide to determine the OLS (optical lever sensitivity, unit = nm/V) of your setup. This constant determines the relationship of change in Z-height of the cantilever and the voltage change recorded on the photodiode.
   (c) Next, move the cantilever away from the surface to avoid any interaction (e.g. electrostatic attraction or repulsion) during the next step.
   (d) Perform a thermal noise calibration to determine the spring constant (unit = N/m). This constant is a material property of the cantilever and can be determined from the resonance frequency of your cantilever. It determines the relationship between the force acting on the cantilever and its bending in the Z-direction.
   (e) *See* **Note 6** for contact-based two-step calibration in air on a Bruker JPK CellHesion® 200 system.

(f) We recommend to run three complete sets of calibration (both OLS and spring constant) while slightly changing the laser position in between. Finally, the mean values for OLS and spring constant should be used. *See* **Note 7** for alternative calibration methods.

8. Functionalize your cantilever so it adheres to the plasma membrane of cells:

(a) Place a clean glass bottom dish (or similar) on the microscope stage and bring the cantilever to within ~500 μm Z-height of the glass surface. (The goal here is to ensure that in the next step the cantilever will be surely submerged in the functionalization solution.)

(b) Remove the AFM head with the AFM cantilever holder and cantilever, apply 100 μl Concanavalin A (ConA) functionalization solution (*see* **Notes 8**, **9**, and **10** for details) to the centre of the dish (it will form a hemispherical drop). In addition, apply 20 μL of the functionalization solution to the cantilever chip to pre-wet the AFM cantilever holder, the chip and the cantilever. Then remount AFM head on the microscope. The cantilever should now be submerged in the functionalization solution. Verify the absence of air bubbles and check for correct submerging by focusing and detecting ConA fibres above and below the cantilever focus (Fig. 5; *see* **Note 11** for avoiding and removing air bubbles).

(c) Incubate for 1 h at 37 °C. Longer incubation times might cause drying of the ConA functionalization solution.

(d) Wash cantilever twice with warm 1x PBS using another clean glass bottom dish.

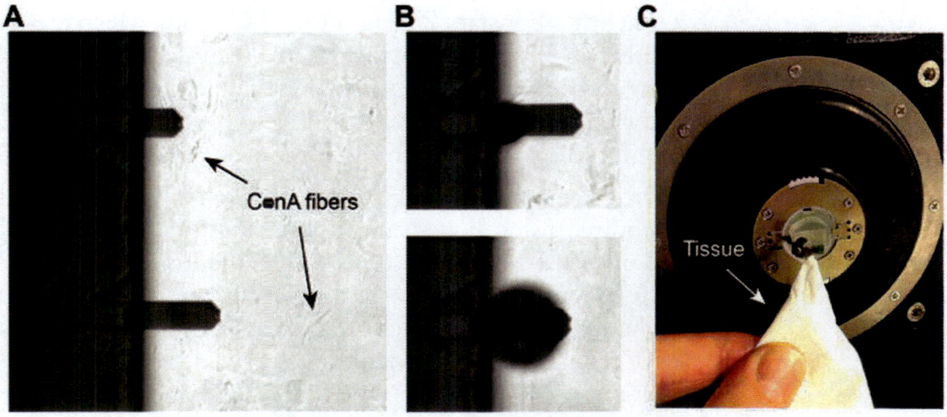

Fig. 5 ConA functionalization. (**a**) DBL-10 cantilever submerged in ConA solution. (**b**) Air bubbles formed at the cantilevers upon submerging in ConA. (**c**) Drying of cantilever holder and chip with cantilevers to remove/burst air bubbles during functionalization procedure

9. Calibrate the OLS of your cantilever in liquid:
   (a) Apply 1 mL warm PBS to the dish, submerge cantilever and wait for equilibration (5 min).
   (b) Run a calibration force curve in PBS on a glass bottom dish to determine the OLS in liquid of your setup (similar to point 7b). Afterwards, enter the spring constant of your cantilever as determined previously during the calibration in air.
10. The AFM setup is now ready for measurements.

3.3 Measurements

3.3.1 Static Tether Pulling

1. Place specimen (e.g. cells in a glass bottom dish) on the microscope and approach the glass surface on an empty region of the dish (e.g. by using the previously described approach function).
2. Wait for thermal equilibration (5 min).
3. In the approached state, locate the tip of the cantilever and mark its position on the screen using either a digital solution such as the software Screen Markers (RollingData) or scotch tape as analogue alternative. (Rationale: If using a high magnification objective, the cantilever tip will not be in focus anymore once the piezo is retracted. When targeting a cell/specimen for measurements, this allows to place the cell/specimen directly below the non-focused cantilever without the need to refocus.)
4. Choose cell/specimen of interest and place below the mark from **step 3**.
5. Run a force curve using the settings listed in Table 1. The cantilever is lowered towards the cell/specimen using a defined velocity until it touches the cell/specimen and reaches a target force setpoint. After a short contact, the cantilever is retracted at a defined distance with a defined velocity. Then the cantilever is held at a constant Z-height for 25 s, during which the formed tether(s) might break, which will be recorded as jumps/steps in the force curve. Tethers breaking after more than 25 s should not be considered, as by then F-actin might have polymerized within the tether, which significantly changes its mechanics.
6. On a Bruker JPK CellHesion® 200 system, the force curve is acquired using the advanced force ramp designer (see Fig. 6, also for the example settings). Note that in order to save time, the extension phase is split into two, an Extend 1 Segment where the cantilever is lowered with a fast velocity until it is near the cell/specimen (this length has to be determined empirically for each cell/specimen) and Extend 2 Segment where the cell/specimen is approached with the target velocity and target force setpoint (as listed in Table 1). *See* **Note 12** for details on TipSaver option.

Table 1
Parameters for static tether pulling

Parameter	Value	Notes
Extension velocity	0.5 μm/s	
Force setpoint	0.1–0.3 nN	To be determined per sample. Modify only, if changing contact time (see below) does not improve tether breaking frequency. Increase if no tethers form, decrease when tethers don't break within 25 s. Try to keep constant during an experiment and between conditions
Contact time	100 ms–10 s	To be determined per sample and experiment session (depends strongly on ConA functionalization quality). Increase if no tethers form, decrease when tethers don't break within 25 s. Usually requires increase over the course of an experiment, as ConA binding efficiency decreases over time
Retraction length	10 μm	Determines tether length. To be kept constant within and between experiments
Retraction velocity	10 μm/s	To be kept constant within and between experiments (If too slow, tether pulling takes too much time and F-actin polymerizes in the tether. If too fast, there might be insufficient adhesion to the cantilever and no tethers will form.)

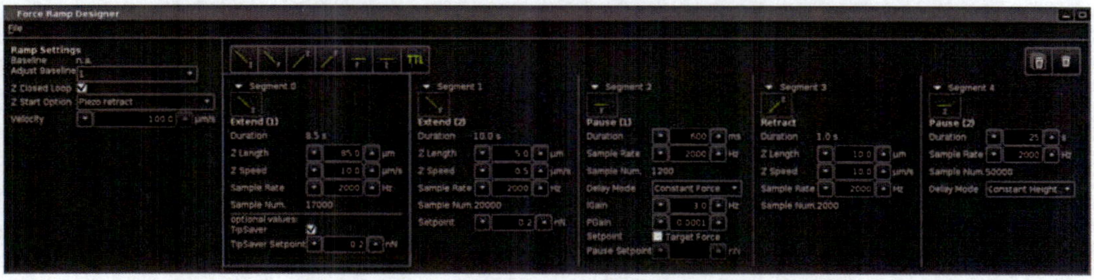

Fig. 6 Advanced force curve settings for static tether pulling using a Bruker JPK CellHesion® 200 system

7. Aim for at least three tethers with high quality per cell/specimen (see Fig. 8d, e for examples of low-quality tethers). Each cell/specimen should undergo a maximum of 12 measurements in order to avoid stress responses.

3.3.2 Dynamic Tether Pulling

1. Follow **steps 1–3** of section Static Tether Pulling.
2. Run a force curve using the settings listed in Table 2. The cantilever is lowered towards the cell/specimen using a defined velocity until it touches the cell/specimen and reaches a target force setpoint. After a short contact, the cantilever is retracted at different velocities for about 80 μm in Z. During this retraction, formed tether(s) might break, which will be recorded as

Table 2
Parameters for dynamic tether pulling

Parameter	Value	Notes
Extension velocity	0.5 μm/s	
Force setpoint	0.1–0.3 nN	To be determined per sample. Modify only, if changing contact time (see below) does not improve tether breaking frequency. Increase if no tethers form, decrease when tethers don't break (within the retraction length or 25 s). Try to keep constant during an experiment and between conditions
Contact time	100 ms–10 s	To be determined per sample and experiment session (depends strongly on ConA functionalization quality). Increase if no tethers form, decrease when tethers don't break. Usually requires increase over the course of an experiment, as ConA binding efficiency decreases over time
Retraction length	80 μm	Determines time window during which tether can break. To be kept constant within and between experiments
Retraction velocity	2/5/10/ 30 μm/s	Use these velocities as starting points for your cell/specimen. Once determined, keep constant within and between experiments

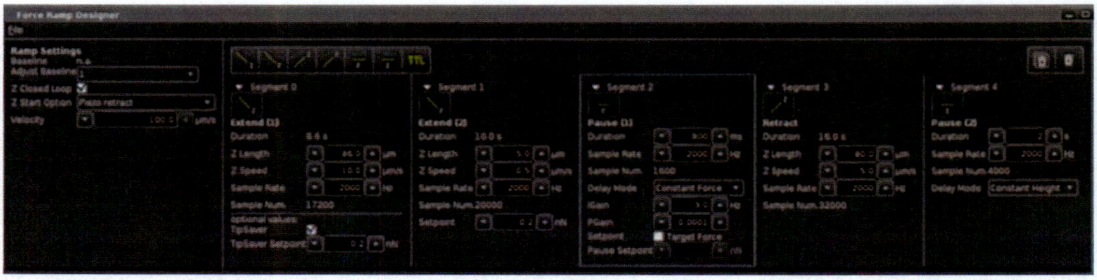

Fig. 7 Advanced force curve settings for dynamic tether pulling using a Bruker JPK CellHesion® 200 system

jumps/steps in the force curve. Consider only tethers that break within 25 s.

8. On a Bruker JPK CellHesion® 200 system, the force curve is acquired using the advanced force ramp designer (see Fig. 7, also for example settings). As for static tether pulling, the extension phase is split into two phases. Important for dynamic tether pulling is to ensure that there is enough Z-range for the AFM piezo to retract at least 80 μm after contacting the cell/ specimen. *See* **Note 12** for details on TipSaver option.

3. Each cell/specimen has to be tested for all the chosen retraction velocities. Aim for at least two tethers with high quality per velocity (see Fig. 8d, e for examples of low-quality tethers). Each cell/specimen should undergo a maximum of 12 measurements in order to avoid stress responses.

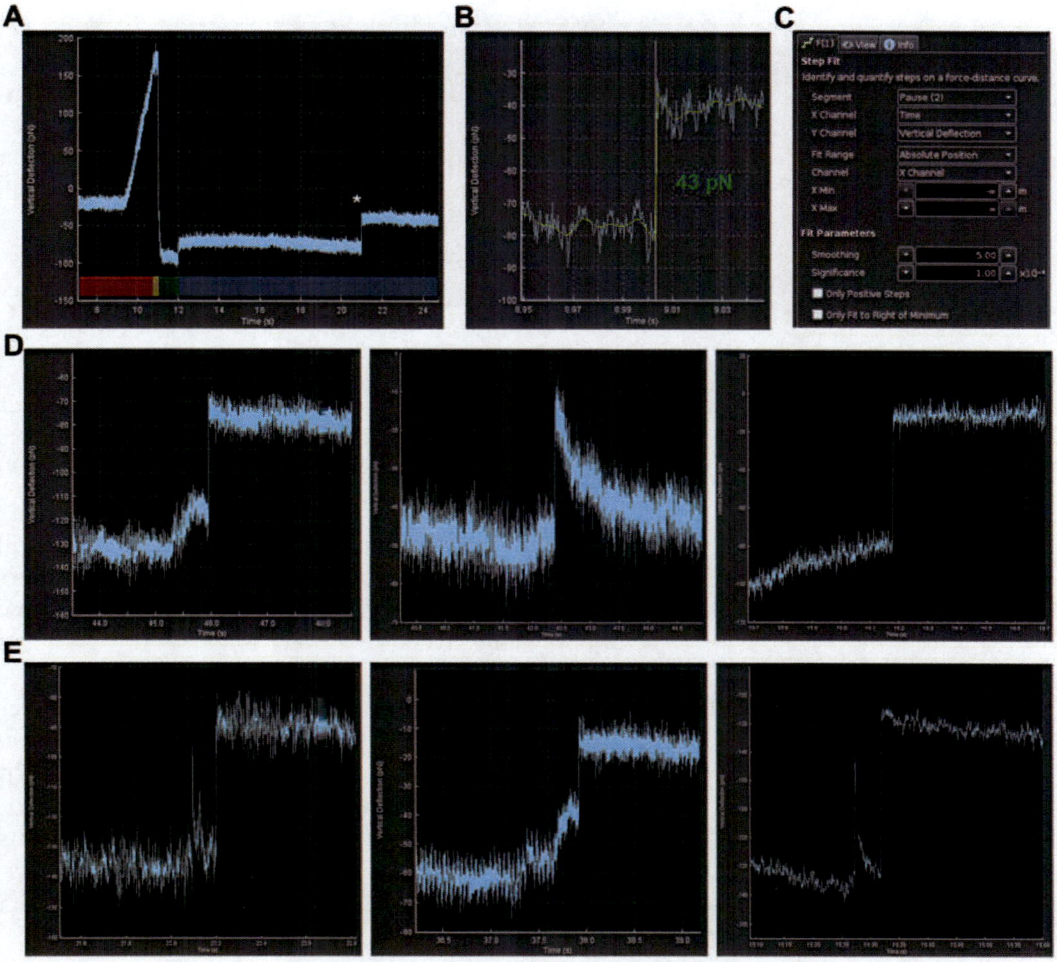

Fig. 8 Force curve processing and tether analysis with the JPD Data Processing software. (**a**) Example curve for static tether pulling. Force curve segments are color-coded: Extension (red), Pause (1) (yellow), Retraction (green), Pause (2) (blue). Asterisk marks breaking tether. (**b**) Zoom on the breaking tether. Green line indicates smoothed data. (**c**) Step-fitting parameters. (**d**) Low-quality tethers with no clear plateau before and after breaking. (**e**) Low-quality tethers with jumps/irregularities just before breaking. Do not accept tethers as shown in **d** and **e**

3.4 Data Analysis

3.4.1 Processing

1. For straightforward processing of the acquired AFM force curves, we recommend the JPK Data Processing software (as described below). However, other AFM processing software or custom analysis with e.g. Matlab (Mathworks) is also possible. In brief, analysis should contain at least a smoothing of the naturally noisy AFM force curve (e.g. Gaussian filter) and a step-fitting algorithm (e.g. from the analysis of in vitro microtubule dynamics [9]) to identify the force difference upon breaking of a tether.

2. A typical force curve for static tether pulling is shown in Fig. 8a. Tethers break in the last phase (Pause).
3. Within the JPK Data Processing software, select the 'Step Fit' function (Fig. 8c). Choose the correct segment of your force curve (i.e. Pause (2)), display 'Time' on the X-channel, and set the significance level of the fitting to 10^{-4}. The software will now check the selected segment of the force curve for steps (i.e. breaking tethers) and quantify their force difference.
4. Important considerations whether to accept a tether as clear/ high quality or discard (see Fig. 8d, e for examples of low-quality tethers):
   (a) For static tether pulling, accept only tethers with a clear plateau in the force value before breaking. This ensures that the lipid flow is equilibrated within the tether and you are not measuring any dynamic effects.
   (b) For dynamic tether pulling, often multiple tethers will break within one force curve (see Fig. 9). Ignore tethers that show jumps/irregularities in the force trace just before breaking. Average all other tethers.

3.4.2 Plotting/Fitting

1. For static tether pulling, we suggest to plot the static tether force F_s instead of calculating apparent membrane tension T (see Eq. 1), unless membrane bending rigidity has been experimentally measured in the different conditions. We propose further to average the static tether force measurements obtained per cell/specimen and then plot individual cell/specimen values (e.g. using a boxplot that includes individual data points, see Fig. 10a).
2. For dynamic tether pulling, we suggest to plot the force-velocity relationship, with average force values at each velocity (see Fig. 10b). Such data can then further be used to fit an appropriate model such as the Brochard-Wyart model ([7], Eqs. 2 and 3). This allows the measurement of MCA by obtaining an MCA parameter α, that is proportional to the density of binders (i.e. the active MCA molecules) and the emerging effective viscosity (Fig. 10c).
3. For both measurements, it is advised to always measure all conditions of interest with the same cantilever and calibrated setup, due to calibration uncertainties and errors. We propose further to only average complete datasets (e.g. where all conditions of interest were measured with the same cantilever and calibrated setup).

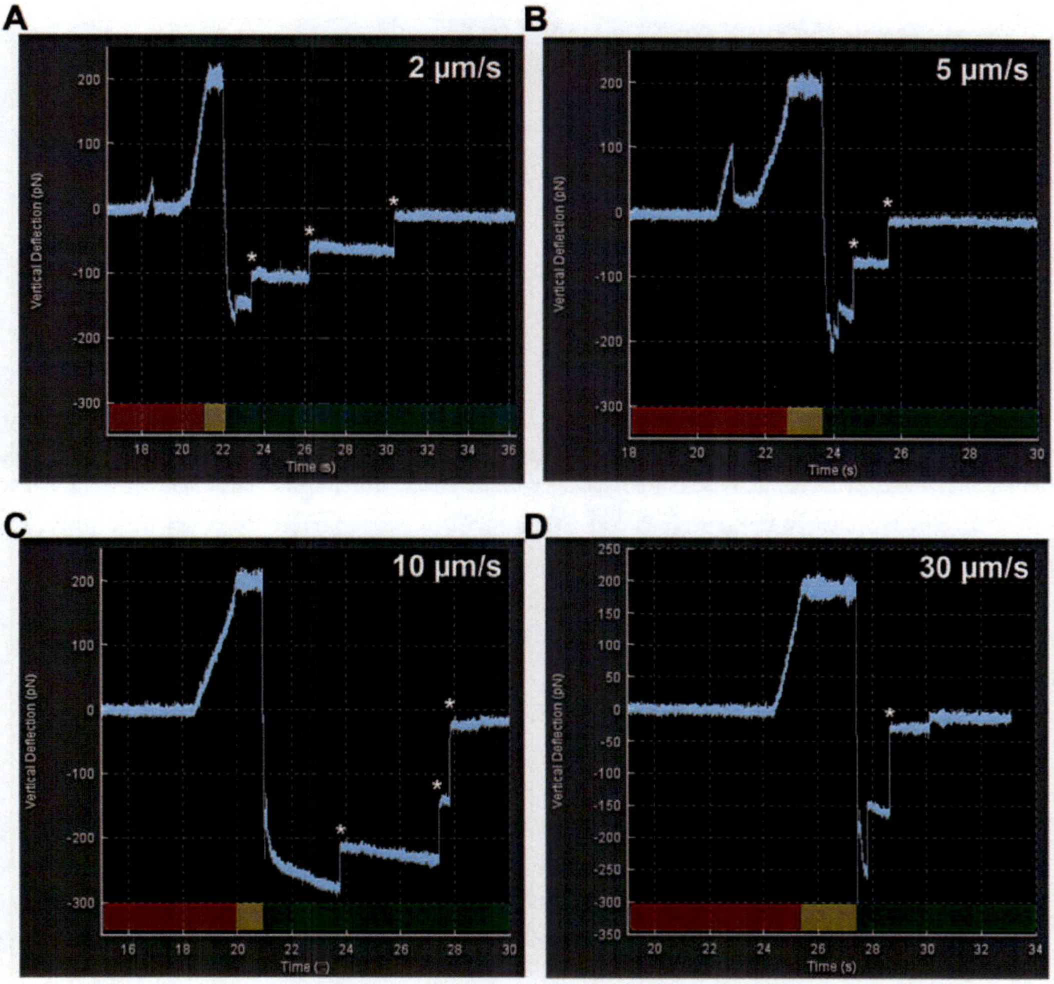

Fig. 9 Example force curves for dynamic tether pulling with increasing retraction velocities (**a**–**d**). Force curve segments are color-coded: Extension (red), Pause (1) (yellow), Retraction (green). Asterisks mark breaking tethers

4 Notes

1. Depending on the AFM system being used, there is a maximal working distance the cantilever holder can penetrate into the cell culture dish/container (see Fig. 11). For the Bruker JPK CellHesion® 200 system, this maximum working distance is about 10.8 mm for 35 mm glass bottom dishes. If the inner depth of the dish/container is >10.8 mm, it cannot be used as the cantilever is not able to reach the cells adhering at the bottom of the dish. Compatible with the Bruker JPK CellHesion® 200 system are e.g. IBIDI µ-Dish low (#80136) or WPI FluoroDish (#FD35–100).

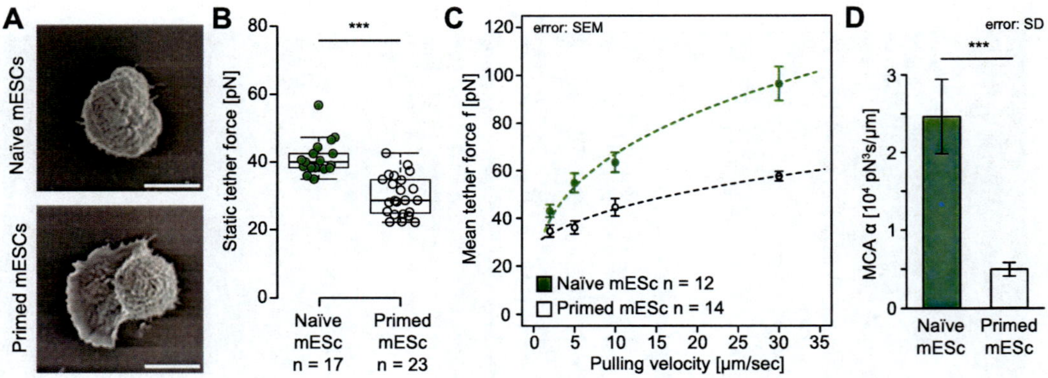

Fig. 10 Example Data: Membrane tension in naïve and primed mouse embryonic stem cells. (**a**) Scanning electron microscopy images of naïve and primed mouse embryonic stem cells. Scale bar: 10 μm. (**b**) Static tether force measurements. (**c**) Results from dynamic tether pulling experiments. Error bars: SEM. (**d**) Fitting the Brochard-Wyart model (see Eqs. 2 and 3) allows to derive mean and standard deviation of the MCA parameter α. n: number of cells analysed. ***: $P < 0.001$. (Data modified from Ref. [13])

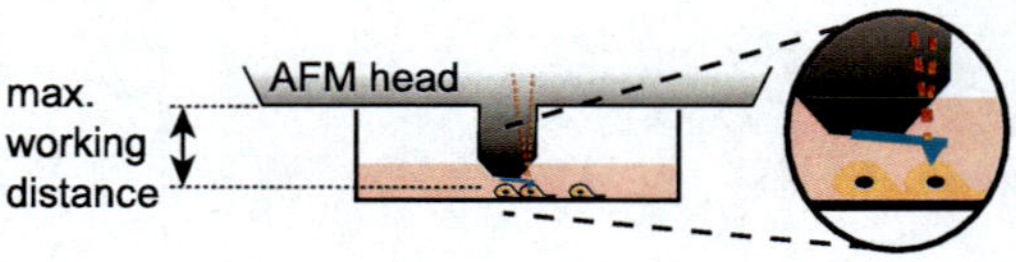

Fig. 11 Schematic of maximum working distance

2. If your medium contains FBS, reduce concentration to 2% (v/v) in the washing step 30–60 min prior to AFM measurements, as FBS neutralizes ConA functionalization over time. High FBS concentrations will significantly limit the number of tethers that can be pulled with a single cantilever.
3. For successful tether pulling with an AFM cantilever, the sample of interest (cells, living tissue, organoids) needs to be adherent. If the cell/specimen is not adhering, natural adhesives such fibronectin, laminin or collagen or artificial adhesives such as Poly-L-Lysine or Cell-Tak Cell & Tissue Adhesive (Corning) can be tested.
4. If you open the heated isolation chamber to physically access the AFM/microscope system, always wait for at least 2–5 min after closure to equilibrate the system enough before performing any measurements.
5. With Z-closed loop being enabled, the piezo movement during the experiment is constantly measured and corrected. This corrects for e.g. any nonlinearity and hysteresis of the piezo during movement and allows for accurate Z positioning as well as for accurate and constant velocity.

6. Contact-based two-step calibration in air on a Bruker JPK CellHesion® 200 system:
   (a) Acquire a force-distance curve on a hard surface (e.g. glass) in air. For OBL-10 cantilevers, use 10 μm/s extension velocity. See example curve in Fig. 12a.
   (b) Open the calibration manager and fit the extension slope in the region of the force-distance curve where the cantilever is in contact with the surface. This will determine the OLS of your setup (Fig. 12b).
   (c) Retract the cantilever at least 200 μm away from the surface.
   (d) Run a 'thermal noise' calibration.
   (e) Fit the resonant peak of the obtained resonant frequency spectrum. Make sure you include the tails of the peak for a correct fit. This will determine the spring constant of your cantilever (Fig. 12c).
   (f) Accepting both the OLS and the spring constant, your setup is now calibrated and ready for measurements.
7. For alternative calibration methods, such as contact-free calibration, see [10].
8. We found ConA concentrations of 2.5–4 mg/ml optimal for tether pulling on the animal cell lines tested (NIH 3 T3 fibroblasts, HL60 leukaemia cells, MDCK cells, mouse embryonic stem cells). If you find tethers not breaking (with lowest possible contact time and contact force for your AFM system), or if you fail to pull tethers, decrease or increase ConA functionalization concentrations, respectively, to determine the optimal concentration for your sample.
9. We found that certain cell lines are highly sensitive to ConA in the medium, displaying signs of cellular toxicity and death (e.g. A549, HeLa Kyoto, A375-M2, A375-P). As some ConA will diffuse from the cantilever into the medium during the time course of the experiment, test first whether you cells of interest tolerate ConA (concentration range: 1–5 μg/ml).
10. The functionalization protocol described above is economical, basic and fast. Other more advanced cantilever functionalization protocols using e.g. biotinylated-ConA or Cell-Tak Cell & Tissue Adhesive (Corning) have been described [11, 12]. These usually resist higher detachment forces (nN range), but have not been tested by our group yet for quantifying membrane tension.
11. When submerging the cantilever in ConA solution for functionalization, air bubbles could be trapped at the cantilever (see Fig. 5b). As they tend to expand over time due to the heat, they

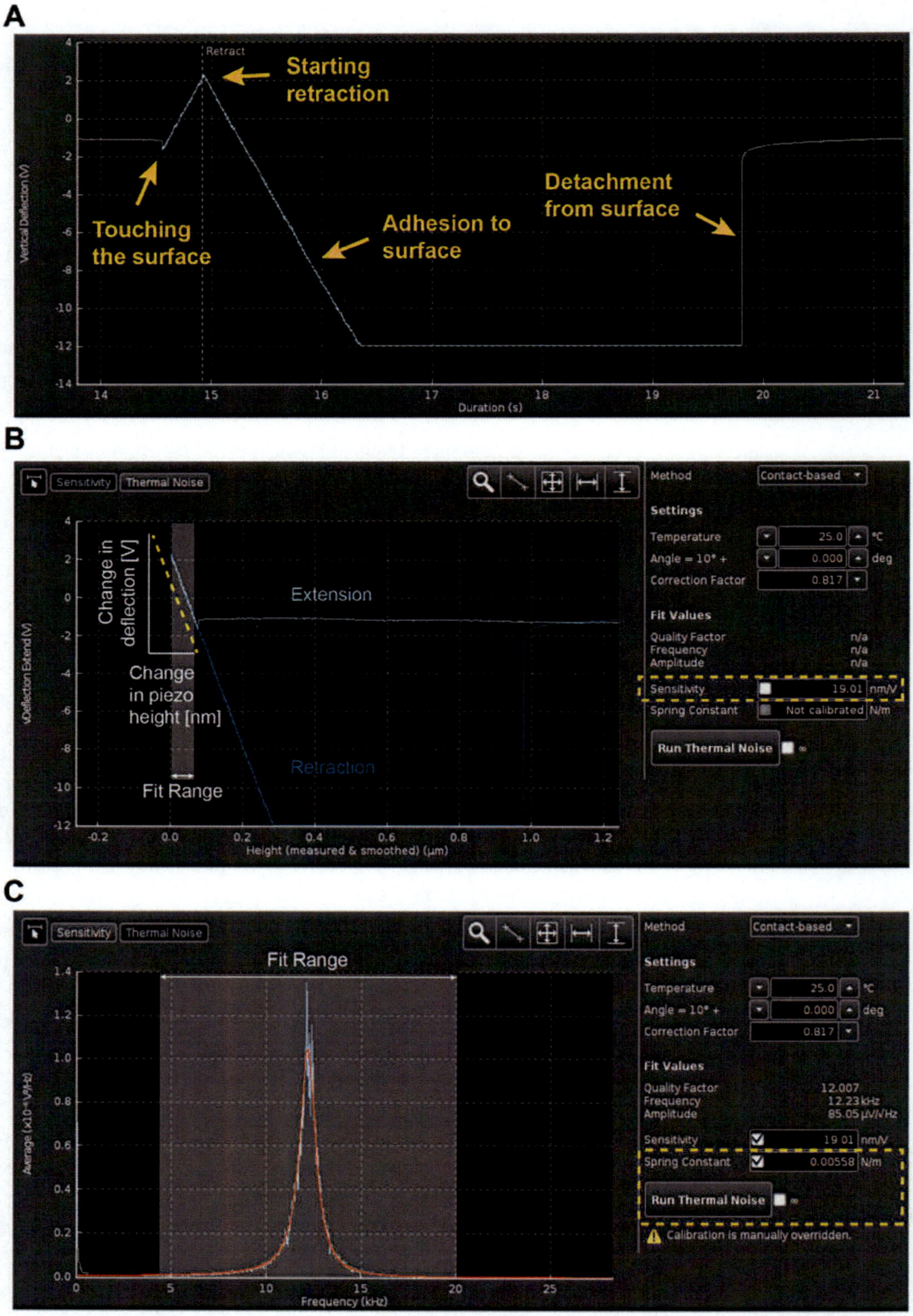

Fig. 12 Two-step calibration on a Bruker JPK CellHesion 200 system. (**a**) Typical force-time curve obtained during contact-based calibration in air. (**b**) The slope of the curve during contact with the substrate (displayed as force-distance curve) determines the OLS (optical lever sensitivity, unit = nm/V) of your setup. (**c**) Determining and fitting the first resonance peak during the Bruker JPK thermal noise measurement determines the spring constant (unit = N/m) of your cantilever

need to be removed, in order to ensure that the cantilever tip is truly being functionalized. To remove them, dry the cantilever holder/chip with cantilevers briefly with a low-lint tissue. To that end, fold tissue tip-shaped and carefully touch the side of the chip to soak up the liquid (see Fig. 5c). Be careful not to touch the cantilever(s). This usually bursts the air bubbles. If necessary, repeat multiple times.

12. If you are using a Bruker JPK CellHesion® 200 system and the advanced force ramp designer, it is advised to enable the Tipsaver option in the segment 0. This prevents damage to the cantilever and the system, as it prevents an ongoing extension in case the cantilever hits an obstacle during the first segment, where it extends with a relatively high velocity. Tipsaver setpoint should be equal or lower than the force setpoint in segment 1.

Acknowledgements

We acknowledge the financial support of the European Molecular Biology Laboratory (EMBL), the Human Frontiers Science Program (HFSP) grant number RGY0073/2018 and the Deutsche Forschungsgemeinschaft (DFG) grant numbers DI 2205/2-1 and DI 2205/3-1 to A.D-M.

We thank Niccolò Banterle, Leanne Strauss, Jia Hui Li, Ewa Maria Sitarska, Sergio Lembo (all EMBL) and Torsten Müller (Bruker) for critically reading the manuscript.

References

1. Diz-Muñoz A, Weiner OD, Fletcher DA (2018) In pursuit of the mechanics that shape cell surfaces. Nat Phys 14(7):648–652
2. Diz-Muñoz A, Fletcher DA, Weiner OD (2013) Use the force: membrane tension as an organizer of cell shape and motility. Trends Cell Biol 23(2):47–53
3. Salbreux G, Charras G, Paluch E (2012) Actin cortex mechanics and cellular morphogenesis. Trends Cell Biol 22(10):536–545
4. Kelkar M, Bohec P, Charras G (2020) Mechanics of the cellular actin cortex: from signalling to shape change. Curr Opin Cell Biol 66:69–78
5. Pontes B, Viana NB, Salgado LT, Farina M, Moura Neto V, Nussenzveig HM (2011) Cell cytoskeleton and tether extraction. Biophys J 101(1):43–52
6. Hochmuth FM, Shao JY, Dai J, Sheetz MP (1996) Deformation and flow of membrane into tethers extracted from neuronal growth cones. Biophys J 70(1):358–369
7. Brochard-Wyart F, Borghi N, Cuvelier D, Nassoy P (2006) Hydrodynamic narrowing of tubes extruded from cells. Proc Natl Acad Sci U S A 103(20):7660–7663
8. Sheetz MP (2001) Cell control by membrane-cytoskeleton adhesion. Nat Rev Mol Cell Biol 2(5):392–396
9. Kerssemakers JWJ, Munteanu EL, Laan L, Noetzel TL, Janson ME, Dogterom M (2006) Assembly dynamics of microtubules at molecular resolution. Nature 442(7103):709–712
10. Sader JE, Chon JWM, Mulvaney P (1999) Calibration of rectangular atomic force microscope cantilevers. Rev Sci Instrum 70(10):3967–3969
11. Friedrichs J, Helenius J, Müller DJ (2010) Quantifying cellular adhesion to extracellular

matrix components by single-cell force spectroscopy. Nat Protoc 5(7):1353–1361

12. Friedrichs J, Werner C, Müller DJ (2013) Quantifying cellular adhesion to covalently immobilized extracellular matrix proteins by single-cell force spectroscopy. Methods Mol Biol 1046:19–37

13. Bergert M, Lembo S, Sharma S, Russo L, Milovanović D, Gretarsson KH, Börmel M, Neveu PA, Hackett JA, Petsalaki E, Diz-Muñoz A (2021) Cell surface mechanics gate embryonic stem cell differentiation. Cell Stem Cell 28(2):209–216.e4

Chapter 4

Characterizing the Biophysical Properties of Adhesive Proteins in Live Cells Using Single-Molecule Atomic Force Microscopy

Andrew Vae Priest and Sanjeevi Sivasankar

Abstract

Cell adhesion proteins play essential roles in the formation, regeneration, and maintenance of tissue. However, the molecular mechanisms by which cells regulate the conformation and binding properties of adhesion proteins are poorly understood. These biophysical properties can be resolved, with single-molecule resolution, using atomic force microscopy (AFM). Here, we outline how AFM force measurements can be used to study the conformation, cytoskeletal linkage, binding strength, and force-dependent bond lifetimes of adhesion proteins in live cells.

Key words AFM force measurements, Catch bonds, Membrane proteins, Cell adhesion, Binding conformation, Inside-out regulation

1 Introduction

Cell adhesion proteins play essential roles in maintaining tissue homeostasis and in facilitating embryogenesis and wound healing; dysregulation of cell adhesion is a characteristic of certain cancers [1] However, the molecular mechanisms by which cells regulate the biophysical properties of adhesive proteins, such as their conformation, adhesion, and binding partners, are difficult to study in live cells. Most technologies that are currently used to study cell adhesion proteins are based on fluorescence imaging which probe adhesion dynamics at the ensemble level. In contrast, the nonequilibrium dynamics of cell adhesion is poorly studied at the single-molecule level.

A powerful technique that can be used to characterize adhesive protein biophysics is atomic force microscopy (AFM). This technique can measure pico-scale forces and distances with applications ranging from topographical imaging to measuring forces between individual molecules. An AFM consists of a force-transducing

Ronen Zaidel-Bar (ed.), *Mechanobiology: Methods and Protocols*,
Methods in Molecular Biology, vol. 2600, https://doi.org/10.1007/978-1-0716-2851-5_4,

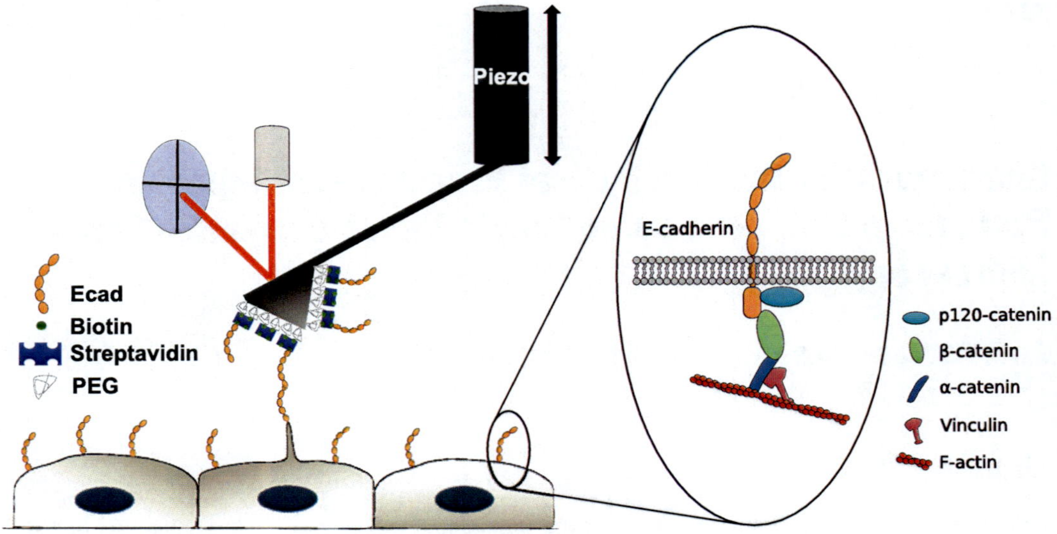

Fig. 1 Live-cell AFM experiment. Ecad monomers were immobilized on an AFM cantilever, and interactions between Ecad on the cantilever and on the apical surface of epithelial cells were measured. *Inset:* The cytoplasmic domain of Ecad couples to the actin cytoskeleton via β-catenin, α-catenin, and vinculin. Ecads also associate with other cytoplasmic proteins including p-120 catenin. (Panels were adapted from Ref. [7])

cantilever that is driven by a piezoelectric actuator. The cantilever bends when force is applied, and the extent of bending is measured using light reflected off the backside of the cantilever onto a position-sensitive photodetector—typically a quadrant photodiode or QPD (Fig. 1). Force is determined by Hooke's Law: $F = -k\,x$, where x is the displacement (the amount of bending) and k is the spring constant or stiffness of the cantilever typically determined using the thermal fluctuation method [2]. Molecular interactions can be explored by biochemically functionalizing cantilevers with a protein of interest and performing AFM force measurements on cell monolayers (Fig. 1). AFM directly provides details about the binding frequencies, unbinding forces, and force-dependent bond lifetimes of interacting molecules.

Here, we describe how AFM force measurements can be performed on the surface of live cells. While this protocol focuses on using AFM to study the biomechanics of E-cadherin (Ecad), an essential transmembrane adhesion protein in epithelial cells, it is important to emphasize that the same approach can be used to study essentially any transmembrane protein. Ecad cell-cell adhesion proteins play key roles in the formation and maintenance of epithelial tissues, in wound healing, and in cancer metastasis [1]. While the extracellular region of Ecad mediates adhesion, the cytoplasmic region of Ecad is coupled to the actin cytoskeleton through various adapter proteins (Fig. 1, *inset*), which reinforce Ecad-based cell adhesion in a force-dependent manner [3–

6]. However, the mechanisms by which cytoplasmic proteins strengthen adhesion are not completely understood at the molecular level. Using single-molecule AFM force measurements, we recently demonstrated that the association of intracellular vinculin to the Ecad cytoplasmic region regulates adhesion by switching Ecad ectodomains from a weak binding to the strongly adhesive conformation [7].

In our experiments, the cantilever of an AFM was decorated with Ecad ectodomains. The AFM cantilever was brought into contact with the apical surface of an epithelial cell, such that a bond was formed between an Ecad on the AFM tip and an Ecad on the cell surface. When the AFM cantilever was withdrawn from the cell surface at a constant velocity, we observed characteristic "signatures" that reported on the cytoskeletal linkage of the Ecad. For example, while pulling on Ecads that are uncoupled from the cytoskeleton resulted in the formation of a membrane tethers (Fig. 2a, b), pulling on strongly interacting Ecads that are coupled to the cytoskeleton resulted in the formation of cyto-tethers (Fig. 2c, d). Finally, if cytoskeleton-bound Ecads were weakly interacting, they exhibited an elastic response as load was applied, but the bond broke before a tether was formed (Fig. 2e, f). Thus, the fraction of membrane tethers vs. cyto-tethers and jumps allowed us to determine the fraction of Ecads that were linked to the underlying cytoskeleton. Using this approach, we were able to demonstrate that in Madin-Darby canine kidney (MDCK) cells, only ~50% of Ecads are linked to the underlying cytoskeleton.

Similarly, an analysis of the length/persistence time of the membrane tether/cyto-tether can be used to determine the conformation of the interacting Ecads. Previous structural studies show that cadherin ectodomains bind in two *trans* conformations: robust strand-swap dimers and weak X-dimers. Cell-free AFM measurements have shown that strand-swap dimers and X-dimers respond to force in distinctly different ways [8–10]. While strand-swap dimers form slip bonds, where the lifetime of the bond decreases with increased application of force (Fig. 3a) [8, 10], X-dimers form catch bonds which strengthen upon force application (Fig. 3b) [8, 9]. In cell-free single-molecule experiments, these characteristic force-dependent lifetimes are determined by subjecting the Ecad-Ecad bond to a constant force and measuring its persistence time [11]. Similarly, in a live-cell AFM force measurements, since both membrane tethers and cyto-tethers largely extend at a constant force, they can be quantitatively analyzed to differentiate bond types for Ecad interactions [7]. For example, this analysis showed that Ecad dimers can exist as either strand-swap dimers or X-dimers on the surface of MDCK cells as indicated by the slip and catch bonds observed from the force-dependent lifetimes (Fig. 3c). Moreover, as has been previously shown in cell-free AFM experiments [8], we were able to confirm that the binding

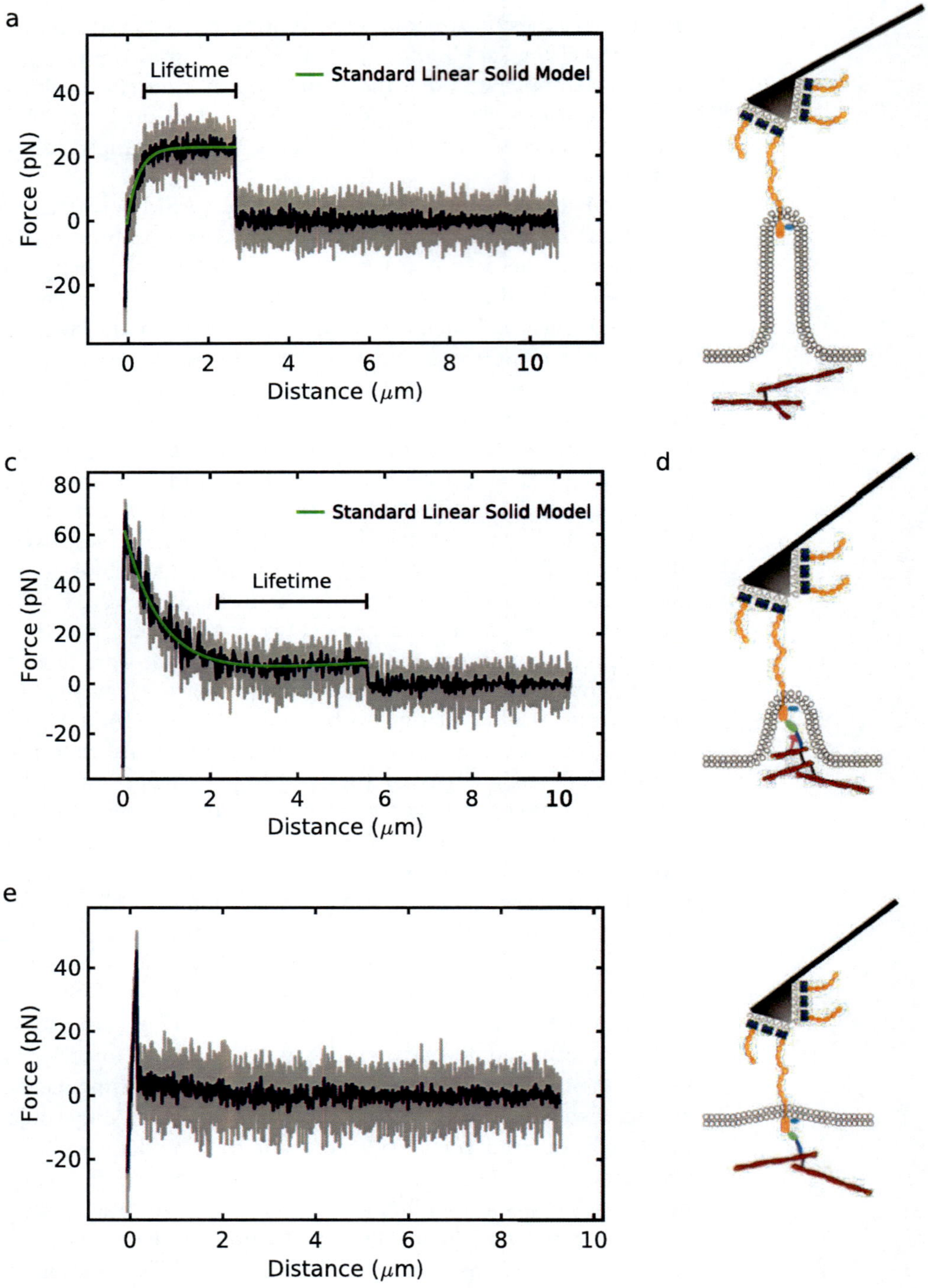

Fig. 2 Force "signatures" used to identify cytoskeletal coupling and strength of *trans* adhesion. (**a**) Typical force curve showing a membrane tether which extends at a constant force. (**b**) Membrane tethers are measured when an Ecad which is not linked to the actin cytoskeleton is pulled. (**c**) Typical force curve showing a cyto-tether with an initial increase in force that viscoelastically relaxes as a membrane tether is formed. (**d**) Cyto-tethers are measured when a robust Ecad trans dimer that is linked to the actin cytoskeleton is pulled. (**e**) Jump events exhibit a linear increase in force and are measured upon pulling (**f**) an Ecad trans

conformation of Ecad on the surface of cells can be restricted to an X-dimer by blocking strand-swapping (Fig. 3d). Finally, we showed, at the single-molecule level, that vinculin association with the Ecad cytoplasmic region allosterically converts weak X-dimers into strong strand-swap dimers and that this process is mediated by myosin II-dependent changes in cytoskeletal tension. These results demonstrated, for the first time, the inside-out regulation of Ecad conformation and established a mechanistic role for vinculin in this process [7].

In these experiments, we increased the likelihood of measuring interactions between single Ecads on the AFM tip and cell surface by decreasing their interaction rate (i.e., *event rate*). Since the probability of forming single bonds follows Poisson statistics, we tuned our event rates to ~10%, which corresponds to ~95% probability of forming a single bond [12]. The event rate in our live-cell AFM experiments was controlled by limiting the surface density of immobilized Ecads on the AFM cantilever tip. Additionally, we performed measurements on the apical surface of cells where the density of Ecads are less than at cell-cell junctions.

The protocol reported in this manuscript details how to perform live-cell AFM measurements on MDCK cells to extract the fraction of Ecad that are coupled to the cytoskeleton and the Ecad binding conformation. Although these instructions are hardware specific, they can be adapted to suit individual needs and can be used to study most membrane proteins on most adherent cell lines.

2 Materials

Prepare all solutions using ultrapure deionized water and analytical grade or better reagents. Prepare and store all reagents at room temperature unless indicated otherwise. Always follow proper safety and waste disposal procedures.

2.1 AFM Equipment

1. Atomic force microscope with environmental chamber
2. CO_2 controller with hydrating bottle and heating source
3. CO_2 cylinder
4. Sonicator
5. Compressed air with inline air filter
6. Vacuum desiccator (*optional*)

Fig. 2 (continued) dimer that is weaker than its cytoskeleton coupling. The green curve is the standard linear solid model fit to membrane tethers and cyto-tethers. Black bars indicate bond lifetimes determined from membrane tethers and cyto-tethers. (Panels were adapted from Ref. [7])

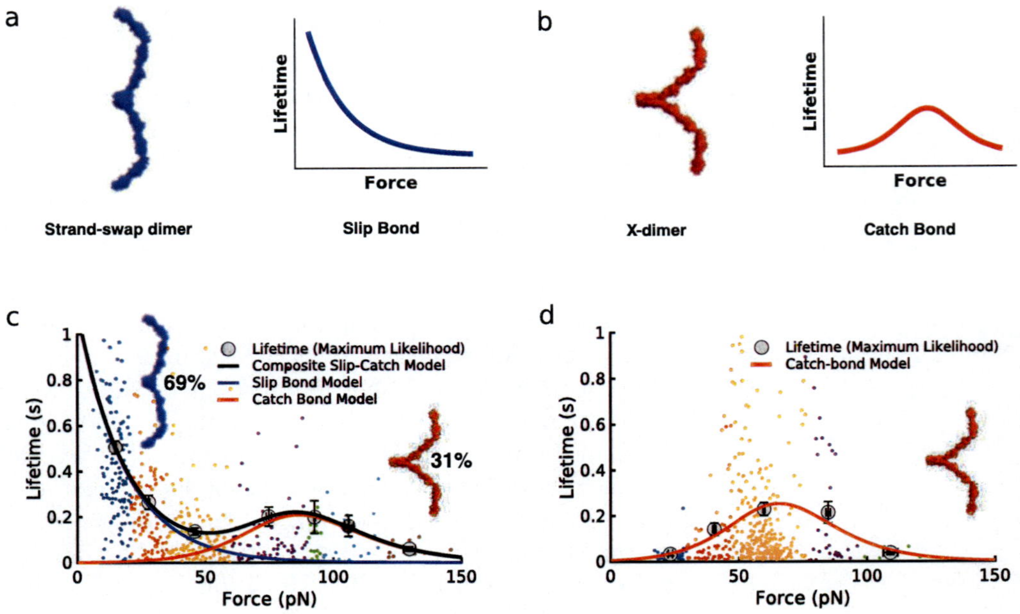

Fig. 3 Conformation-dependent Ecad biomechanics. (**a**) Ecad strand-swap dimers form slip bonds (blue), while (**b**) X-dimers form catch bonds (red). Structures were generated from the crystal structure of mouse Ecad (PDB ID code 3Q2V); the X-dimer was formed by alignment to an X-dimer crystal structure (PDB ID code 3LNH). (**c**) Force-dependent bond lifetimes for Ecad on MDCK cells show that Ecad forms both strand-swap dimers and X-dimers. (**d**) Ecad on MDCK cells only form X-dimers when strand-swapping is blocked using free tryptophan. (Panels were adapted from Ref. [7])

2.2 Cell Culture

1. Madin-Darby canine kidney (MDCK) cells. Store at −80 °C or lower.
2. Cell culture media: DMEM (low glucose, with sodium pyruvate) containing 10% fetal bovine serum (FBS), 1% kanamycin, and 1% penicillin/streptomycin. Store at 4 °C.
3. AFM experiment buffer: same as cell culture media but without 10% FBS. Store at 4 °C.
4. Trypsin. Store at 4 °C.

2.3 AFM Sample Preparation

1. Acid-resistant fine-tipped tweezers.
2. Parafilm.
3. Piranha solution: Mix a solution of 3 parts sulfuric acid and 1 part hydrogen peroxide. All preparation should occur, and waste should be stored in a fume hood (*see* **Notes 1** and **2**).
4. Acetone (HPLC grade). Store in flammable safety cabinet.
5. Silane: APTES (3-aminopropyltriethoxysilane). Store in flammable safety cabinet (*see* **Note 3**).
6. Tris/Ca^{2+} buffer: 10 mM Tris, 100 mM NaCl, 10 mM KCl, 2.5 mM $CaCl_2$, pH 7.4.

7. Pegylation buffer: Dissolve 100 mM sodium bicarbonate ($NaHCO_3$) in a stock solution of 600 mM potassium sulfate (K_2SO_4), pH 8 (*see* **Note 4**).
8. m-PEG-SVA 5 k: Solid should be store at −20 °C.
9. Biotin-PEG-SVA 5 k: Solid should be stored at −20 °C.
10. Sulfo-NHS-acetate (sulfosuccinimidyl acetate): Solid should be stored at −20 °C.
11. 0.1 mg/mL Streptavidin in Tris/Ca^{2+} buffer. Aliquots stored at −20 °C (*see* **Note 5**).
12. 80 μM Biotin in Tris/Ca^{2+} buffer. Store at 4 °C.
13. Recombinant biotinylated E-cadherin protein as reported [8, 10, 13, 14]. Store at −80 °C.

3 Methods

3.1 Tip Blunting and Spring Constant Calculation

These steps should be performed in air before chemical functionalization of the AFM cantilevers.

1. Mount cantilever and bare coverslip on the AFM.
2. Scan the cantilever across the coverslip in a 20 um × 20 um area while maintaining a pressing force of about 1 nN.
3. Determine the sensitivity of the cantilever by performing force spectroscopy measurements with a pressing voltage of 3 V (or enough voltage to observe a linear relation between *distance moved by cantilever* and *deflection*).
4. With the cantilever far away from the surface, measure the power spectral density (PSD) of the cantilever oscillations.
5. Fit the peak in the PSD, at the resonance frequency of the cantilever, to a Lorentzian function, and calculate the area under the curve.
6. Calculate the spring constant using the equipartition theorem: $k_B T = k \langle x \rangle^2$, where k is the spring constant of the cantilever, $\langle x \rangle^2$ is area under the curve, k_B is Boltzmann's constant, and T is the temperature.

3.2 Cantilever Functionalization

For all of the following steps except **step 1**, cover the beakers containing cantilevers with aluminum foil to avoid contamination.

1. Using acid-resistant tweezers, place AFM cantilevers in piranha solution. Leave overnight (*see* **Notes 6** and 7).
2. Carefully transfer the cantilevers to new beakers with ultrapure deionized water.
3. Transfer the cantilevers to a new beaker with HPLC grade acetone (repeat this step once).

4. Mix 2% (vol/vol) of APTES (3-aminopropyltriethoxysilane) in acetone and incubate cantilevers for 30 min (*see* **Note 8**).
5. Transfer the cantilevers to another beaker with HPLC grade acetone (repeat this step once).
6. Prepare cloud-point PEG solution. Dissolve m-PEG-SVA (and biotin-PEG-SVA) in pegylation buffer to a final concentration of 100 mg/mL (*see* **Note 9**).
7. Spin down PEG solution. Taking only the clear supernatant, mix the m-PEG and biotin-PEG so the mixture contains 10% biotin-PEG (*see* **Note 10**).
8. Pipette droplets of the 10% biotin-PEG solution on Parafilm and incubate cantilevers in the dark for 4 h at room temperature or overnight at 4 °C (*see* **Notes 11** and **12**).
9. Place cantilevers into a small beaker containing ultrapure deionized water, dipping in and out of the water several times to rinse (*see* **Note 13**).
10. Place cantilevers into a small, dry beaker and allow to air dry.
11. Prepare sulfo-NHS-acetate solution. Dissolve sulfo-NHS-acetate in deionized water to a final concentration of 10 mg/mL (*see* **Notes 14** and **15**).
12. Place droplets on Parafilm and incubate the cantilevers for 30 min at room temperature.
13. Transfer cantilevers to a beaker with ultrapure deionized water (dip to rinse). Then place in an empty, clean beaker to dry.
14. After the cantilevers are dry, they can be immediately used (Subheading 3.3) or stored in a vacuum desiccator for up to 7 days for later use.

3.3 Immobilization of Biotinylated Ecad on AFM Cantilevers

For all steps in this section, refer to **Notes 11** and **16**.

1. Pipette 100 μL droplet of 1 mg/mL BSA on Parafilm and place a cantilever in droplet. Leave at room temperature overnight (*see* **Note 17**).
2. Place the cantilever into 50 μL droplets of 0.1 mg/mL streptavidin for 30 min.
3. Wash off unbound streptavidin by dipping the cantilever in three successive droplets of Tris/Ca^{2+} buffer.
4. Next, incubate the cantilever with 200 nM biotinylated Ecad for 45 min.
5. Block remaining biotin-binding sites on streptavidin by incubating the cantilever with 80 μM free biotin for 20 min.
6. Finally, as in **step 3** above, wash cantilevers with Tris/Ca^{2+} buffer, and store in fresh Tris/Ca^{2+} buffer until you are ready to start the experiment.

3.4 Cell Culture

1. Remove cells from −80 °C.
2. Warm with hands or water bath.
3. Put cells and 3–5 mL of cell culture media in a 15 mL centrifuge tube and centrifuge at 2500 rpm for 3 min.
4. Decant the media, leaving the pellet.
5. Add 5 mL of cell culture media to the pellet, mix well but gently, and pipette contents into a T-75 cell culture flask containing 10 mL of cell culture media.
6. Place in cell culture incubator at 37 °C with 5% CO_2 (*see* **Note 18**).
7. Check the flask under the microscope daily. When cells are 70–100% confluent, they are ready to be cultured on glass coverslips for the AFM experiment.

3.5 Cell Culture on Glass Coverslips

1. Clean glass coverslips with ethanol and water or autoclave. Store coverslips in a ceramic rack inside a beaker in ultrapure deionized water or air that is covered with aluminum foil.
2. Pour media from T-75 flask into an appropriate waste container. Add 1–2 mL of trypsin to the flask and tip flask to coat the surface.
3. Discard trypsin and replace with 5 mL of fresh trypsin and store in incubator for approximately 15 min (*see* **Note 19**).
4. While the cells are incubating in trypsin, prepare a six-well plate by adding cell culture media and a cleaned coverslip to each well as needed.
5. After the cells have detached from the flask surface, neutralize the trypsin by pipetting the cells into a 15 mL centrifuge tube containing 5 mL of cell culture media.
6. Centrifuge at 2500 rpm for 3 min.
7. Decant the supernatant and add 3 mL of cell culture media. Pipette gently to mix.
8. Place 50–200 μL of cells (this is more than 1×10^6 cells for confluent MDCK cells from T-75 flask) on a cleaned coverslip in a six-well plate so they will be confluent in 2–3 days (*see* **Note 20**). The remainder of the cells can be discarded or seeded on a new flask for subsequent experiments.

3.6 AFM Experiment

These instructions are specific to a Keysight 5500 AFM. Therefore, not all of these instructions may apply if your system is different and would need to be adapted as necessary.

1. Calibrate CO_2 sensor – with the environmental chamber open, turn on the CO_2 sensor to calibrate the baseline. At this point, no CO_2 should be flowing into the chamber.

2. In a cell culture hood, load the cell-coated coverslips onto the sample holder. Wash with experiment buffer (FBS-free low--glucose DMEM) five times (*see* **Notes 21** and **22**).
3. Load the sample holder onto the AFM (*see* **Note 23**). Load the functionalized cantilever onto the AFM scanner. Perform closed-loop calibration as per the system instructions.
4. Turn on the laser and align the laser to the tip of the cantilever by adjusting the mirrors (*see* **Note 24**). Center the reflection on the quadrant photodiode (QPD).
5. If necessary, wait and let the system thermally equilibrate. The system is thermally stable when the voltage deflection output from the QPD is constant (not drifting).
6. A typical measurement consists of approaching the surface, stopping at a voltage set point corresponding to the force the cantilever presses with on the cell surface, waiting a predefined time (typically 0.1 s) to allow the molecules to interact, then retracting the tip away from the surface at a constant velocity (*see* **Notes 25** and **26**).
7. Perform measurements in an x-y grid with a 20 μm area with each subsequent measurement on the opposite x-y coordinate. For example, start the first measurement at the bottom left of the measurement area; then move to the top right and back to the bottom left but one position over. Continue alternating x-y positions while snaking through the 20 μm measurement area (*see* **Notes 27** and **29**).
8. After all measurements are taken, replace the coverslip with a bare coverslip, and measure the sensitivity of the cantilever in an aqueous environment (preferably the experiment buffer). This sensitivity will be used to convert the raw deflection voltage to force: $F = S \times k \times V$, where F is the force, S is the sensitivity, k is the spring constant (calculated in Subheading 3.1), and V is the deflection voltage readout of the QPD.
9. Finally, remove the cantilever and coverslip, clean the scanner and sample holder, and shutdown the AFM hardware/software.

3.7 Data Analysis

Analysis can be performed with your choice of software or programming language. The process for data analysis is force curve categorization, fitting to the standard linear solid model, filter events based on fit error, and extracting relevant statistics (event rates, unbinding forces, and bond lifetimes).

1. Preprocess the force curves. Perform a baseline correction by fitting to the portion of the force curve that is farthest from the surface to a line and subtracting the fitted values from the force.

2. Correct the distance values by subtracting the tip bending, distance $= x - F/k$ where x is the piezo distance, F is the force, and k is the spring constant of the cantilever.
3. Categorize force curves depending on the interaction type (i.e., membrane tether, cyto-tether, jump, multiple interactions, nonspecific, and no-interaction).
4. Fit membrane tethers and cyto-tethers to the standard linear solid model [15]:

$$F(x) = (F_0 + k_t\, x_0 - v\, \mu_c)e^{-k_c \frac{(x-x_0)}{v\, \mu_c}} + k_t\, x + v\, \mu_c,$$

where F_0 is the initial force, k_t is the stiffness of the tether, x_0 is the position, v is the AFM pulling velocity, μ_c is the viscosity of the cell, and k_c is the stiffness of the cell. Fit jumps to an elastic model (linear fit).
5. Filter events based on root mean squared errors (RMSE) of the fits, and remove data for which any of the fitting parameters are stuck at the bounds. Use the filtered events to determine event rates.
6. Record the unbinding forces and determine the bond lifetimes of tethers from the time constant in the standard linear solid model (*see* **Note 30**).
7. Plot the force-dependent lifetimes (as in Fig. 3c, d), and fit to a molecular bond type (slip or catch) [11, 16, 17].

4 Notes

1. Sulfuric acid and hydrogen peroxide stock should be stored in an acid cabinet.
2. Do not put piranha waste in a closed container; use a vented or loose cap. Piranha solution releases gas so a closed container will pressurize and explode. Piranha is also hygroscopic (i.e., it absorbs water from the air), so the volume of the solution will increase over time.
3. APTES is moisture sensitive so storage under inert gas is recommended.
4. Upon mixing, pH of the solution will be at pH 8. However, this solution must be prepared each time as the pH will change over time. A stock solution of potassium sulfate may be used, but sodium bicarbonate will need to be dissolved in the potassium sulfate solution each preparation.
5. To avoid freeze/thaw cycles, store 0.5 mg/mL streptavidin in 20 μL aliquots at −20 °C, and use one vial each experiment by diluting to 100 μL.

6. Wear proper PPE. Work with piranha in a fume hood, and DO NOT mix piranha with any organic solvents.
7. Take care when removing the cantilevers from the piranha solutions as they may float to the top of the solution on a gas bubble. If the cantilevers have a low spring constant, they may break during this step, so it is best to check them under a microscope to see if cantilevers are broken before proceeding.
8. Using a Teflon beaker is best during this step as the APTES will react with glass beakers.
9. It is best to work with PEG aliquots of 20–40 mg and use a fresh aliquot each functionalization prep.
10. The amount of biotin-PEG determines the surface density of Ecad, which in turn determines the event rate. It is important to keep the event rate low to ensure that single-molecule interactions are being measured with high probability. Poisson statistics predicts that the probability of forming a single bond in a given measurement is given by:

$$P(n;\lambda) = \frac{\lambda^n}{n!} e^{-\lambda}$$

where λ is the event rate and n is the number of bonds. To determine the probability that an interaction is a single bond, we use the conditional probability:

$$P(n|n > 0) = \frac{P(n > 0|n) * P(n)}{P(n > 0)} = \frac{P(n)}{1 - P(0)}$$

where $n = 1$ for a single bond.
11. It is convenient to use an empty pipette tip box. Add deionized water to the bottom of the box to decrease evaporation, and cut a piece of Parafilm and use the back of the tweezers to press on the edges to fix it to the tip box rack. The tip box lid can then be closed to retain humidity and also keep the surfaces from becoming contaminated.
12. We use a Styrofoam container to cover the tip box, so the PEG reaction occurs in the dark.
13. There may still be a water bead on the cantilever. You may wait for it to dry, or it can be carefully pipetted off the top of the cantilever or removed by dabbing with Kimwipe.
14. Sulfo-NHS-acetate should be prepared fresh each time. It is best to work with small aliquots (<10 mg).
15. This step caps the unreacted amine groups of silane molecules with acetate to reduce nonspecific interactions between the cantilever and cell membrane.
16. It can be helpful to use a second pipette tip box and replace the Parafilm as needed.

17. More than one tip can be placed in the 100 μL droplet.
18. If you are not using flasks with vented caps, open the flask a quarter turn to allow the CO_2 to enter the flask.
19. The time it takes for the cells to detach from the surface will depend on the cell lines used and how adherent they are to the surface. Keep checking under the microscope to see when the cells have detached from the surface. They will round up and float in the cell culture media.
20. To account for variations in seeding density and cell growth rates, prepare backup coverslips with different dilutions of cells.
21. The washing helps to remove dead cells that are stacked on top of the monolayer but take care not to flush the cells off the coverslip.
22. If the cells weakly adhere to the surface or keep getting washed off, you may want to culture the cells on a collagen-coated coverslip for increased adhesion to the surface.
23. Adding a tray of deionized water inside the environmental chamber during the experiment reduces evaporation of the experiment buffer.
24. The reflection of the laser should be round and symmetric as seen on the frosted screen of the scanner. Consult your AFM user manual for more information.
25. When using an uncoated cantilever, there may be a distance-dependent oscillation in each measurement. This occurs due to interference between the laser reflecting off the cantilever and the laser reflecting off the coverslip. If possible, adjust the angle between the coverslip and AFM scanner to reduce this interference. If this is not possible, you may be restricted to using a metallic coated cantilever at the cost of thermal stability.
26. We perform measurements at a range of retraction velocities (3, 5, 7, 9, 11 μm/s) to vary the unbinding force; the greater the velocity, the greater the unbinding force [18]. While in principle these measurements can be performed at any velocity, there are issues at low and high velocities. At low velocities (<1 μm/s), the cantilever may be in contact with the cell surface significantly longer than at higher velocities which may increase nonspecific interactions. At high velocities (>15 μm/s), the difference between unbinding forces due to retraction velocity is negligible [18].
27. The Keysight 5500 AFM can be controlled with Python, MATLAB, LabView, or C/C++.
28. We perform the AFM measurements by alternating the x and y positions to reduce the possibility of piezo depolarization.

29. We perform measurements at 35 positions in x and y corresponding to $35 \times 35 = 1225$ measurements per experiment. For a 20 μm area, the minimum distance between measurements is more than 500 nm.
30. Since we withdraw the AFM cantilever at a constant velocity, the standard linear solid model can be converted to time directly (using $x = vt$). Therefore, the time constant is $\tau = \frac{\mu_c}{k_c}$, and the lifetime of the bond is $t - \tau$, where t is the full tether time.

Acknowledgments

This research was supported by the National Institute of General Medical Sciences of the National Institutes of Health (R01GM121885 and R01GM133880).

References

1. Mendonsa AM, Na T-Y, Gumbiner BM (2018) E-cadherin in contact inhibition and cancer. Oncogene 37(35):4769–4780
2. Hutter JL, Bechhoefer J (1993) Calibration of atomic-force microscope tips. Rev Sci Instrum 64(7):1868–1873
3. le Duc Q, Shi Q, Blonk I, Sonnenberg A, Wang N, Leckband D, de Rooij J (2010) Vinculin potentiates E-cadherin mechanosensing and is recruited to actin-anchored sites within adherens junctions in a myosin II-dependent manner. J Cell Biol 189(7):1107–1115
4. Choi H-J, Pokutta S, Cadwell GW, Bobkov AA, Bankston LA, Liddington RC, Weis WI (2012) αE-catenin is an autoinhibited molecule that coactivates vinculin. Proc Natl Acad Sci U S A 109(22):8576–8581
5. Yao M, Qiu W, Liu R, Efremov AK, Cong P, Seddiki R, Payre M, Lim CT, Ladoux B, Mège R-M (2014) Force-dependent conformational switch of α-catenin controls vinculin binding. Nat Commun 5:4525
6. Barry AK, Tabdili H, Muhamed I, Wu J, Shashikanth N, Gomez GA, Gottardi CJ, De Rooij J, Wang N, Leckband DE (2014) α-Catenin cytomechanics–role in cadherin-dependent adhesion and mechanotransduction. J Cell Sci 127(8):1779–1791
7. Koirala R, Priest AV, Yen C-F, Cheah JS, Pannekoek W-J, Gloerich M, Yamada S, Sivasankar S (2021) Inside-out regulation of E-cadherin conformation and adhesion. Proc Natl Acad Sci U S A 118(30):e2104090118
8. Rakshit S, Zhang Y, Manibog K, Shafraz O, Sivasankar S (2012) Ideal, catch, and slip bonds in cadherin adhesion. Proc Natl Acad Sci U S A 109(46):18815–18820
9. Manibog K, Li H, Rakshit S, Sivasankar S (2014) Resolving the molecular mechanism of cadherin catch bond formation. Nat Commun 5:3941
10. Manibog K, Sankar K, Kim S-A, Zhang Y, Jernigan RL, Sivasankar S (2016) Molecular determinants of cadherin ideal bond formation: conformation-dependent unbinding on a multidimensional landscape. Proc Natl Acad Sci U S A 13(139):E5711–E5720
11. Rakshit S, Sivasankar S (2014) Biomechanics of cell adhesion: how force regulates the lifetime of adhesive bonds at the single molecule level. Phys Chem Chem Phys 16(6):2211–2223
12. Tees DF, Waugh RE, Hammer DA (2001) A microcantilever device to assess the effect of force on the lifetime of selectin-carbohydrate bonds. Biophys J 80(2):668–682
13. Zhang Y, Sivasankar S, Nelson WJ, Chu S (2009) Resolving cadherin interactions and binding cooperativity at the single molecule level. Proc Natl Acad Sci U S A 106(1): 109–114
14. Shafraz O, Rübsam M, Stahley SN, Caldara AL, Kowalczyk AP, Niessen CM, Sivasankar S (2018) E-cadherin binds to desmoglein to facilitate desmosome assembly. elife 7:e37629
15. Schmitz J, Benoit M, Gottschalk K-E (2008) The viscoelasticity of membrane tethers and its

importance for cell adhesion. Biophys J 95(3): 1448–1459

16. Bell GI (1978) Models for specific adhesion of cells to cells. Science 200(4342):618–627
17. Chakrabarti S, Hinczewski M, Thirumalai D (2014) Plasticity of hydrogen bond networks regulates mechanochemistry of cell adhesion complexes. Proc Natl Acad Sci U S A 111(25):9048–9053
18. Krieg M, Helenius J, Heisenberg CP, Muller DJ (2008) A bond for a lifetime: employing membrane nanotubes from living cells to determine receptor–ligand kinetics. Angew Chem Int Ed 47(50):9775–9777

Part II

Exposing Cells to External Mechanical Forces

Chapter 5

Application of Shear Stress to Endothelial Cells Using a Parallel Plate Flow Chamber

Jason Sedlak and Alisa Clyne

Abstract

Endothelial cell response to shear stress from the flowing blood plays an important role in diseases such as atherosclerosis. It is therefore essential to study how different flow regimes, including both atheroprotective laminar flow and atheropromoting disturbed flow, impact endothelial cell function. Shear stress effects can be studied in vitro using a variety of techniques, each with their own advantages and disadvantages. In this chapter, we describe how to use a parallel plate flow chamber to study the impact of both laminar and disturbed flow on endothelial cell monolayers.

Key words Endothelial cells, Shear stress, Atherosclerosis, Flow chamber

1 Introduction

Endothelial cells lie at the interface between the flowing blood and the tissue, and as such, they are constantly exposed to shear stress [1]. Arterial shear stress magnitude depends on flow rate, vessel radius, and blood viscosity. Straight, unbranched arterial sections are exposed to laminar flow and high, unidirectional, shear stress (10–70 dynes/cm^2). Endothelial cells in these arterial regions are elongated with their long axis and actin stress fibers aligned parallel to the blood flow direction [2–4]. Laminar flow-adapted endothelial cells are quiescent, downregulate atherogenic gene expression, and release nitric oxide, a vasodilator that is also important in preventing vascular disease through regulation of thrombosis, inflammation, dilation, and remodeling [5–8].

Endothelial cells that are located at arterial branches and curvatures are exposed to disturbed flow, including recirculation and flow reversal, and experience a lower net shear stress (<4 dynes/cm^2) [9–12]. Endothelial cells in disturbed flow have a rounder morphology [2, 3], with activation of atherogenic genes, [10, 13–16] Additionally, ECs in disturbed flow have reduced nitric oxide

Ronen Zaidel-Bar (ed.), *Mechanobiology: Methods and Protocols*,
Methods in Molecular Biology, vol. 2600, https://doi.org/10.1007/978-1-0716-2851-5_5,

production [17, 18] and are typically linked to the development of atherosclerosis [18, 19].

To study endothelial cell response to both laminar and disturbed flow, cells are often exposed to flow in vitro in a parallel plate flow chamber, cone-and-plate viscometer, parallel disk viscometer, orbital shaker, or capillary tube [1]. In a parallel plate flow chamber, fluid flows between two parallel plates, with cells cultured on the lower plate. A gasket creates the space between the two plates. The applied shear stress can be controlled by manipulating the gasket thickness, volume flow rate, fluid viscosity, or chamber geometry. Either pressure- or gravity-driven flow can be used to create the required pressure differential to drive flow [20]. The small gap size between the two plates makes the flow highly laminar, and in most cases, it is difficult to create large regions of disturbed flow.

However, we recently created a parallel plate flow chamber that induces large regions of disturbed flow by turning the flow via a system of repeating baffle protrusions [21]. In this chapter, we present methods for how to culture and seed the endothelial cells, set up a parallel plate flow chamber, and analyze the cells for alignment.

2 Materials

2.1 Collagen Coating and Cell Seeding

1. 10 mg/mL stock rat tail type I collagen solution
2. 1X phosphate buffered saline (PBS)
3. Trypsin
4. Endothelial Basal Medium-2 (EBM-2)
5. Endothelial Growth Medium-2 (EGM-2)
6. Fetal bovine serum (FBS)
7. 25 × 75 mm polystyrene microscope slides
8. Bacteria cell culture dish

2.2 Parallel Plate Flow Chamber Set Up

1. GlycoTech parallel plate flow chamber
2. Glycotech silicone gaskets (0.127 mm thickness): laminar or disturbed flow versions
3. Peristaltic pump (e.g., Ismatec Reglo Digital 8-Roller)
4. MS/CA Click-n-Go cassettes
5. PharMed BPT three-stop tubing
6. HelixMark platinum-cured silicone tubing (1.5875 mm ID)
7. Silastic tubing
8. 50 mL sterile bottle or beaker

9. 10 ml sterile syringe
10. Bubble trap (Stovall Flow Cell)
11. Valve with a small mouth (tube) adapter
12. Y-connector
13. Sterile forceps
14. Kimwipes
15. 70% ethanol

2.3 Endothelial Cell Flow Adaptation Analysis

1. 4% paraformaldehyde (PFA)
2. Phosphate-buffered saline (PBS)
3. 0.2% v/v Triton X-100 in PBS (10 mL per sample)
4. 1% w/v bovine serum albumin (BSA) in PBS
5. Hoechst 33258 (bis-Benzimide)
6. Rhodamine phalloidin
7. Aluminum foil
8. 50:50 v/v of glycerol in PBS
9. Parafilm
10. 100 mm Petri dishes (1 per sample)
11. 12 mm square glass coverslips (3–5 per sample)

3 Methods

3.1 Collagen-Coated Microscope Slide Preparation

1. Dilute collagen solution 1:1000 to a final concentration of 10 μg/mL. Create 1 mL of collagen solution per sample.
2. Place a polystyrene microscope slide in a 100 mm cell culture dish. Add 1 mL of 10 μg/mL collagen solution to each microscope slide. Use a pipet tip to evenly spread the collagen solution over the entire slide (*see* **Note 1**).
3. Carefully transfer the slide with collagen solution to the incubator, making sure that the collagen solution remains on the slide. Incubate for 3 h at 37 °C.
4. Aspirate collagen solution and rinse the slide with PBS. Store the collagen-coated slide in a sterile location until use.

3.2 Endothelial Cell Seeding

1. Culture endothelial cells to >80% confluency in EGM-2 in a tissue culture dish or flask.
2. Wash endothelial cells with 10 mL warm PBS. Detach endothelial cells from the dish or flask by adding 2 mL 0.05% Trypsin-EDTA. Monitor the cells by phase-contrast microscopy. When the cells are balled up but remain attached to the dish, they are ready to be removed.

3. Aspirate trypsin out of the dish without disturbing the cells. Add 8 mL of EGM-2 to the dish. Remove cells from the dish by repeatedly pipetting the medium across the entire dish.
4. Alternatively, add 5 mL EGM-2 to the cells to neutralize the trypsin. Pipette the cell-media mixture up and down to remove cells from the dish and divide cell clusters. Pipette the cell suspension into a 15 mL conical tube. Centrifuge at 1200 × g for 3 min. Aspirate the supernatant and resuspend the cell pellet in 10 mL EGM-2.
5. Count endothelial cells using Trypan blue to determine the number of viable cells. Create a 100,000–500,000 cells/mL solution, depending on your endothelial cell type.
6. Add 1 mL of cell solution to the microscope slide. Rock the microscope slide to carefully spread the cell solution over the entire microscope slide. If necessary, use a pipet tip to spread the cell solution to completely cover the microscope slide.
7. Transfer the slide with cell solution to the incubator, making sure that the cell solution remains on the slide. Incubate for 1 h at 37 °C.
8. Add 10 mL EGM-2 and incubate at 37 °C for 2 days until >80% confluent (*see* **Note 2**).

3.3 Parallel Plate Flow Chamber Setup (Fig. 1)

1. Spray all materials with 70% ethanol and bring into a biosafety cabinet to maintain sterility.
2. Add 50 mL EBM-2 into the sterile 50 mL bottle or beaker. Keep the media warm in a 37 °C water bath until use. 1.25 mL 1 M HEPES can be added to the media to maintain pH (*see* **Note 3**).
3. Lay the Glycotech silicone gasket (with the knobs facing up) on a Kimwipe, and ensure that there are no folds in the gasket. Spray the gasket with 70% ethanol (*see* **Note 4**).
4. Place the Glycotech flow chamber (either laminar or disturbed flow design, with the knobs facing down) onto the gasket and press firmly (Fig. 2). Lift off flow chamber and ensure that the gasket remains attached to the flow chamber. If the gasket comes off the flow chamber, repeat **steps 3** and **4**.
5. Remove any creases in the gasket, and make sure the gasket does not cover the flow inlet or outlet. Place the flow chamber with the gasket side up on the bench (*see* **Note 5**).
6. Extract 10 mL of media into the sterile 10 mL syringe. Attach a tube adapter valve on the end of the syringe. Then attach the tube adapter valve to one end of the three-stop tubing.
7. Connect the free end of the three-stop tubing to a short length of silicone tubing using a linear adapter. Then connect the free end of the silicone tubing to the positive side of the bubble trap, and open the bubble trap valve.

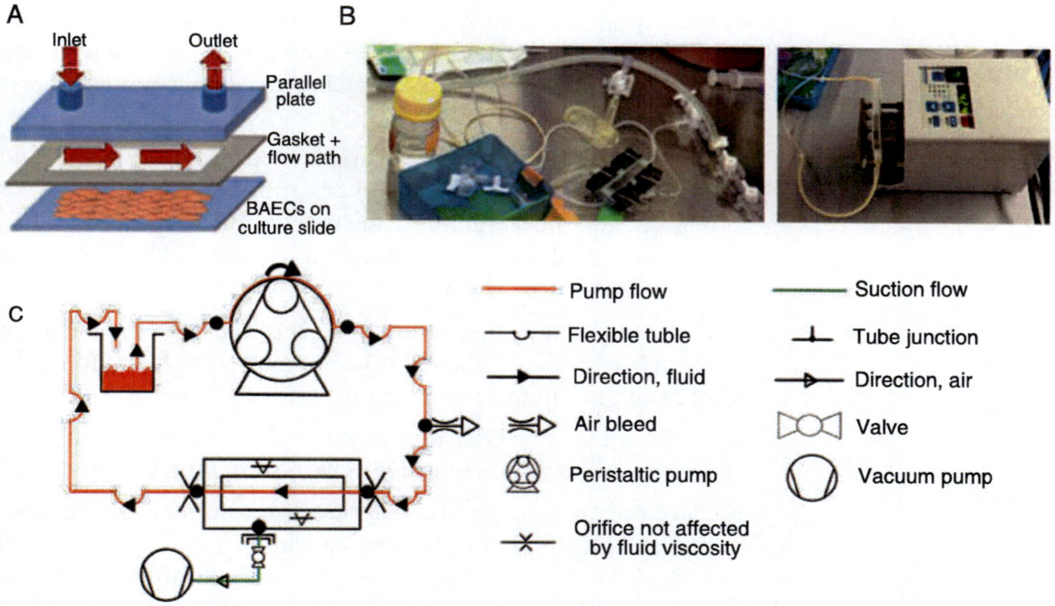

Fig. 1 General parallel plate apparatus components and assembly. (**a**) Schematic of gasket-defined flow over the culture ECs. (**b**, left) Basic circuit assembly including (clockwise) bubble trap, vacuum manifold, flow chamber with slide held on top, media reservoir. (**b**, right) Fluid flow driven by adaption to peristaltic pump. (**c**) Schematic of flow circuit with legend. Fluid pumped clockwise

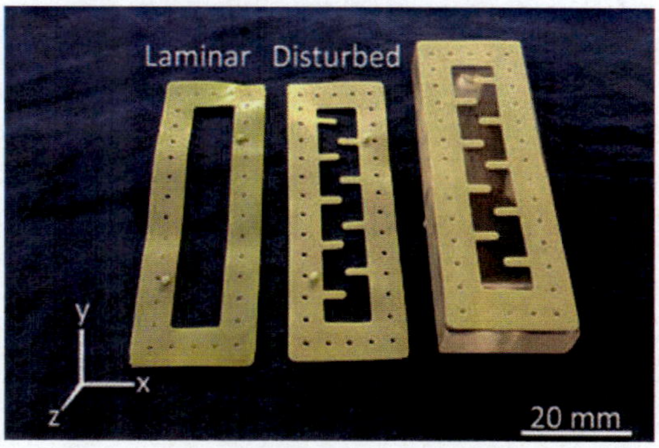

Fig. 2 Parallel plate flow chamber gaskets. (Left) Standard laminar flow and (Right) disturbed flow parallel plate flow chamber gaskets

8. Connect a short length of silicone tubing to the negative end of the bubble trap. Then connect the free end of the silicone tubing to the flow chamber inlet, keeping the gasket side of the flow chamber facing up.
9. Connect a longer length of silicone tubing to the flow chamber outlet. Place the free end of the silicone tubing into the media bottle. The end of the tubing does not need to be immersed in the media.

10. Checkpoint: You should have a media-filled syringe → three-stop tubing → silicone tubing → bubble trap → silicone tubing → flow chamber with gasket facing up → silicone tubing → media bottle.
11. Depress the syringe plunger to push media into the tubing and the bubble trap. When the media in the bubble trap rises halfway up the column, close the bubble trap valve. Then continue to dispense the media until media fully covers the parallel plate flow chamber.
12. Disconnect the syringe and attach the free end of the dialysis tubing to a longer silicone tube. Immerse the free end of the silicone tubing into the media in the bottle.
13. Retrieve the cell-seeded microscope slide from the incubator, and remove it from culture dish with sterile forceps.
14. Place the microscope slide cell side down on the flow chamber (gasket side). Start by placing one end of the slide on one end of the flow chamber, and then slowly lower the other end of the slide onto the flow chamber to prevent bubble formation. It is normal for some media to spill out. You may also want to add clamps on the sides (Fig. 1b) to provide compression to hold the microscope slide in place (*see* **Note 6**).
15. Attach the Silastic tubing to the flow chamber vacuum ports (knobs on the side). Use a Y-connector adapter to connect the tubes coming out of the two vacuum ports to each other.
16. Turn on the vacuum. Once the vacuum is steady, then connect the free end of the Y adapter to the vacuum tube.
17. After the system has stabilized for 30 s gently nudge the glass slide. The slide should be firmly in place and should not move. Do not push too hard or the slide will move, potentially killing cells (*see* **Note 7**).
18. Transfer the complete circuit to a 37 °C incubator.
19. Place the peristaltic pump either adjacent to the incubator or inside the incubator. Attach the three-stop tubing.
20. Fit the three-stop tubing into the MS/CA Click-n-Go cassette, and then attach the cassette onto the pump. Set the pump flow rate to the desired value (*see* **Note 8**). Turn on the pump and open the bubble trap valve. Once the bubble trap is three-quarters full, close the valve again.
21. Check the flow chamber for bubbles or leaks for the first 30 min. If no bubbles or leaks are observed, turn the flow chamber over so that the flat side (microscope slide) is down (*see* **Note 9**).

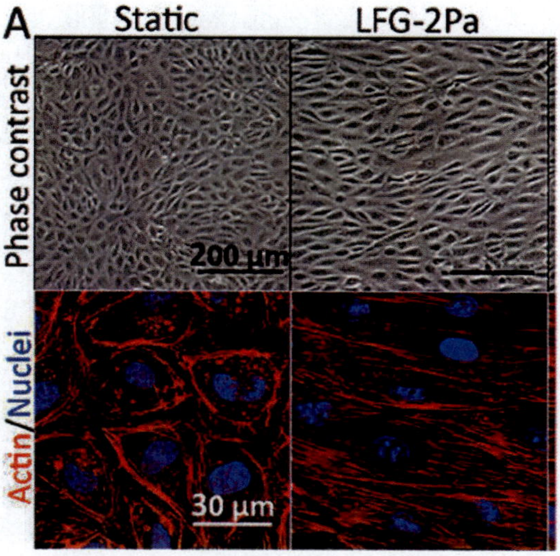

Fig. 3 Endothelial cells aligned in the flow direction in response to shear stress. (Top) Phase-contrast images (10×) of endothelial cells after 36 h of flow in the parallel plate flow chamber fitted with the laminar flow gasket. Endothelial cells in static conditions are shown as a control. (Bottom) Confocal microscopy images (63×) showing actin fibers (rhodamine phalloidin, red) and nuclei (bisbenzimide, blue)

22. Expose the endothelial cells to flow for the prescribed time. Signaling cascades and nitric oxide can be measured within minutes, while cell alignment takes 24–48 h.

3.4 Endothelial Cell Adaptation Analysis via Immunofluorescence (Fig. 3)

1. Turn off the pump and move the flow chamber apparatus into a biosafety cabinet. Disconnect the vacuum tubing from the flow chamber, and carefully remove the slide from the flow chamber. Place the slide with the cells facing up on a strip of Parafilm in a 100 mm Petri dish. Gently wash the cells with chilled PBS.
2. Add 1 mL of 4% paraformaldehyde (room temperature) to cover the slide. Incubate at 4 °C for 15 min.
3. Aspirate the paraformaldehyde and wash the slide two times with chilled PBS. At this point, the slide can be covered with PBS and stored at 4 °C for several days prior to moving on to the next step.
4. Permeabilize the cells by adding 10 mL 0.2% v/v Triton X-100 in PBS. Incubate for 15 min at room temperature. Wash the samples two times in PBS (*see* **Note 10**).
5. Block nonspecific binding by incubating the samples in 10 mL of 1% w/v BSA in PBS for 1 hour at room temperature. Wash the samples two times in PBS.
6. Create 300 µL of labeling solution per sample. For example, if labeling for nuclei and actin, create a 1:2000 dilution of

Hoechst and a 1:100 dilution of rhodamine phalloidin in 1% w/v BSA in PBS (1:2000 dilution).

7. Place a sheet of Parafilm on the bottom of a 100 mm Petri dish (1 per sample), which is placed on top of a sheet of aluminum foil. Place 300 μL of labeling solution in droplets on the Parafilm, with the droplets sufficiently spaced out so as to cover the entire glass slide. Place the glass slide, cell side down, onto the droplets of labeling solution using the same method as in 3.3.14. If fluid spills out from under the glass slide and touches the top surface, some or all of the labeling solution may be wicked out from under the glass slide. In that situation, lift the slide off the labeling solution, wipe the slide dry on the top and sides with a KimWipe (taking care not to touch the cells on the bottom slide surface), and repeat the slide placement on the labeling solution.
8. Cover each dish and wrap it in the aluminum foil to protect it from the light. Incubate the samples with the labeling solution for one hour at room temperature.
9. Wash the samples two times in PBS, continuing to protect the samples from light.
10. Samples can be imaged as is using an upright fluorescent microscope. If using an inverted fluorescent microscope, mount the samples by adding 10 μL drops of 50:50 v/v of glycerol in PBS to the cell side of the glass microscope slide in locations where you would like to image the cells. Carefully place 12 mm glass coverslips onto the drops. Use clear nail polish to seal the edges of the coverslips.
11. Store the slides at 4 °C until imaged (*see* **Note 11**).

3.5 Cleaning and Sterilizing the Flow Chamber

1. Aspirate all of the media from the flow circuit.
2. Wash all components with 10% bleach solution, followed by multiple rinses in deionized water (*see* **Note 12**).
3. Autoclave all components that can be autoclaved, including the media bottle, metal tools, glass components, and tubing.

4 Notes

1. Glass microscope slides can also be used for the parallel plate flow chamber setup. We found that the polystyrene slides were better for cell adhesion and alignment.
2. It is essential that the cells are not fully confluent or overconfluent. If the cells are too confluent, they will not align in response to flow.
3. Maintaining media pH at 7.4 can be challenging. If the cells are not surviving or aligning in flow, check the media pH. HEPES

can be added as an additional buffer, or media pH can be adjusted using an acid or base.

4. Either the laminar or disturbed flow gasket can be used here. Thicker gaskets can also be used. However, the volume flow rate will need to be adjusted to create the desired shear stress.
5. The gasket must be smooth and without creases. Otherwise, you will not be able to hold the vacuum and keep the parallel plate flow chamber sealed.
6. Ensure that there are no bubbles in the parallel plate flow chamber or the tubing leading up to the chamber. Bubbles will shear the endothelial cells off the surface if they go through the flow chamber.
7. You can also check at this point to make sure that the vacuum is not aspirating media out of the parallel plate flow chamber. If you see continuous media flow into the vacuum tubing, you have a leak and you should disassemble and then reassemble the flow chamber.
8. The flow rate to achieve the desired shear stress (τ) comes from the infinite parallel plate solution of the Navier Stokes equations.

$$\tau = \frac{6\mu Q}{bh^2}$$

where μ is viscosity, Q is volume flow rate, b is chamber width, and h is chamber height.

9. You can mark the starting media level on the media bottle at the beginning of the experiment to help identify fluid loss due to leaks. When first using the parallel plate flow chamber, we start with short-term experiments (1–6 h) and then move up to the longer times (12–48 h).
10. Methanol can also be used for simultaneous fixation and permeabilization. The best fixation method depends on the antibody used.
11. Many antibodies are best imaged the same day. Actin and nuclei are relatively stable and can be imaged several days after fixation.
12. 70% ethanol or UV light can also be used to sterilize parallel flow chamber components.

Acknowledgments

This work was support by NIH grants (1R01HL140239, DK102107-01) and NSF grants (1916997, 0846751).

References

1. Chiu J-J, Chien S (2011) Effects of disturbed flow on vascular endothelium: pathophysiological basis and clinical perspectives. Physiol Rev 91:327–387. https://doi.org/10.1152/physrev.00047.2009
2. Langille BL, Adamson SL (1981) Relationship between blood flow direction and endothelial cell orientation at arterial branch sites in rabbits and mice. Circ Res 48:481–488
3. Nerem RM, Levesque MJ, Cornhill JF (1981) Vascular endothelial morphology as an indicator of the pattern of blood flow. J Biomech Eng 103:172–176
4. Wong AJ, Pollard TD, Herman IM (1983) Actin filament stress fibers in vascular endothelial cells in vivo, vol 219. Science, New York, pp 867–869
5. Berk BC (2008) Atheroprotective signaling mechanisms activated by steady laminar flow in endothelial cells. Circulation 117:1082–1089. https://doi.org/10.1161/circulationaha.107.720730
6. Federici M et al (2002) Insulin-dependent activation of endothelial nitric oxide synthase is impaired by O-linked glycosylation modification of signaling proteins in human coronary endothelial cells. Circulation 106:466–472. https://doi.org/10.1161/01.CIR.0000023043.02648.51
7. Li YS, Haga JH, Chien S (2005) Molecular basis of the effects of shear stress on vascular endothelial cells. J Biomech 38:1949–1971. https://doi.org/10.1016/j.jbiomech.2004.09.030
8. Traub O, Berk BC (1998) Laminar shear stress: mechanisms by which endothelial cells transduce an atheroprotective force. Arterioscler Thromb Vasc Biol 18:677–685
9. Cunningham KS, Gotlieb AI (2005) The role of shear stress in the pathogenesis of atherosclerosis. Lab Investig J Techn Methods Pathol 85:9–23. https://doi.org/10.1038/labinvest.3700215
10. Malek AM, Gibbons GH, Dzau VJ, Izumo S (1993) Fluid shear stress differentially modulates expression of genes encoding basic fibroblast growth factor and platelet-derived growth factor B chain in vascular endothelium. J Clin Invest 92:2013–2021. https://doi.org/10.1172/jci116796
11. Asakura T, Karino T (1990) Flow patterns and spatial distribution of atherosclerotic lesions in human coronary arteries. Circ Res 66:1045–1066
12. Bharadvaj BK, Mabon RF, Giddens DP (1982) Steady flow in a model of the human carotid bifurcation. Part I--flow visualization. J Biomech 15:349–362
13. Hsiai TK et al (2001) Pulsatile flow regulates monocyte adhesion to oxidized lipid-induced endothelial cells. Arterioscler Thromb Vasc Biol 21:1770–1776
14. Hsiai TK et al (2003) Monocyte recruitment to endothelial cells in response to oscillatory shear stress. FASEB J 17:1648–1657. https://doi.org/10.1096/fj.02-1064com
15. Huang C-C et al (2015) Glycolytic inhibitor 2-deoxyglucose simultaneously targets cancer and endothelial cells to suppress neuroblastoma growth in mice. Dis Models Mech 8: 1247. https://doi.org/10.1242/dmm.021667
16. Wilcox JN, Smith KM, Williams LT, Schwartz SM, Gordon D (1988) Platelet-derived growth factor mRNA detection in human atherosclerotic plaques by in situ hybridization. J Clin Invest 82:1134–1143. https://doi.org/10.1172/jci113671
17. Noris M et al (1995) Nitric oxide synthesis by cultured endothelial cells is modulated by flow conditions. Circ Res 76:536–543
18. Yetik-Anacak G, Catravas JD (2006) Nitric oxide and the endothelium: history and impact on cardiovascular disease. Vasc Pharmacol 45: 268–276. https://doi.org/10.1016/j.vph.2006.08.002
19. Ku DN, Giddens DP, Zarins CK, Glagov S (1985) Pulsatile flow and atherosclerosis in the human carotid bifurcation. Positive correlation between plaque location and low oscillating shear stress. Arteriosclerosis (Dallas, Tex) 5:293–302
20. Davis CA et al (2015) Device-based in vitro techniques for mechanical stimulation of vascular cells: a review. J Biomech Eng 137: 040801. https://doi.org/10.1115/1.4029016
21. Sedlak JM, Clyne AM (2020) A modified parallel plate flow chamber to study local endothelial response to recirculating disturbed flow. J Biomech Eng 142. https://doi.org/10.1115/1.4044899

Chapter 6

Cell Stretcher Assay to Analyze Mechanoresponses to Cyclic Stretching

Robin Püllen, Jens Konrad, Bernd Hoffmann, and Rudolf Merkel

Abstract

In their natural environment, most cells and tissues are continuously exposed to cyclic mechanical strain. Sensing these stimuli by mechanosensory proteins and subsequent conversion into a variety of biological responses (referred to as mechanotransduction) are key processes for tissue homeostasis, survival, and differentiation. Perturbations of underlying signaling pathways lead to severe diseases in vivo (Urciuoli E, Peruzzi B, Int J Mol Sci 21(24). https://doi.org/10.3390/ijms21249426, (2020)). In addition, cellular mechanoresponses to cyclic stretching of an isolated single cell differ from those of a cell monolayer, network, or even three-dimensional tissue. Since these processes depend on various physical and biological parameters, the development of a precise, well-characterized, and highly reproducible but also easily tunable stretcher assay is indispensable. Here, we describe the fabrication of defined elastic substrates and their application in cyclic stretching of cultured cells in a custom-made cell stretcher device. We focus on the detailed description of the system and provide a possibility for mechanoresponse characterization, using the analysis of actin stress fiber orientation as exemplary mechanoresponse to cyclic stretching of adherent cells.

Key words Mechanobiology, Cell stretcher, PDMS, Cyclic stretching, Mechanoresponse, Uniaxial, Elastic substrate, Mechanosensing

1 Introduction

Cells within their native environment are exposed to a variety of both physiological (e.g., movement and muscle contraction or pumping blood flow) or pathological (e.g., brain trauma, chafing, and other injuries) mechanical stresses. During the last two decades, it has become increasingly clear that cells not only sense these external mechanical stimuli but also respond to them through various mechanisms (reviewed in [1]), both in the short and long term (see, e.g., [2, 3]). Yet, we are just beginning to understand the underlying molecular pathways and processes as well as inter-cooperative effects of closely connected cells in an integrated tissue (see, e.g., [4–6]). To add another layer of complexity, specific mechanoresponses differ not only between cell types but also

Ronen Zaidel-Bar (ed.), *Mechanobiology: Methods and Protocols*,
Methods in Molecular Biology, vol. 2600, https://doi.org/10.1007/978-1-0716-2851-5_6,

depend on parameters such as cell density, strain profile, amplitude, and frequency [7–9]. Therefore, an increasing number of in vitro analysis tools and techniques have been implemented to explore stretch-induced mechanoresponses (reviewed in [10, 11]). Uniaxial, biaxial, or equibiaxial stretching of cells on an elastic silicone rubber by pneumatic or electromagnetic actuation is the most common stretching concept in mechanobiological research. In contrast to multiaxial stretching, uniaxial strain profiles have the advantage of an intuitive and more direct correlation between mechanical impact and biological response and also simulate strain parameters most often found in vivo. While pneumatic actuation requires low technical investment and is more suitable for equibiaxial stretching, electromagnetic actuation generally provides higher precision and scalability as well as more experimental flexibility. Both principles are implemented in commercially available cell stretcher systems (e.g., STREX, Osaka, Japan or FlexCell, Burlington, NC, USA). However, these systems and associated consumables are costly and barely adaptable to individual needs. The majority of cell stretchers found in the literature are therefore custom-made systems designed for specific experimental requirements and budgets and are only suitable for a narrow range of applications [12, 13]. Here, we present the principle of our cell stretcher assay based on a simple example of actin stress fiber alignment and reorientation upon uniaxial, cyclic stretching with our custom-made, easily tunable stretcher system for multiple elastomer stretching in parallel. The cytoskeletal reorientation response away from the direction of uniaxial strain is a universal and vital process described for many cell types [14–16]. The fabrication of our well-characterized elastic substrates and the subsequent cultivation of cells on them are spatiotemporally separated from the stretching device and the actual stretching assay. Thus, parallel sample preparation under different conditions or treatments and multiple post-processing options are applicable. This facilitates a versatile approach and is transferable to a constantly increasing number of research questions related to stretch-induced mechanoresponses. Here, we describe our cell stretcher design and its benefits in detail and refer to customization options at related steps throughout the protocol.

2 Materials

Unless otherwise stated, warm all solutions that come in direct contact with living cells in a water bath to 37 °C prior to use. Prepare all buffers using deionized, ultrapure water, and store reagents according to manufacturer's instructions.

2.1 A7r5 Smooth Muscle Cell Culture

1. Smooth muscle cell line A7r5 (*see* **Note 1**).
2. Growth medium: Add 10% (v/v) fetal bovine serum of premium quality, 100 units/mL penicillin, 0.1 mg/mL streptomycin, and 2 mM L-alanine-L-glutamine dipeptide to Dulbecco's Modified Eagle's Medium (DMEM with 1 g/L glucose and additional pyruvate but without phenol red and glutamine).
3. Phosphate buffered saline (PBS): Solve 155.2 mM NaCl, 1.06 mM KH_2PO_4, and 2.97 mM $Na_2HPO_4 \times 12\ H_2O$ in water and adjust pH with 2 M NaOH to 7.4. Sterilize by autoclaving and check pH (7.3–7.5) and osmolality (290–320 mOsm/kg).
4. Trypsin-EDTA solution (0.05%).
5. Standard T75 culture flask.
6. Biological safety cabinet.
7. Humidified cell culture incubator at 37 °C and 5% CO_2.

2.2 Elastomer Substrates

1. Two-part, platinum-catalyzed polydimethylsiloxane (PDMS) elastomer Sylgard™ 184 Kit (*see* **Note 2**).
2. Suitable and inert gloves, spatulas, and container or weighing pan (*see* **Note 3**).
3. Laboratory balance.
4. Desiccator with connected vacuum pump.
5. Oven with planar surface set at 60 °C.
6. Positive displacement pipette for viscous media with disposable tips.
7. Chamber molds (Fig. 1) with polished surface (*see* **Note 4**).
8. Isopropanol for cleaning and mold disassembly.

2.3 Preparation of Elastomer Chambers

1. Chamber holder with frame (Fig. 2) and matching screwdriver for Torx T20 (*see* **Note 5**).
2. Calibration block for 80 mm (so-called zero position) and 81.5 mm pin distance (so-called 1.5 mm pre-stretch position).
3. Isopropanol in a beaker to briefly sterilize the substrates.
4. Box for handling and transportation of mounted and sterilized chambers.
5. Coating solution: Solve 20 μg/mL human fibronectin (*see* **Note 6**) in PBS.

2.4 Cell Stretching and Analysis of Cytoskeletal Mechanoresponse

1. 6× Cell stretcher system (Fig. 3 and more information at www.fz-juelich.de/ibi/ibi-2) for uniaxial stretching assays (*see* **Note 7**).

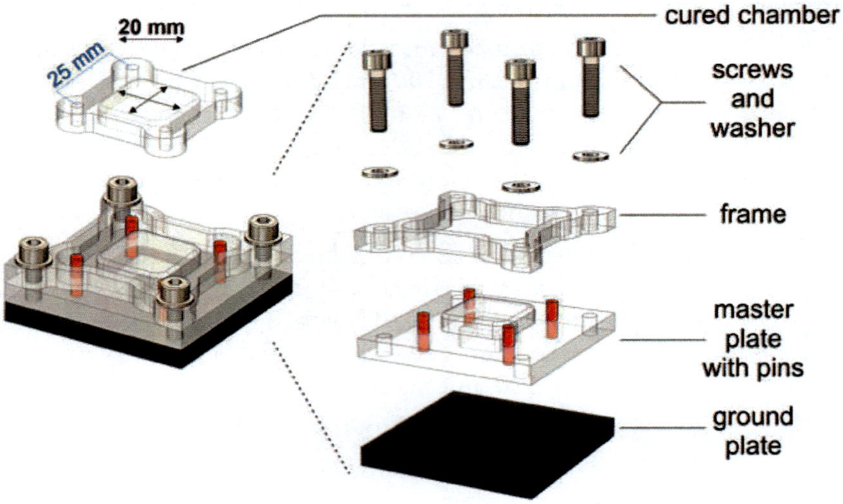

Fig. 1 Mold for casting of PDMS-based elastomer substrates for cell stretcher assays. The three-part system guarantees an easy assembly and disassembly and therefore fabrication of elastomer chambers with reproducible geometry and reliable mechanical properties. Cured chamber molds exhibit an inner area of 20 × 20 mm encircled by a 5 mm rim

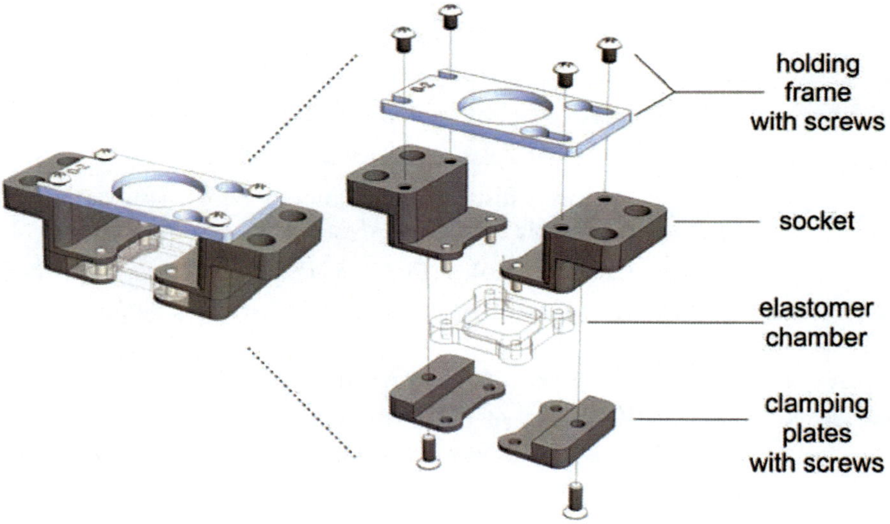

Fig. 2 Holder design for immobilization and stabilization of individual elastomer substrates. The elastomer chambers are clamped between detachable plates of defined thickness and two sockets on each side. Together, plates and sockets form the chamber holder. The sockets physically connect the elastomer chamber to the cell stretcher, while the variable holding frame on top enables to lock the chamber at any position and thus permits working outside the cell stretcher

2. For optimal preservation of cytoskeletal structures, prepare cytoskeleton buffer (CB): Solve 5 mM ethylene glycol-bis (2-aminoethylether)-*N,N,N′,N′*-tetraacetic acid (EGTA), 5 mM D-Glucose, 5 mM 2-(*N*-morpholino)ethanesulfonic

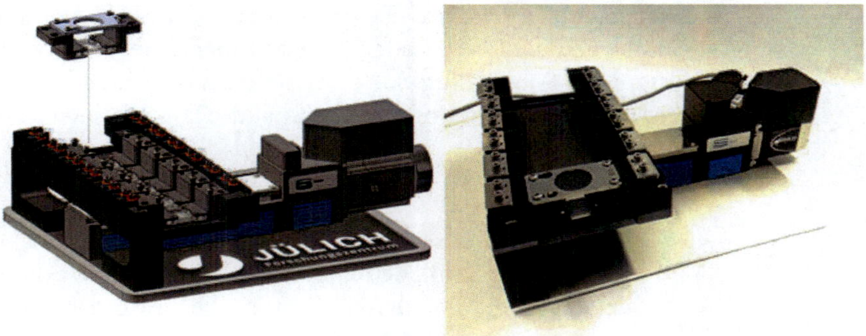

Fig. 3 Cell stretcher setup for simultaneous, uniaxial stretching of six elastomer chambers. With a total size of 350 × 280 × 100 mm, the instrument fits into any standard CO_2 incubator. The specifications of the installed linear stage offer a wide range of possible applications. For this, the motor is connected to a computer via a thin cable and precisely controlled at all times. Setup is indicated as CAD image (left) and actual system (right)

acid (MES), 5 mM $MgCl_2$, 150 mM NaCl in deionized water. Adjust pH to 6.1 with 5 M NaOH solution. Depending on established immunofluorescence protocols, alternative buffers like PBS can be used as well.

3. Fixative: 3.7% formaldehyde in CB (*see* **Note 8**).
4. Quenching solution: 100 mM glycine in CB.
5. Glass cutter.
6. Trimmed (approx. 44 × 16 mm) glass or plastic slide and Parafilm® to cover the elastomer chamber.
7. 1% (v/v) Octoxynol-9 in CB.
8. Phalloidin-Atto 488 conjugate or similar.
9. Longitudinal-trimmed (approx. 76 × 13 mm) standard microscopy slides.
10. Razor blade and/or scalpel.
11. Aqueous mounting medium (e.g., Fluoromount-G™ or Prolong™ Glass).

3 Methods

Handling of fluid and curing of PDMS should ideally be performed in a separate room or working space without typical latex gloves nearby to avoid contamination with potential catalyst poisons (*see* **Note 3**). After sterilization of fabricated elastomer substrates and for all manipulations on cells, working under aseptic conditions with sterile solutions is highly recommended.

3.1 Cell Culture and Passaging

A7r5 cells (*see* **Note 1**) are routinely cultured in T75 tissue culture flasks at 37 °C and 5% CO_2 until they reach 80–90% confluency (approximately every 2–3 days) and subcultured in a ratio of 1:2 or 1:3 in new flasks or seeded on elastomer chambers for stretching assays (*see* Subheading 3.4). Dissociation of adhered A7r5 cells on plastic for passaging or experimental use is performed as follows:

1. Aspirate medium from a T75 flask with 80–90% confluent cells and rinse with PBS.
2. Aspirate PBS and add 2 mL Trypsin-EDTA (0.05%) solution. Rock gently and incubate for 5–7 min until cells begin to detach from the surface. Tap at the bottom of the flask to mechanically detach remaining cells.
3. Add 6 mL growth medium and transfer the suspension in a centrifuge tube.
4. Centrifuge at 180 × g for 3 min, aspirate the supernatant, and resuspend the cell pellet in 2 mL growth medium. Either split the suspension for subculturing in a new T75 flask with 12 mL growth medium or determine the concentration for further use in cell stretching assays.

3.2 Fabrication of Elastomer Chambers

1. Assemble and gently clean molds with isopropanol and dust-free paper tissue.
2. Mix the can containing the base material well by shaking upside down before each use.
3. Weigh 40 parts of base and add 1 part of curing agent as precisely as possible (*see* **Note 9**). Prepare at least 4 g per elastomer chamber.
4. Mix thoroughly with a spatula for at least 3–5 min and change stirring direction several times to ensure homogeneous elastomer properties.
5. Degas fluid PDMS under mild vacuum in a desiccator (*see* **Note 10**).
6. Slowly pipette 3.8 mL of the fluid PDMS mixture into each chamber mold. Let the viscous PDMS spread evenly in the mold on a flat surface, and remove any remaining air with a standard pipette or needle.
7. Cure the PDMS at 60 °C for 16 h.
8. Disassemble chamber molds by unscrewing and slightly lifting the outer frame (*see* **Note 11**).
9. Store silicone chambers protected from light and dust at room temperature (*see* **Note 12**).

3.3 Mounting of Elastomer Substrates and Surface Coating

1. Assemble chamber holders and use the calibration block to adjust and fix the holding frame to 1.5 mm pre-stretch position (*see* **Note 13**).

2. Wipe holder parts, screw driver, and transport box with ethanol, and let them dry in a sterile biosafety cabinet for several minutes.
3. Place the prepared elastomer chambers and a beaker with isopropanol in the biosafety cabinet, and put a holder upside down next to it.
4. Briefly dip each chamber in isopropanol (*see* **Note 14**), shake off remaining solvent, and mount the chamber using the four pins of the holder sockets. The open side of the box-shape chamber should point toward the floor or the upper frame. Add a clamping plate with a screw on both sides and tighten the screws. Due to their defined spacing, the clamping plates enable mounting with reproducible contact pressure.
5. Place the mounted chambers overnight in an incubator at 37 ° C to allow all remaining solvent to evaporate completely and to let the chambers equilibrate in the humidified environment.
6. After overnight evaporation of remaining solvent, add 500 μl coating solution to every chamber surface, and spread the solution by moving in an L-shape. Incubate at 37 °C for at least 1 h (*see* **Note 15**).
7. The chambers are now ready for cell seeding or optional pre-experimental treatment.

3.4 Cell Seeding and Pre-incubation

1. Cells are harvested from the T-flask as described (*see* Subheading 3.1), and the suspension is adjusted to 24,000 A7r5 cells/mL for a single-cell assay. Used cell numbers depend on specific research question and cell type.
2. Remove the coating solution from the elastomer chamber, and immediately add 500 μL cell suspension (*see* **Note 16**). Avoid surface drying. Swirl the chamber horizontally in an L-shape several times to evenly distribute the cells (*see* **Note 17**).
3. Cover the top side of the elastomer chamber with a trimmed glass or plastic slide and a piece of Parafilm® to minimize medium evaporation and the risk of contaminations.
4. Place the chambers back in a humidified incubator at 37 °C, and let the A7r5 cells adhere for 16 h or overnight. Adhesion times can vary and depend on cell type and coating.
5. After proper cell adhesion, remove old culture medium and nonadhered cells, and add the same volume of fresh medium (*see* **Note 18**). Cover the chamber again with a slide and a piece of Parafilm®.

3.5 Cell Stretcher Assay

1. Adjust the cell stretcher system to zero position (equal to 80 mm pin distance), and set the pre-stretch in the cell stretching software tool to 1.5 mm (*see* **Note 19**).

2. Mount up to six chamber holders in the cell stretching device, and remove the holding frames.
3. Place the prepared cell stretching device and additional non-stretch controls in a humidified incubator, and program the desired stretch protocol in the cell stretching software (*see* **Note 20**).

   For an exemplary cyclic (trapezoid) stretching of 20% at a frequency of 300 mHz, this would result in:

   Amplitude: 4.3 mm.

   Velocity (equal stretch and release): 3.44 mm/s.

   Dwell time (equal stretch and release): 0.417 s
4. Click on "Run OK" to activate the stretching protocol.

3.6 Fixation and Staining of the Actin Cytoskeleton

1. At desired termination time (e.g., here after 30 min and 4 h), click on "Release Halt" to stop the cyclic stretching in pre-stretch position. Reinstall holding frames on the corresponding samples, and remove them from the stretching device. Subsequently, click again on "Run OK" to continue the stretching program for the remaining samples (*see* **Note 21**).
2. For fixation: Aspirate the cell culture medium from the chambers, and add 500 μL 3.7% formaldehyde in CB.
3. Incubate for 15 min at 37 °C, remove fixative, and wash twice with 100 mM glycine in CB.
4. Wash the chambers twice with CB.
5. Aspirate buffer and permeabilize the cells with 1% (v/v) Octoxynol-9 in CB at room temperature for 10 min.
6. Rinse three times with CB.
7. Aspirate buffer and add 300 μL CB with 1 μL Phalloidin-Atto 488 (Stock 20 μM). Incubate for 1 h protected from light (*see* **Note 22**).
8. Press a longitudinal-trimmed glass slide under the bottom of every elastomer chamber to stabilize the sample and ensure later microscopy (*see* **Note 23**).
9. Aspirate the staining solution, wash twice with CB, and add 20 μL of aqueous mounting medium in the center of the chamber (*see* **Note 24**).
10. Put a circular coverslip (Ø 15 mm, #1) on top, and allow mounting medium to cure according to the manufacturer's instructions. Avoid trapping air bubbles under the coverslip.
11. Use a sharp razor blade or scalpel to cut the elastomer chamber along the edges of the slide, and remove the side walls. The sample is then ready for imaging and further analysis by widefield and confocal light microscopy applications with appropriate filter settings (Fig. 4, *see* **Notes 25**).

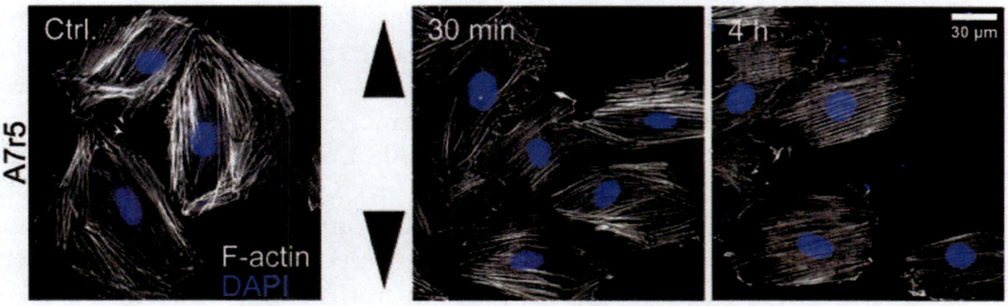

Fig. 4 Cyclic stretch-induced actin fiber reorientation of A7r5 smooth muscle cells. The cells were uniaxially stretched by 20% at a repeat frequency of 300 mHz for indicated times on elastomeric substrates. After fixation, F-actin (grayscale) and nuclei (blue) were fluorescently stained. Upon stretch, cell body and actin fibers align in a time-dependent manner perpendicular to the direction of stretch (indicated by bold arrowheads). Images kindly provided by L. Lövenich, IBI-2: Mechanobiology, Forschungszentrum Jülich. Actin reorientation analysis can be performed as previously described [7, 9]

4 Notes

1. The smooth muscle cell line A7r5 was intensively used by our group to investigate actin stress fibers and related mechanically induced autophagosome formation [9] and serves here just as an example for an adhesive cell type. Besides, we extensively used different primary cell types (e.g., neurons and astrocytes, fibroblasts, or keratinocytes [4, 5, 7]) and a variety of cell lines in single cell as well as monolayer and even microtissue analysis under mechanical strain. Coating of elastomer substrates as well as cell seeding and culture conditions can be easily adapted to established requirements and/or cell types (*see also* **Notes 7, 15–18**).
2. Sylgard™ 184 is a well-characterized and widely used standard elastomer system. It can be used as a linear elastic material for up to 50% strain in a wide range of applications (in terms of coating material, strain amplitude, substrate stiffness, etc.). For certain experimental questions and applications (e.g., very high strain amplitude), other PDMS-based elastomer systems with different attributes may be required and are already established in our group (*see also* **Notes 9, 13, 23**). Important properties for fabrication of silicone substrates by two-component systems like pot life, curing time, and temperature depend on the elastomer type and may vary.
3. Cross-linking by addition curing is sensitive to catalyst poisons like sulfur or tin compounds. These are frequently found in consumables like medical latex gloves. As suitable option, latex- and silicone-free cleanroom gloves can be used.

4. The molds function as negative templates for a standard replica technique and are build up as a three-part construct for easy assembly and disassembly (Fig. 1). The design leads to elastomer substrates (called chambers) with a well-defined geometry (patent number 20120208229). Here, the inner thin part covers 20 × 20 mm and has a polished surface to generate a plane and smooth cell adhesion area without any topographical cues for cells. For specific purposes, however, a certain pattern can be applied (e.g., microstructures or micromolds). This rectangular area with a thin bottom (approx. 500 μm thickness) is surrounded by a 5 × 5-mm-thick wall that stabilizes the chamber and ensures a reliable strain profile. In contrast to other substrate designs for cyclic cell stretching, the strain is linearly applied in an almost fully uniaxial direction. This means that the strain in x-direction is barely accompanied by lateral compression in y-direction (e.g., $\varepsilon_{xx} = 0.17$, $\varepsilon_{yy} = -0.024$, and $\varepsilon_{xx} = 0.28$, $\varepsilon_{yy} = -0.033$ for 370 kPa stiff elastomer chambers), but by a change in material thickness, that is not recognized by the cells. Moreover, strain levels are almost constant over the whole area. Consequently, the experimental conditions are precisely determined and reduce the result's spread. Additionally, the box-shaped chamber design allows for spatiotemporally unlimited cell cultivation. Thus, parallel experiments under equal conditions but with different cells/incubation times/pattern/drugs/read-outs, etc. are easy to implement.
5. The holder connects the moving stretcher with the elastomeric chambers. It must prevent slippage of the elastomeric substrate under load. A slight contact pressure is therefore necessary and is defined by the choice of clamping plates with different thicknesses. This results in a defined gap between clamping plate and socket and thus in a defined compression of the elastomer chamber that is clamped in this gap (Fig. 2).

   In order to enable cell seeding, cultivation, and postprocessing of multiple samples spatiotemporally separated from the stretching device, the elastomer substrate must be stabilized and fixed in position whenever detached from the stretcher. For this, the bright metal frame on top of the chamber holder (Fig. 2) can be locked by four screws in variable positions. After installation of the holder in the cell stretcher, the metal frame must be removed before stretching starts. All parts are resistant to solvents and fully autoclavable. They can be used independently of the cell stretching device in humidified CO_2 incubators.
6. Fibronectin works well for most cell lines and primary cells as standard coating of hydrophobic silicone surfaces by physisorption. Nevertheless, other coating materials (e.g., poly-L-lysine or artificial ECM like Matrigel®) have been used successfully

for certain cell types as well. However, pure type 1 collagen seems to adhere poorly on hydrophobic silicone chambers.

7. The stretching device can be used for up to six chamber holders in parallel and fits in every standard CO_2 cell culture incubator (Fig. 3). In this setup, a stepper motor drives a precise linear translation stage (LTP-100-DC-R, Steinmeyer Mechatronik, Dresden, Germany) and performs the precise uniaxial movement at one side of the installed chambers, while the other side remains fixed. A custom-made software tool (*see also* **Note 20**) enables the accurate spatiotemporal control of various parameters (e.g., stretch frequency, amplitude, and time interval between stretch cycles).

   Based on this principle, we developed further cell stretchers for specific applications in mechanobiology. In recent work, systems for live cell imaging [16] and atomic force microscopy [17] but also long-term stretching and tensile testing of living tissue samples [18] have been implemented. For more information about the different systems, visit the IBI-2 website at www.fz-juelich.de/ibi/ibi-2.

8. A 10% formalin has proven to be useful for staining of many antigens by secondary immunofluorescence. However, methanol fixation for 10 min at −20 °C might be better for some epitopes and is also applicable on elastomers. After the methanol fixation, the chambers should additionally be stored at RT for at least 1 h to allow the methanol to evaporate from the elastomer.

9. For example, weigh 40 g of base and 1 g of curing agent to the third decimal place. The 40 + 1 ratio results in an elastomer stiffness of approximately 50 kPa for the current batch of Sylgard™ 184 Kits with the Poisson's ratio being approximately 0.5. By using a specific ratio for every two-component system, a defined Young's modulus can be adjusted. These mixing ratios must be (re)calibrated for every new batch by indentation as described in [19].

10. Degas until any air trapped in the mixture is removed. Vent briefly if mixture starts to foam over, and do not degas longer than 30 min or until larger bubbles appear. This may interfere with the curing process.

11. Soft PDMS-based elastomers are sticky and sometimes difficult to remove from the mold. Adding a few drops of isopropanol around the edges and twisting the chamber will aid in separation. Use isopropanol sparingly as it can partially diffuse into silicone and damage the cells if partially remaining (*see* **Note 14**).

12. Petri dishes are useful to store prepared elastomer chambers. Chambers can be stored for several months but no longer than a year post fabrication.

    For certain applications (e.g., traction force microscopy or calibration of strain profile), fluorescently labeled microparticles can be covalently coupled on the elastomer surface at this point [17, 20].

13. Application of a certain, basic strain (so-called pre-stretch) as starting position before adding any kind of cells is necessary to avoid potential sagging of the chamber, which is most likely a result of clamping and added medium. Soft elastomer chambers (like 50 kPa used here) can be mounted directly in 1.5 mm pre-stretch position. Stiffer elastomer chambers may require higher contact pressure to prevent slipping during the experiment and therefore a higher pre-stretch (e.g., 3 mm for approximately 400 kPa stiff chambers). To apply a pre-stretch of more than 1.5 mm (≙7.5%), the chambers are mounted in zero position (use calibration block to adjust holders to 80 mm pin distance) and then moved to the desired pre-stretch position using the stretcher system (*see also* **Notes 7, 19**). All following experimental data about amplitude, cell adhesion area, and other parameter are related to the pre-stretched chamber as initial position. Conversely, also negative strain (compression) can be applied.

14. Immerse the substrates for 5–10 s but no longer for sterilization. Isopropanol can diffuse into cross-linked PDMS and enter the medium during an experiment with living cells. Do not use ethanol. It diffuses even faster into the silicone.

15. PDMS-based silicone substrates are hydrophobic. Be sure to evenly distribute the solution by swirling several times. Other coating materials (*see* **Note 6**) possibly require slightly different concentrations and incubation times. However, always use buffer with osmolalities in the range of 150–320 mOsm/kg instead of dH_2O as solvent.

16. Volumes between 500 and 800 μL are appropriate to cover the cells during experiments but avoid spillage when moving chambers from cabinet to incubator.

17. Especially for higher cell densities, repetitive swirling during the initial adhesion phase (e.g., three times during the first hour) improves the formation of homogeneous cell layers.

18. Depending on the research question, further manipulations can be applied in parallel at this point of the protocol. We performed several experiments with medium containing toxins or drugs or initiated differentiation of primary cells. In addition, the delivery of plasmid DNA or siRNA with liposomal or fusogenic transfection systems or viral transduction assays from

various suppliers are possible. Protocols may need to be adjusted to account for the hydrophobic nature of PDMS-based silicone.

19. For stretcher adjustment to zero position, an empty chamber holder (set to zero with the calibration block) or a corresponding metal plate with defined distance can be used. To move the stretcher precisely and set the effective stretcher position as new zero position, use the options under the tab "Setup Position." The desired pre-stretch (in mm) is then preset by the user under the tab "Stretcher Control."
20. The precise movement of stepper motor and connected linear translation stage are spatiotemporally controlled by an end-user-friendly, LabVIEW-based software tool (Version 2.0 FTDI – LabVIEW 2019, 04.11.2021), which enables the individual setting of various spatiotemporal parameters under the tab "Stretcher Control" (e.g., amplitude, velocity of stretch and release term, dwell time in stretch and release position, number of cycles, etc.). Consequently, a multitude of load types and load shapes with high reproducibility and accuracy can be realized. For simulated sinusoidal stretching, we typically use isosceles trapezoids with suitably chosen holding periods as load shape.
21. Many mechanosensitive processes are time and often strain and/or frequency dependent. Therefore, the optimal time for termination of the cell stretcher assay is variable and highly dependent on the studied processes and post-processing type of the sample. Stretch cycles can be interrupted at any time in stretch or release position, and the relevant samples can then be quickly removed and fixed or processed otherwise. Subsequently, the stretching program can be continued for the remaining samples. If stretching for all samples is terminated at the same time and/or rapid fixation is needed, fixing is also possible while installed in the stretcher.

    In addition to fixation, live cell analysis (e.g., FRAP analysis, atomic force microscopy, or traction force microscopy) and harvesting of cells and lysates for proteomics, transcriptomics, and epigenomics are feasible. Typical routines and protocols are easily transferable and usually work well on elastomeric substrates.
22. Shaking on a 2D laboratory rocker improves staining efficiency. Of course, secondary immunofluorescence according to established protocols can be performed instead of or in parallel to the described phalloidin-conjugate staining.
23. Soft Sylgard™ 184 is sticky and inherently adheres to glass slides. For other elastomer types or mixing ratios, additional

glue or freshly prepared soft PDMS on a glass slide may be required.

24. All aqueous mounting media are applicable. Good results were obtained with the nonhardening Fluoromount-G™ and the hardening mounting medium Prolong™ Glass.
25. The cyclic stretch-induced reorientation of the filamentous actin cytoskeleton (Fig. 4) can be further analyzed via image processing routines based on a 2D structure tensor approach as previously described [7]. This quantification of the orientation angles from gray value gradients has been used extensively for different cell types and research questions [4, 9].

Acknowledgments

We thank Werner Hürttlen (IBI-TAE, Forschungszentrum Jülich, Germany) for the LabVIEW-based programming of the cell stretcher control software tool. We additionally thank Wolfgang Rubner (former member of the IBI-2, Forschungszentrum Jülich, Germany) for the early stretcher designs.

References

1. Muhamed I, Chowdhury F, Maruthamuthu V (2017) Biophysical tools to study cellular Mechanotransduction. Bioengineering (Basel) 4(1). https://doi.org/10.3390/bioengineering4010012
2. Nava MM, Miroshnikova YA, Biggs LC, Whitefield DB, Metge F, Boucas J, Vihinen H, Jokitalo E, Li X, Garcia Arcos JM, Hoffmann B, Merkel R, Niessen CM, Dahl KN, Wickstrom SA (2020) Heterochromatin-driven nuclear softening protects the genome against mechanical stress-induced damage. Cell 181(4):800–817 e822. https://doi.org/10.1016/j.cell.2020.03.052
3. Walters B, Turner PA, Rolauffs B, Hart ML, Stegemann JP (2021) Controlled growth factor delivery and cyclic stretch induces a smooth muscle cell-like phenotype in adipose-derived stem cells. Cells 10(11). https://doi.org/10.3390/cells10113123
4. Noethel B, Ramms L, Dreissen G, Hoffmann M, Springer R, Rubsam M, Ziegler WH, Niessen CM, Merkel R, Hoffmann B (2018) Transition of responsive mechanosensitive elements from focal adhesions to adherens junctions on epithelial differentiation. Mol Biol Cell 29(19):2317–2325. https://doi.org/10.1091/mbc.E17-06-0387
5. Abraham JA, Blaschke S, Tarazi S, Dreissen G, Vay SU, Schroeter M, Fink GR, Merkel R, Rueger MA, Hoffmann B (2021) NSCs under strain-Unraveling the Mechanoprotective role of differentiating astrocytes in a cyclically stretched Coculture with differentiating neurons. Front Cell Neurosci 15:706585. https://doi.org/10.3389/fncel.2021.706585
6. Gogele C, Hoffmann C, Konrad J, Merkel R, Schwarz S, Tohidnezhad M, Hoffmann B, Schulze-Tanzil GG (2021) Cyclically stretched ACL fibroblasts emigrating from spheroids adapt their cytoskeleton and ligament-related expression profile. Cell Tissue Res 384(3):675–690. https://doi.org/10.1007/s00441-021-03416-9
7. Faust U, Hampe N, Rubner W, Kirchgessner N, Safran S, Hoffmann B, Merkel R (2011) Cyclic stress at mHz frequencies aligns fibroblasts in direction of zero strain. PLoS One 6(12):e28963. https://doi.org/10.1371/journal.pone.0028963
8. Jungbauer S, Gao H, Spatz JP, Kemkemer R (2008) Two characteristic regimes in frequency-dependent dynamic reorientation of fibroblasts on cyclically stretched substrates. Biophys J 95(7):3470–3478. https://doi.org/10.1529/biophysj.107.128611

9. Lovenich L, Dreissen G, Hoffmann C, Konrad J, Springer R, Hohfeld J, Merkel R, Hoffmann B (2021) Strain-induced mechanoresponse depends on cell contractility and BAG3-mediated autophagy. Mol Biol Cell 32(20):ar9. https://doi.org/10.1091/mbc.E21-05-0254
10. Kamble H, Barton MJ, Jun M, Park S, Nguyen NT (2016) Cell stretching devices as research tools: engineering and biological considerations. Lab Chip 16(17):3193–3203. https://doi.org/10.1039/c6lc00607h
11. Monemian Esfahani A, Rosenbohm J, Reddy K, Jin X, Bouzid T, Riehl B, Kim E, Lim JY, Yang R (2019) Tissue regeneration from mechanical stretching of cell-cell adhesion. Tissue Eng Part C Methods 25(11): 631–640. https://doi.org/10.1089/ten.TEC.2019.0098
12. Boulter E, Tissot FS, Dilly J, Pisano S, Feral CC (2020) Cyclic uniaxial mechanical stretching of cells using a LEGO((R)) parts-based mechanical stretcher system. J Cell Sci 133(1). https://doi.org/10.1242/jcs.234666
13. Lien JC, Wang YL (2021) Cyclic stretching-induced epithelial cell reorientation is driven by microtubule-modulated transverse extension during the relaxation phase. Sci Rep 11(1):14803. https://doi.org/10.1038/s41598-021-93987-y
14. Roshanzadeh A, Nguyen TT, Nguyen KD, Kim DS, Lee BK, Lee DW, Kim ES (2020) Mechanoadaptive organization of stress fiber subtypes in epithelial cells under cyclic stretches and stretch release. Sci Rep 10(1):18684. https://doi.org/10.1038/s41598-020-75791-2
15. Abraham JA, Linnartz C, Dreissen G, Springer R, Blaschke S, Rueger MA, Fink GR, Hoffmann B, Merkel R (2019) Directing neuronal outgrowth and network formation of rat cortical neurons by cyclic substrate stretch. Langmuir 35(23):7423–7431. https://doi.org/10.1021/acs.langmuir.8b02003
16. Zielinski A, Linnartz C, Pleschka C, Dreissen G, Springer R, Merkel R, Hoffmann B (2018) Reorientation dynamics and structural interdependencies of actin, microtubules and intermediate filaments upon cyclic stretch application. Cytoskeleton (Hoboken) 75(9): 385–394. https://doi.org/10.1002/cm.21470
17. Ahrens D, Rubner W, Springer R, Hampe N, Gehlen J, Magin TM, Hoffmann B, Merkel R (2019) A combined AFM and lateral stretch device enables microindentation analyses of living cells at high strains. Methods Protoc 2(2). https://doi.org/10.3390/mps2020043
18. Pullen R, Konrad J, Merkel R, Hoffmann B (2021) Skin under strain: from epithelial model tissues to adult epithelia. Cell 10(7). https://doi.org/10.3390/cells10071834
19. Ulbricht A, Eppler FJ, Tapia VE, van der Ven PF, Hampe N, Hersch N, Vakeel P, Stadel D, Haas A, Saftig P, Behrends C, Furst DO, Volkmer R, Hoffmann B, Kolanus W, Hohfeld J (2013) Cellular mechanotransduction relies on tension-induced and chaperone-assisted autophagy. Curr Biol 23(5):430–435. https://doi.org/10.1016/j.cub.2013.01.064
20. Gaiko-Shcherbak A, Eschenbruch J, Kronenberg NM, Teske M, Wolters B, Springer R, Gather MC, Merkel R, Hoffmann B, Noetzel E (2021) Cell force-driven basement membrane disruption fuels EGF- and stiffness-induced invasive cell dissemination from benign breast gland acini. Int J Mol Sci 22(8). https://doi.org/10.3390/ijms22083962

Chapter 7

Two-Point Optical Manipulation of Cell Junctions in the Early Epithelium of the *Drosophila* Embryo

Kenji Nishizawa, Claire Chardès, Raphaël Clément, and Pierre-François Lenne

Abstract

Laser manipulation is widely used to study mechanics from the molecular to the tissue scale. We implemented optical tweezers to directly manipulate single cell-cell junctions in a developing tissue. We further extended the approach to two-point laser manipulation to enable extensive remodeling of cell-cell junctions. Here, we describe two-point laser manipulation and its implementation to probe the mechanics of cell junctions in the *Drosophila* embryo.

Key words Optical tweezers, Tissue mechanics, Force measurements, Epithelia, Morphogenesis

1 Introduction

Tissue shaping is determined by how forces are applied by and to cells, and how cells respond to these forces. There are several ways to probe cell mechanics in tissues [1, 2], but they are often indirect and inferred from the analysis of the molecular content (e.g., cytoskeletal constituents) or from cell shapes. Such approaches require information on the relationship between molecular content and force application or cell shape and cell mechanical response. A more direct way to probe cell mechanics consists in applying controlled forces and measure the resulting cell and tissue deformation. Optical forces offer the unique possibility to manipulate structures locally, enabling, for example, manipulation of cell junctions [3]. By locally deflecting single-cell junction, single-point manipulation can provide direct measurements of tension and rheological properties [4]. We extended this approach to two-point manipulation, which enables to exert forces in various configurations. Here, we describe the method and its application to the early *Drosophila* embryo.

Ronen Zaidel-Bar (ed.), *Mechanobiology: Methods and Protocols*,
Methods in Molecular Biology, vol. 2600, https://doi.org/10.1007/978-1-0716-2851-5_7,

2 Materials

2.1 Safety Measures

1. Light-protecting goggles (1070 nm IR laser).
2. A sign saying "IR laser in use" to warn people who may enter the room.

2.2 Lasers and Calibration

1. 1070 nm laser for optical trapping.
2. 488 and 561 nm lasers for illumination.
3. Red fluorescent beads (r = 500 nm).

2.3 Optical Path

1. A mechanical shutter with opening/closing time below 1 ms.
2. A telescope containing two lenses with IR coating. The width of the IR laser is expanded 2.7 times here (Fig. 1).
3. IR-coated two mirrors to change the direction of the path of the laser. At least two mirrors are necessary for the alignment of the laser path.
4. A dual-axis galvanometric mirrors. Laser focus is controlled in the x-y plane by changing the angle of the mirrors. The laser focus is split into two points by moving quickly between two angles. Details are explained below.
5. A periscope to bring the laser beam into the microscope.

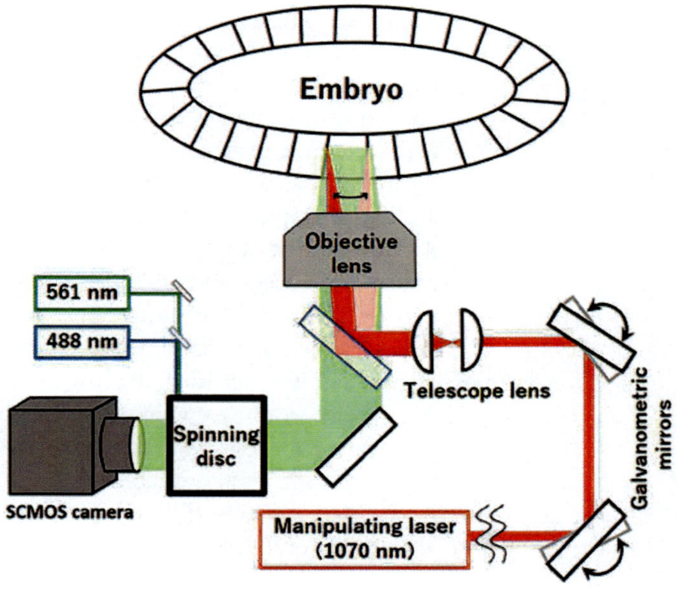

Fig. 1 Setup for two-point optical manipulation. Cell junctions are manipulated by the IR laser (1070 nm). The laser focus position is controlled by two galvanometric mirrors (x-y plane)

6. A dichroic mirror for IR light reflection and visible light transmission.
7. A sample stage on which an embryo is placed. The laser is focused on the junction of the embryo.
8. A commercial spinning disc coupled to an inverted microscope for fast fluorescence imaging.

2.4 Drosophila Embryo Sample

1. Agarose plate with a paste of yeast.
2. Sieve with pore size ~100 μm.
3. Bleach solution: 50% bleach (2.6% sodium hypochlorite).
4. Cover glass with agar pad, coverslip, and coverslip glue.
5. *Drosophila* strain with fluorescently labelled cell-cell junctions.

3 Methods

3.1 Calibration of the Optical Trap Position

The position of the laser focus is moved in the x-y plane by changing the angles of the dual-axis galvanometric mirrors. This requires prior calibration of the position of the laser focus with respect to the voltage signals controlling the dual-axis galvanometric mirrors.

3.2 Recording of a Calibration Movie

1. Put a drop of a low-concentration solution of red fluorescent beads ($r = 500$ nm) on a coverslip, and place the coverslip on the sample holder. Beads in solution are easily trapped by the laser beam. Make sure that the concentration is low enough so that only one bead is trapped at a time.
2. Adjust the z position to find the particles close to the coverslip. This z-position will be within a few μm of the position where the experiments will be carried out in the embryo.
3. Turn on the image acquisition software.
4. Select the parameters for image acquisition: exposure time, gain, time between two images, number of images.
5. Turn on the infrared laser (IR laser, 1070 nm) and set the power to 1 W. The IR laser is used to trap a bead but also here to image it, as it is able to excite the red fluorescence by two-photon excitation.
6. The galvanometers and the shutter are controlled via a national instrument card. The angles of the galvanometer mirrors correspond to voltages. We designed a program in C++ (OTC program) with the Qt library to control the card that sends the voltages to the mirrors and the shutter. The National Instrument card can also record the voltages and the camera trigger, which we will then use to match the mirror positions to the trap position in the image. *We can provide the code that generates this program on request.* Control the angle of dual-axis

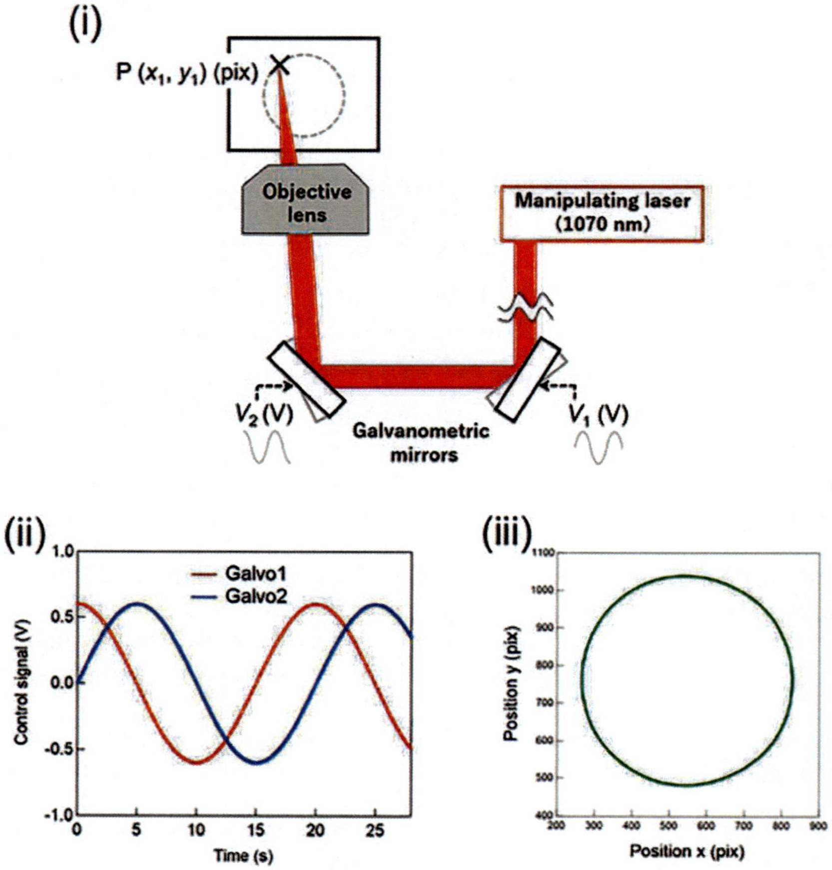

Fig. 2 Calibration of the optical trap position and control voltages of galvanometers

galvanometric mirrors to make the laser focus trace a circle in *x*-*y* plane (Fig. 2). The control signals are two sine waves of amplitude *V1* and *V2* (volts) and out of phase with each other by $\pi/2$ (rad) and 10-second period (to draw the circle slowly). Figure 3 shows the interface of the optical tweezers control (OTC) program.

7. Start the image acquisition (1000 images with a high frame rate, e.g., 50 fps).
8. Run the software that opens the IR laser shutter and controls the dual-axis galvanometric mirrors.
9. Wait that the trapped bead completes at least two full circles, and stop the optical tweezers.
10. Stop the image acquisition. Note that a tiff movie and a text file with control voltages of galvanometers are created.

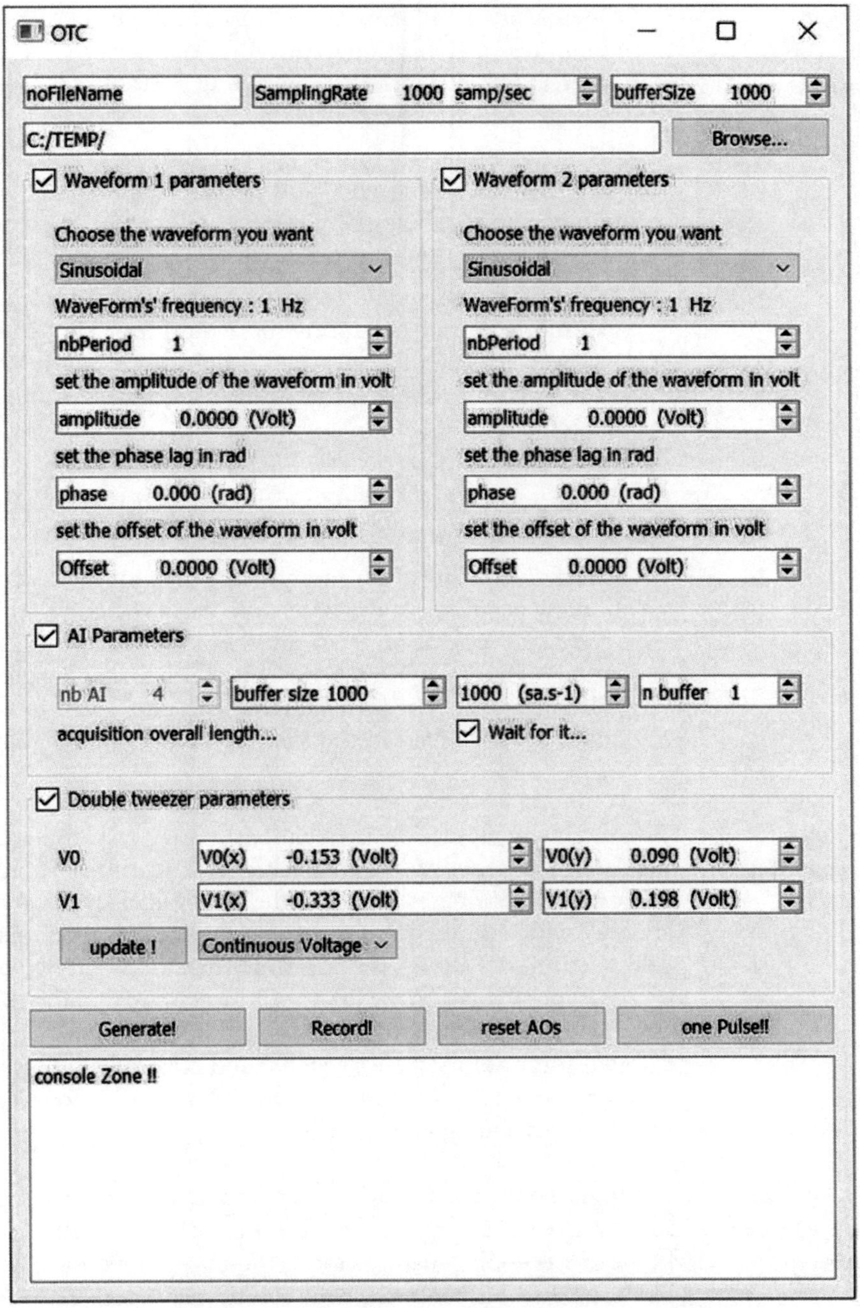

Fig. 3 Interface of the optical tweezers control (OTC) program

3.3 Position Interpolation with Image Analysis

1. Open Matlab and launch the "position2tension.m" script that computes the table converting the control voltages of galvanometers (V_1, V_2) into the optical trap positions (x_1, y_1) (Matlab script provided on request).
2. Select a calibration movie (several movies can be used, but generally one is enough), and provide the first and last image numbers in the dialog box corresponding to a sequence with a clear trajectory and a good signal-to-noise ratio.
3. Select the corresponding text file containing the voltages of the galvanometer mirrors during the acquisition. The script detects the bead for each frame with a subpixel resolution by fitting a 2D Gaussian to the bead. It displays a graph showing the bead trajectory.
4. Select the circular path by mouse clicks around it. If necessary, eliminate by clicking any poorly detected points (outliers) to only keep the points that define the circle.
5. Check the quality of the interpolated map for the laser position: If many positions of the bead describing the circle are missing, repeat the operation with another calibration movie with a better signal-to-noise ratio or/and with a slower speed of the galvanometers. These images and the interpolation values are automatically saved in the same folder as the image, with the name convertFct.mat.

3.4 Two-Point Optical Manipulation

Two-point manipulation is obtained by very fast displacement of the laser focus between two positions. The signal to adjust the mirror angles switches between the two values every 4×10^{-3} s (Fig. 4), which is much shorter than the viscous relaxation time of a cell junction ~1 s (Ref. 4). Equal time share between two points $\mathbf{P}_1$ (x_1, y_1) and $\mathbf{P}_2$ (x_2, y_2) leads to a power split of 50% [5]. If the distance between the two points is within 20 μm, the positions of the two focal points are very stable and controlled. This is a long enough distance to manipulate two different junctions belonging to each of two neighboring cells.

3.5 Selecting the Optical Trap Position on the Screen

In single-point manipulation, the motorized stage is moved to superimpose the target junction's midpoint with the laser focus [4]. For two-point manipulation, the two points must be defined by the user according to the configuration of junctions. We use a file that records the positions of the two traps to communicate between the acquisition software (Micro-Manager [6] in our case) and OT program that controls the tweezers.

1. Select in the image the two positions you want to manipulate. In Micro-Manager, we select the two positions with the tool multipoint selection, we visualize the positions with Analyze>Measure (After selecting only Centroid in Analyze>Set Measurements...)

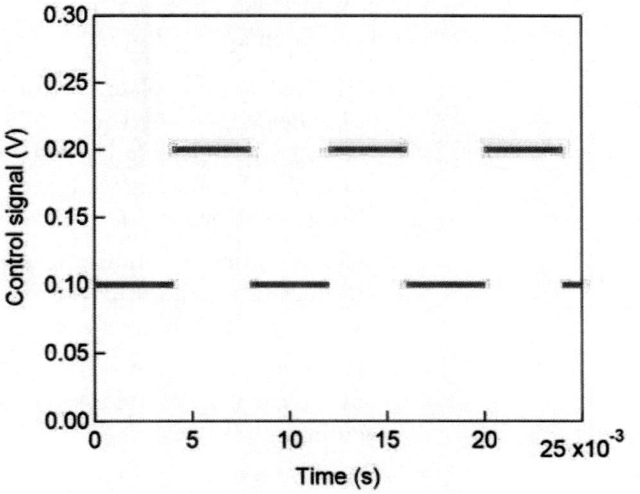

Fig. 4 Control signal sent to the galvanometers for two-point manipulation. In this example for galvanometer 1, the angle alternates between two positions every 4 msec

2. Save the selected positions $\mathbf{P}_1(x, y)$ (pix) as a text file.
3. Run OTC program. The values of the control voltage are selected from the calibration file (obtained in *2.6*) to minimize $|\mathbf{P}_1\text{-}\mathbf{C}|$ where $\mathbf{C}(x, y)$ is the position in the calibration file. This is performed for the two positions.
4. Select the amplitude to be imposed on the laser movement. The two laser foci are moved at constant speed from the initial positions. Select parallel or antiparallel movement.
5. Start the image acquisition (with a fast frame rate, such as 10 fps).
6. Open the shutter of the optical tweezers, and run OTC program which will start sending voltages to the galvanometers and record the galvanometer voltages, the status of the shutter (5 is open, 0 is closed), and trigger of the camera in a text file.
7. After the desired time, close the shutter of the IR laser in OTC program.
8. Stop the image acquisition. Check the acquired movie and the corresponding text file giving the galvanometer voltages.

3.6 Sample Preparation

We use the early epithelium of the *Drosophila* embryo as a model system. Experiments were carried out after cellularization and before germ-band extension. Tissue movement is not too fast at this stage. At later stages when tissue movements are fast, such as during germ-band extension, the feedback tracking system of motorized stage is required (details are in 2.11). Experiments were carried out at the adherens cell junctions close to the apical side, where the adhesion molecule E-cadherin is highly enriched and can be imaged using fluorescently labeled lines.

1. Put the agarose plate with a paste of yeast in a fly cage of *Drosophila* incubated at 25 °C, and collect *Drosophila* embryos after 1 h. *Drosophila* lay eggs on the paste of yeast. Keep the plate 2–3 h after egg laying to wait until they are growing just before cellularization.
2. Collet embryos from the yeast on the plate. Dissolve the yeast by adding water and pass the liquid through a sieve (pore size should be slightly smaller than the embryo ~100 μm).
3. Remove the chorion of the embryos [7]. Soak the embryos in 50% bleach (2.6% sodium hypochlorite) for 50 s, and wash the embryos with water.
4. Put the washed embryos on an agar pad with a brush. If they are still covered by chorion, remove it by a needle.
5. Select embryos which are just after cellularization, and align them with a needle or a moist brush. Put the embryos as the target part becomes top.
6. Add glue on one side of the cover glass, and let it dry for 20 s. Attach the side of the cover glass to the embryos on the agar pad. Here, the target part becomes the bottom.
7. Cover the embryos with 200 dpi oil to avoid drying.
8. Put the cover slide on the stage of the microscope.

3.7 Optical Manipulation of Junctions in the Epithelium of the Drosophila Embryo

To manipulate the junction stably, the focus of the laser is moved so that it pushes outward from slightly inside the cell boundary rather than over the top of the cell boundary.

1. Turn on IR laser (1070 nm) and illumination lasers (488 nm laser and 561 nm) ~ 30 min before experiments for their power stabilizing.
2. Select the junction to target in the tissue (Fig. 5i).
3. Move the sample so that the midpoint of the target junction is close to the center of the image before selecting the positions to manipulate on the image, as the laser focus is precisely controlled within a limited range (Fig. 5ii).
4. Select the positions you want to manipulate in the image. Save these positions as a text file (Fig. 5iii).
5. Start the movie acquisition (with a fast frame rate ~ 5 fps) (Fig. 5iv).
6. Select the amplitude **A** to be imposed on the laser movement.
7. When you set the amplitude **A** to control the galvanometers, laser focus is moved at a constant speed from the initial location $\mathbf{P}_1$ (x_1+, y_1) + $\mathbf{A}^* t/T$, where **A** is the amplitude of the displacements of laser foci, T is the time period for the laser focus to

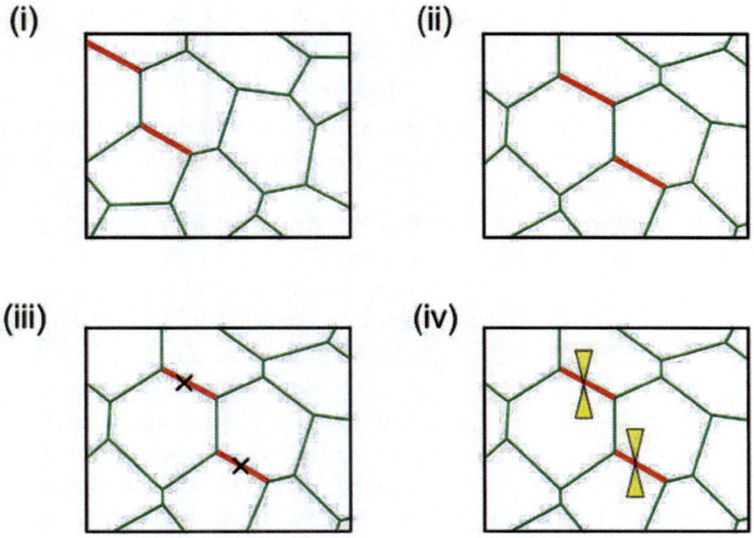

Fig. 5 Steps for two-point optical manipulation in an embryo. (i) Find the junctions to target. (ii) Move the stage to locate the junction close to the center of the image. (iii) Select the positions for manipulation. (iv) The two laser traps are moved to manipulate the two (red) junction

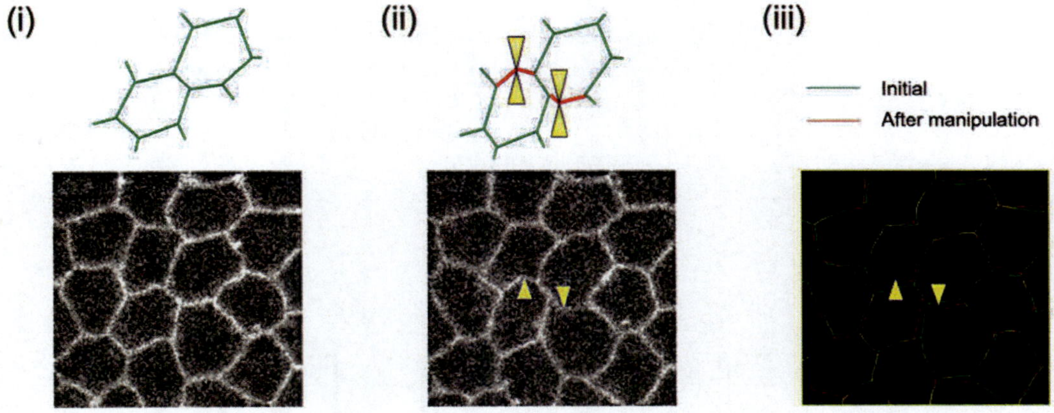

Fig. 6 Applying optical forces to two different cell junctions in an embryo
Images of cells (i) before and (ii) after manipulation. (iii) Both of the two cell junctions were observed to bend after optical force application. The yellow arrows point to junction points that are manipulated by the laser

move, and t is the elapsed time since the manipulation started. The details about how to manage the positions of the two laser foci can be found in (*2.10*).

8. Start the image acquisition (with a fast frame rate, such as 10 fps).
9. Open the shutter of the optical tweezers. When the optical forces are successfully trapping cell junctions, one observes that cell junctions bend (Fig. 6).

10. After the desired time, switch off the trap.
11. Stop the image acquisition. Check the acquired movie and the corresponding text file giving the galvanometer voltages.
12. For efficient manipulation of the junctions, it is important to pay attention to certain points (*see* **Notes**).

3.8 Extending/Shrinking the Junction

1. To extend a junction, the two laser foci are moved in opposite directions and away from each other. Acquisition must start as soon as possible after selecting the points to be manipulated in order to avoid drift (mainly due to tissue movement). To achieve this, it is important that the laser focus be controlled quickly and automatically without requiring time-consuming manual settings. After the initial laser foci $\mathbf{P}_1(x_1, y_1)$ and $\mathbf{P}_2(x_2, y_2)$ are selected, they are automatically moved to the direction away from each other. For example, for horizontal movement, laser foci $\mathbf{P}_1$ (x_1, y_1) and $\mathbf{P}_2$ (x_2, y_2) are moved to $\mathbf{P}_1$ $(x_1 + A, y_1)$ and $\mathbf{P}_2$ $(x_2\text{-}A, y_2)$ if $x_1 > x_2$, or $\mathbf{P}_1$ $(x_1\text{-}A, y_1)$ and $\mathbf{P}_2$ $(x_2 + A, y_2)$ if $x_1 < x_2$ (Fig. 7i). The directions in which the laser foci move are automatically determined according to the initial positioning of the two points.
2. For junction shrinkage, laser foci are moved toward each other (Fig. 7ii).
3. If you would like to approach the two laser foci to the same position gradually, both laser foci are moved at a constant speed to the middle point between two points $\mathbf{P}_m$ $((x_1 + x_2)/2, (y_1 + y_2)/2)$ after the initial laser foci $\mathbf{P}_1$ (x_1, y_1) and $\mathbf{P}_2$ (x_2, y_2) are selected (Fig. 7iii).

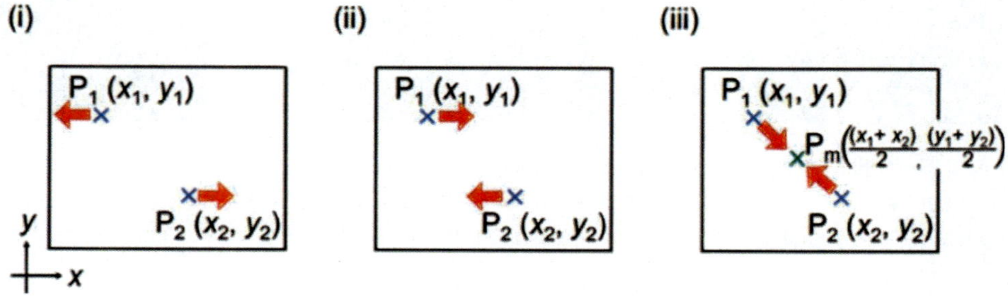

Fig. 7 Controlling the two laser focus positions. The two laser foci are moved along parallel directions away or toward each other

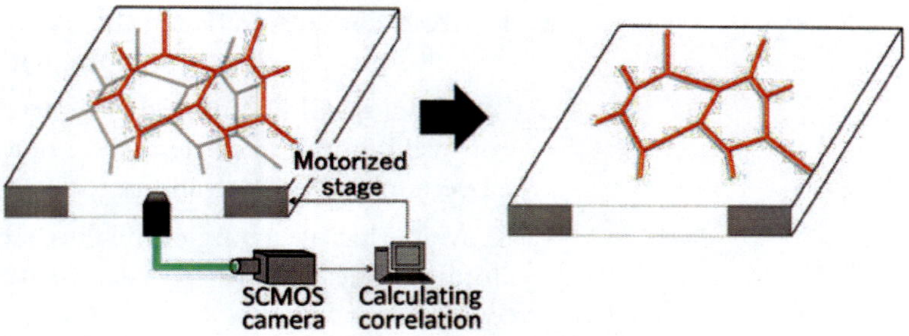

Fig. 8 Tracking tissue movement by image correlation. Tissue movement (translation here) is analyzed by calculating correlation between two images and corrected by moving the motorized stage

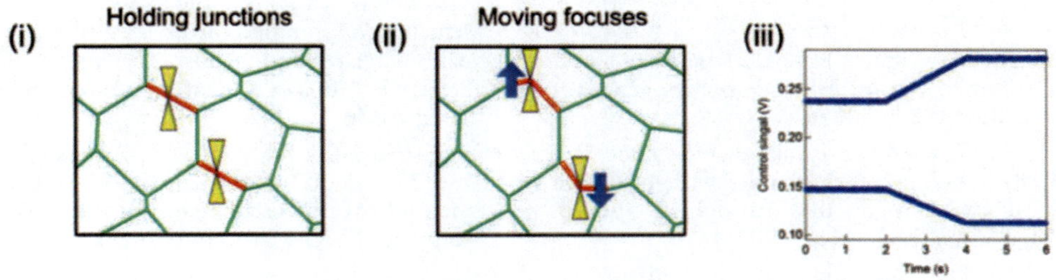

Fig. 9 Efficient manipulation of junctions. For efficient manipulation of junctions, laser foci hold the junctions first (for 2 s) and then are moved

3.9 Feedback Control for Motorized Stage

In a developing tissue, junction positions are not static: they fluctuate, move, and rearrange. Moreover, the cells can respond to the applied forces and change their shape. Therefore, we introduce a feedback control to maintain the target junction at the same position so that the manipulation is well controlled. We calculate correlation between two images to correct potential drifts during the acquisition (Fig. 8).

1. Run the macro in Micro-Manage that calculate the displacement of the tissue by correlation.
2. The motorized stage is then moved to correct the calculated drift. The target area is moved to the center of the imaging window.

4 Notes

1. Before moving laser foci, it is important to hold the junction well by laser manipulation. A few seconds after (2 s) the laser is turned on and junctions are trapped, laser foci are moved. Laser foci are kept at $\mathbf{P}_1$ (x_1, y_1) and $\mathbf{P}_2$ (x_2, y_2) in the first t_m seconds. After t_m seconds, laser foci are moved in constant speed from the initial place $\mathbf{P}_1$ $(x_1+, y_1) + \mathbf{A}^*(t-t_m)/\mathrm{T}$ and $\mathbf{P}_2$ $(x_2+, y_2)-\mathbf{A}^*(t- t_m)/T$ (Fig. 9).

2. To avoid curvature effects, the experiments are performed on cells close to the coverslip, in the flat part of the embryo.
3. Laser foci should be moved close to the center of the image. If not, calibration becomes less precise due to large angular movements of the galvanometers.
4. Be aware that bleaching of the fluorescence can be fast under IR illumination. Thus, it is easier to work with a stable fluorescence marker.

References

1. Rauzi M, Lenne P-F, Lecuit T (2010) Planar polarized actomyosin contractile flows control epithelial junction remodelling. Nature 468(7327):1110–1114
2. Ishihara S, Sugimura K (2012) Bayesian inference of force dynamics during morphogenesis. J Theor Biol 313:201–211
3. Chardès C, Clement R, Blanc O, Lenne P-F (2018) Probing cell mechanics with bead-free optical tweezers in the drosophila embryo. JoVE J Visual Exp 141:e57900
4. Bambardekar K, Clément R, Blanc O, Chardès C, Lenne P-F (2015) Direct laser manipulation reveals the mechanics of cell contacts in vivo. Proc Natl Acad Sci 112(5): 1416–1421
5. Visscher K, Gross SP, Block SM (1996) Construction of multiple-beam optical traps with nanometer-resolution position sensing. IEEE J Selected Topics Quantum Electron 2(4): 1066–1076
6. Edelstein AD, Tsuchida MA, Amodaj N, Pinkard H, Vale RD, Stuurman N (2014) Advanced methods of microscope control using Manager software. J Biol Methods 1(2):e11
7. Cavey M, Lecuit T (2008) Imaging cellular and molecular dynamics in live embryos using fluorescent proteins. In: *Drosophila*. Springer, pp 219–238

Part III

Quantitative Characterization of Cell Mechanosensing

Chapter 8

A Microfluidic-Like System (MLS) to Grow, Image, and Quantitatively Characterize Rigidity Sensing by Plant's Roots and Root Hair Cells

David Pereira, Thomas Alline, Gaurav Singh, Marie-Edith Chaboute, and Atef Asnacios

Abstract

Plant's roots grow in soils of different rigidities. Understanding how the stiffness of the surrounding environment impacts growth and cell fate of roots and root hair cells is an important and open question. Here, we describe a simple method to setup a microfluidic-like system (MLS) to tackle this question. This system enables to grow plantlets during weeks in microfluidic chips filled with gels of controlled stiffness and to image them under a microscope from a few minutes up to a few days. Furthermore, MLS keeps the numerous benefits of microfluidic chips, such as high-resolution imaging, precise control of the geometry of growth, and standardization of the measurements. In sum, MLS enables one to quantitatively test, even on long time scales, the effect of the rigidity and the geometry of the environment on the growth of roots and root hair cells, including mechanotransduction to the nucleus.

Key words Mechanosensing, Rigidity, Microfluidics, Plants, *Arabidopsis thaliana*, Nucleus, Live imaging

1 Introduction

Understanding the effect of mechanical constraints on the growth of plants roots [1] has been widely investigated but less has been done on root hair cells (RHC). In particular, how the rigidity of the surrounding environment impacts the growth and cell fate of roots and RHC is an open question. Growing and imaging biological systems such as plant roots and root hair cells for a long period of time require microfluidic tools [2–4]. Microfluidics has transformed the microscale scientific field [5]. Nowadays, it is extensively used in scientific fields related to chemistry [6], physics [7], and biology [8]. Among many others, this tool has the advantage to standardize the measurement by increasing the reproducibility and to reduce the costs by minimizing the quantity of product used for

Ronen Zaidel-Bar (ed.), *Mechanobiology: Methods and Protocols*,
Methods in Molecular Biology, vol. 2600, https://doi.org/10.1007/978-1-0716-2851-5_8,

the experiments. Nevertheless, it can be heavy going to set up this kind of technique for inexperienced people because microfluidics relies on chemical physics, fluid dynamics, and soft condense matter. Also, it requires many devices which are not available in most laboratories such as flow or pressure controllers, chips made of elastomers or glass, tubing, and connectors. Besides, the latter can lead to contamination and leaks. Consequently, it reduces the experimental efficiency, increases the costs, and can lead to a waste of time. Finally, and obviously, in order to study the effect of rigidity on the growth of roots and RHC, one needs to grow plants in a solid medium of controlled stiffness, which cannot be achieved with standard microfluidic chips.

For all these reasons, we developed a microfluidic-like system (MLS) which enables one to grow plantlets' roots and RHC in standardized conditions with controlled geometry and stiffness and to image their growth, as well as nucleus shape and dynamics, with high resolution. The simplicity and efficiency of the MLS make it usable in many laboratories by most people without any training course in microfluidics. To overcome the main problems explained above, we modified classical microfluidic chips to get them filled with solid culture media, namely, agar gels with embedded regular plant culture media (termed "solid medium" in the following). An outline of the protocol is presented in Fig. 1. This trick enables us to use the chip to monitor plantlet's growth on long culture times and to test the effect of rigidity on growth by varying the agar concentration in the solid medium. To do so, after the bonding step, the chip and the surrounding chamber are filled with prewarmed melted solid medium, before it recovers its solid state after cooling down at room temperature (Fig. 2). The solid medium around the chip serves as a reservoir of liquid and nutrients for the plantlet. The MLS can be easily filled with different agar concentrations (typically 0.5%w/v, 1% w/v, and 1.5% w/v).

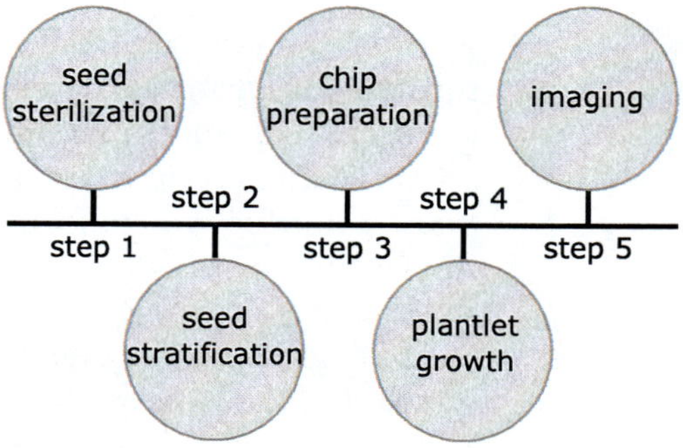

Fig. 1 Protocol timeline

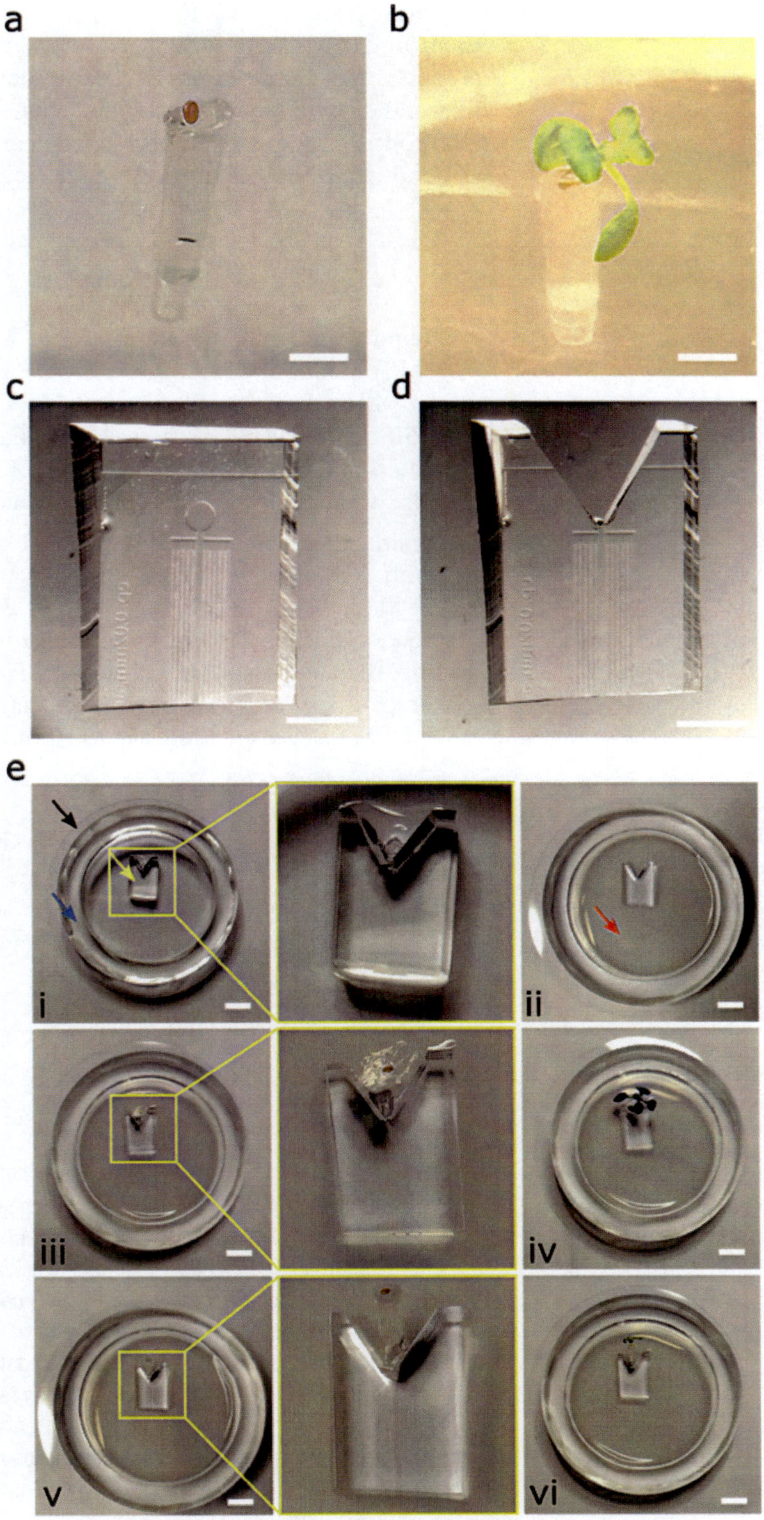

Fig. 2 (**a**) Slice of pipette tip filled with solid media with an *Arabidopsis thaliana* seed at the top of it. Scale bar 1 mm. (**b**) An *Arabidopsis thaliana* plantlet on a tip

This system enables to grow plantlets during weeks and image them under a microscope from a few minutes up to a few days (Fig. 3). Furthermore, MLS keeps the numerous benefits that microfluidic chips give, for instance, high-resolution imaging, precise control of the geometry of the growth environment, and standardization of the measurements (Fig. 4).

2 Materials

2.1 Culture Media

1. Liquid culture medium: ½ Murashige and Skoog (MS) medium, pH 5.7. Add in a 1 L clean glass bottle 1.8 g of MS including vitamins and MES, 3.75 g of sucrose, and 750 mL of distilled water. Use a magnetic stirrer to mix the solution; then adjust the pH to 5.7 using a NaOH solution. Use an autoclave to sterilize the liquid medium.
2. Solid culture medium: ½ Murashige and Skoog (MS) medium with agar 1% w/v, pH 5.7. Add in a 1 L clean glass bottle 1.8 g of MS including vitamins and MES, 3.75 g of sucrose, 7.5 g of agar, and 750 mL of distilled water. Use a magnetic stirrer to mix the solution; then adjust the pH using a NaOH solution. Prepare also 750 mL of solid media with agar 0.5% w/v and 1.5% w/v. Use an autoclave to sterilize the solid medium.

2.2 Microfluidic Chips

1. A mold with the desired design.
2. PDMS chip: Sylgard-184 elastomer. Mix 10:1.

2.3 Chip Bonding

1. Plasma cleaner.
2. 50 mm round coverslip (*see* **Note 1**).
3. 5 cm Petri dish (*see* **Note 1**).
4. PDMS wall structure (blue arrow, Fig. 2i; *see* also **Note 2**).

Fig. 2 (continued) after ten days of growth. Scale bar 1 mm. (**c**) PDMS chip with sinusoidal channels. Scale bar, 2 mm. (**d**) PDMS chip with a V-shape incision. Scale bar, 2 mm. (**e**) 50 mm glass coverslip (black arrow) bonded to a PDMS chip (yellow arrow) and wall (blue arrow). The PDMS chip is filled with solid media (i). The space between the chip and the wall is filled with solid media to serve as a reservoir (red arrow) (ii). A hole is made in the agar at the entrance of the chip to place a seed (iii). After ten days, the plantlet has reached a four-leaf stage (iv). A tip filled with solid media and a seed place on the top of it is placed at the entrance of the chip (v). After tens of days, a plantlet has grown in the tip (iv). Scale bar, 0.5 cm

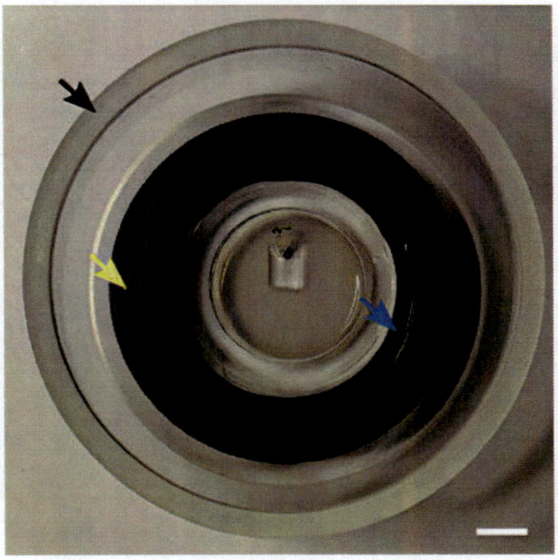

Fig. 3 Microscope holder for the MLS (yellow arrow). A lid is placed on top of the whole system (black arrow), and the channel between the holder and the PDMS wall (blue arrow) is filled with water to limit evaporation. Scale bar, 1 cm

2.4 Seedling Preparation

1. Sterilization solution 1: 70% ethanol solution with 0.05% of Tween 20. Can be stored at room temperature.
2. Sterilization solution 2: 70% ethanol solution. Can be stored at room temperature.
3. Sterilization solution 3: 96% ethanol solution. Can be stored at room temperature.
4. Stratification solution: liquid culture medium.
5. Stratification on tip: tips filled with solid media: Slice a small piece of a 10 μL pipette tip, use a 200 μL pipette tip to fill the small piece with warm solid media. Petri dish filled with solid media.
6. Microporous sealing tape.

2.5 Manipulation

1. Bunsen burner or laminar flow hood.
2. Scalpel.
3. Razor blade.
4. Tweezers.

3 Methods

The timeline gives an overview of the different steps of the protocol (Fig. 1). As a general comment, **steps 1, 2,** and **3** can be done on the same day.

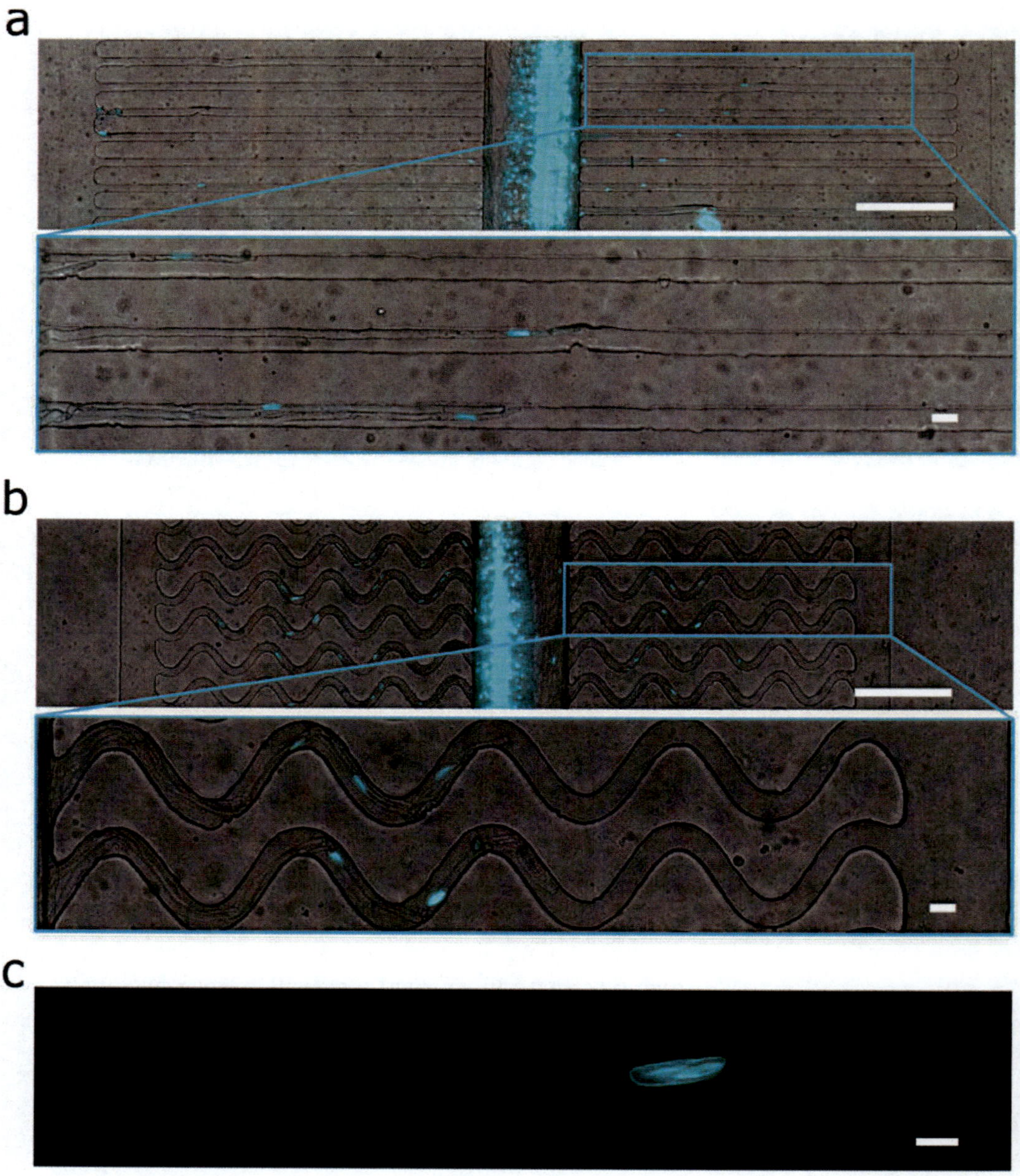

Fig. 4 (**a**) *Arabidopsis Thaliana* plantlet root expressing the nuclear envelope marker SUN2-RFP (light blue) growing inside a PDMS chip with straight lateral channels. Images were taken with a 20X objective. Every light blue structure represents a nucleus. Scale bar, 250 μm; lines, 25 μm. (**b**) *Arabidopsis thaliana* plantlet root expressing SUN1-GFP (light blue) growing inside a PDMS chip with sinusoidal lateral channels. Scale bar, 250 μm; sinus, 25 μm. (**c**) Single root hair cell image with a 60X objective. Scale bar, 10 μm

3.1 Seed Sterilization

The procedure should be done under a laminar flow hood and can be done at least one month before the rest of the protocol.

1. Take few *Arabidopsis thaliana* seeds into a 1.5 mL Eppendorf tube, and add 500 μL of a 70% ethanol solution with 0.05% Triton X100. Resuspend the seeds in the solution, and keep them under rotatory shaking for 10 min.
2. In order to avoid losing too many seeds, remove carefully the ethanol solution using a small pipette tip (p200). Then, add 500 μL of a 70% ethanol solution, and keep under rotatory shaking for 10 min.
3. Remove carefully the ethanol solution, and add 500 μL of 96% ethanol solution. Wait 30 s, and then remove the ethanol using a small pipette tip (p200). Leave the Eppendorf tube open under a laminar flow hood until the remaining ethanol has evaporated (*see* **Note 3**). Close the Eppendorf tube, and store at room temperature in a dark and dry room. The sterilized seeds can be stored and used for a month.

3.2 Seed Stratification

There are three stratification methods: the stratification in liquid media, the stratification on solid media, or the stratification on tip. One can choose one among the three methods presented below.

The stratification can be done in liquid media 3 days before seeding (stratification in liquid media) or directly on the chip on the day of the chip preparation (stratification on solid media or on tip).

1. For the stratification in liquid media, take the required amount of *Arabidopsis thaliana* seeds into a 1.5 mL Eppendorf tube, and add 500 μL of liquid media. Keep at 4 °C in the dark for 3 days.
2. For the stratification on solid media, use the microfluidic chip prepared at the Subheading 3.3. Place a seed at the entrance of the microfluidic chip. Keep at 4 °C in the dark for 3 days (Fig. 2a).
3. For the stratification on tip, first slice a small piece of a 10 μL tip; then use a 200 μL tip to fill it with prewarmed melted solid medium. Then take a tip filled with solid media, and stick it into a Petri dish filled with solid media. Drop a seed on the top of the tip (Fig. 2a–b). Keep at 4 °C in the dark for 3 days.

3.3 Chip Preparation

1. Retrieve the PDMS microfluidic chip from the mold.
2. Use a razor blade or a scalpel to make an incision with a V-shape at the entrance of the PDMS chip (Fig. 2c–d). Make sure that the incision is perpendicular to the chip plane. The bottom of the V-shape should coincide with the beginning of the central channel (Fig. 2d).

3.4 Chip Bonding and Solid Media Filling

1. Take a coverslip, a PDMS wall structure (blue arrow, Fig. 2i; but this is not mandatory, *see* **Note 2**) and a microfluidic chip. Place everything in the plasma cleaner for 45 s (*see* **Note 4**). First, take the coverslip off the plasma cleaner, bond the wall structure on the coverslip, and then bond the microfluidic chip in the center of the coverslip (*see* **Note 5**) (Fig. 2).
2. Few seconds after the bonding step, using a pipette tip, inject melted agar medium inside the microfluidic chip at the bottom of the V-shape (*see* **Note 6, 7,** and **8**). Place the bonded chip filled with solid media in a plastic container (Petri dish).
3. Repeat **steps 1** and **2** if more than one chip is necessary.
4. Add few milliliters of prewarmed melted medium (the one used in **step 3.4.2**) inside the PDMS wall. Make sure that the melted agar medium does not overflow the PDMS wall. Leave the whole system under the hood with the lid open. To avoid any condensation, make sure that the solid medium is at room temperature before putting the lid on top of the Petri dish. Wrap the Petri dish with a Parafilm sealing film. The sealed Petri dish can be stored at 4 °C for a few weeks.

3.5 Plant Seeding

The procedure should be done under a laminar flow hood.

1. Remove a small quantity of agar on the V-shape with a thin scalpel (Fig. 2eiii).
2. If stratification has been done in liquid media (*see* 3.2.1), take the seeds from the Eppendorf tube kept at 4 °C. Using tweezers, place one seed at the bottom of the V-shape of the PDMS chip (Fig. 2eiii).
3. Seal the box with a microporous film.
4. Repeat **steps 1** and **2**, if more than one chip is necessary.

3.6 Plant Seeding with the Tip Technique

If stratification have been done with the tip technique:

1. Take the Petri dish with the tip at 4 °C, and put it horizontally in the incubator.
2. Wait few days for the seed to germinate on the tip in the incubator (*see* **Note 9**).
3. Take the Petri dish with the tip, and put it under a laminar flow hood.
4. Use tweezers to take a tip with a plantlet for which the root has grown inside the tip (*see* **Note 10**). Stick the tip with a 45° tilt at the bottom of the V-shape (Fig. 2ev). Seal and place the system back into the incubator. When the root has reached the central channel of the microfluidic chip, take it for the imaging step.

3.7 Plantlet Culture

1. Place the sealed Petri dish in the incubator with a 45° tilt. The gravity and the V-shape will help the root to enter inside the microfluidic device. The plantlets can be kept in the incubator for a few weeks.
2. When the root starts to enter into the central channel and the RHC in the lateral channels, the system is ready to be imaged.

3.8 Video Microscopy

1. Take the coverslip with the plantlet and the PDMS chip from the incubator. Mount the system on the appropriate microscope stage holder. Add hundreds of microliter of liquid media on the top of the system, place above a chamber to limit evaporation (Fig. 3) (*see* **Note 11**).
2. The roots and RHC can be imaged from a few minutes up to a few days (Fig. 4). A 4x to 60x objective can be used.

4 Notes

1. One can choose the size of the coverslip according to the microscope stage holder. The Petri dish size should be chosen accordingly.
2. The PDMS wall is not mandatory for short-term imaging (few hours). This PDMS wall can be handmade with a simple Petri dish used as a mold and a large punch to make a hole. The mold can also be obtained by 3D printing or made in a workshop from drilling a Plexiglas item. The PDMS wall serves as a container for the surrounding solid media. It is also important because it creates a channel between the experimental chamber and itself (Fig. 3, blue arrow). This channel can be filled with water to limit evaporation during long-term imaging (from a few hours to a few days).
3. To accelerate the process of evaporation, use tweezers to separate the seeds on one side of the Eppendorf tube. Make sure that all the ethanol has evaporated before storing the seeds. The remaining ethanol would damage the seeds.
4. Make sure that the chip channels face up. This ensure a good treatment of the surface and an efficient bonding with the coverslip.
5. The whole **step (3.3)** must be done under a laminar flow hood or close to a Bunsen burner. Most contaminations occur during this step.
6. This step is the most important of the protocol. The media has to be injected at the latest 45 s after bonding the chip to the coverslip. To be sure to fill the small lateral channels with solid media (from 0.5%w/v to 1.5 w/v agar), one has to keep the media warm, for instance, by using an incubator (especially for

high agar concentrations). This is not necessary if the filling step is done quickly after melting the media (10 to 20 min). Also, to avoid solidification of the gel during the filling process, a huge pipette tip (as compared to the volume of the microchannels!) and volume of melted media (at least 200 μL) should be used (especially for high agar concentrations). Indeed, when the tip size decreases, the surface-to-volume ratio increases; therefore, the cooling rate increases, and as a consequence, the solidification of the gel happens faster.

7. The plasma treatment temporarily increases the hydrophilicity of the PDMS and the glass coverslip surfaces. The injection of warm solid media must be done quickly after (less than a minute) the surface treatment in the plasma cleaner to ensure an adequate filling of the chip (small lateral channels).

8. For chips with large channels (width superior to 100 μm), it is not necessary to fill the channels with solid media right after the bonding step. In that case, it is better to first sterilize the chips after the bonding step by using an autoclave, and then fill them with melted solid medium under the laminar flow hood.

9. The tip technique allows to prepare many tips and seeds, and then, after the seed germination, the experimentalist can select the seed that has germinated and put it in the microfluidic chip. It avoids wasting time and products by preparing chips for seeds that may not germinate. But it increases the risk of contaminations as the Petri dish is manipulated outside of the laminar flow hood and many tips are in the same Petri dish. If one still wants to use the tip stratification technique but without risking more contaminations, one can directly take the Petri dish with the tip at 4 °C after 3 days of stratification and go directly to **step 3.6.4** with the tip and the seed.

10. For this step, select plantlets for which the root has not yet exited the tip; otherwise, the root will be damaged and not usable.

11. The chamber can be as simple as a lid from a Petri dish. One should be careful to effectively seal the chamber to avoid any evaporation (for long-term imaging).

References

1. Kolb E, Legué V, Kolb E, Legué V, Interactions MBPR (2017) Physical root-soil Interactions to cite this version: HAL id: HAL-01609984 physical root-soil. Interactions 14
2. Grossmann G et al (2011) The Rootchip: an integrated microfluidic chip for plant science. Plant Cell 23:4234–4240
3. Meier M, Lucchetta EM, Ismagilov RF (2010) Chemical stimulation of the Arabidopsis thaliana root using multi-laminar flow on a microfluidic chip. Lab Chip 10:2147–2153
4. Parashar A, Pandey S (2011) Plant-in-chip: microfluidic system for studying root growth and pathogenic interactions in Arabidopsis. Appl Phys Lett 98

5. Whitesides GM (2006) The origins and the future of microfluidics. Nature 442:368–373
6. Weibel DB, Whitesides GM (2006) Applications of microfluidics in chemical biology. Curr Opin Chem Biol 10:584–591
7. Trégouët C, Salez T, Monteux C, Reyssat M (2019) Microfluidic probing of the complex interfacial RHCeology of multilayer capsules. Soft Matter 15:2782–2790
8. Singh G et al (2021) Real-time tracking of root hair nucleus morphodynamics using a microfluidic approach. Plant J:303–313. https://doi.org/10.1111/tpj.15511

Chapter 9

Photoresponsive Hydrogels for Studying Mechanotransduction of Cells

Keun-Young Park, David J. Odde, and Mark D. Distefano

Abstract

Hydrogels are important platform materials for in vitro cellular studies. Mechanistic studies on durotaxis, the directional movement of a cell affected by a spatial gradient of stiffness of the underlying substrate, requires materials such as polyacrylamide, polyethylene glycol, or PDMS, in which the stiffness can be controlled in a spatiotemporal manner. Here, we describe the synthesis of an *o*-nitrobenzyl-based photocleavable cross-linker and its incorporation into a polyacrylamide hydrogel to render it photoresponsive. Precise control over the physical properties of the gel allows observation of glioblastoma durotaxis under surface stiffness conditions relevant to the actual brain environment.

Key words Photocleavable cross-linker, *o*-nitrobenzyl, Photoresponsive hydrogel, Polyacrylamide, Stiffness pattern, ECM micropatterning, Fibronectin, U-251 glioblastoma, Mechanotransduction, Durotaxis

1 Introduction

Glioblastoma multiforme is the most severe brain cancer with poor prognosis [1]. The extremely low recovery rate is, in large part, due to the tendency of glioma cells to rapidly migrate to the surrounding tissue [2]. There are many factors known to affect the migration of cells, and one that is particularly influenced by the stiffness gradient of the surrounding tissue is known as durotaxis. Hydrogel-based materials have made it possible to perform in vitro tracking of cancer cells, which has contributed to the discovery of durotaxis [3]. Since then, various methods have been developed to create gels exhibiting various stiffness features enabling the study of this mechanoresponsive behavior. Some examples exploit the diffusion of two prepolymer solutions [4], tilted-superposition of two hydrogels [5], freeze-thaw-induced cross-linking of polyvinyl alcohol [6], or toehold-mediated strand displacement of DNA [7] to create various stiffness gradients and

Ronen Zaidel-Bar (ed.), *Mechanobiology: Methods and Protocols*,
Methods in Molecular Biology, vol. 2600, https://doi.org/10.1007/978-1-0716-2851-5_9,

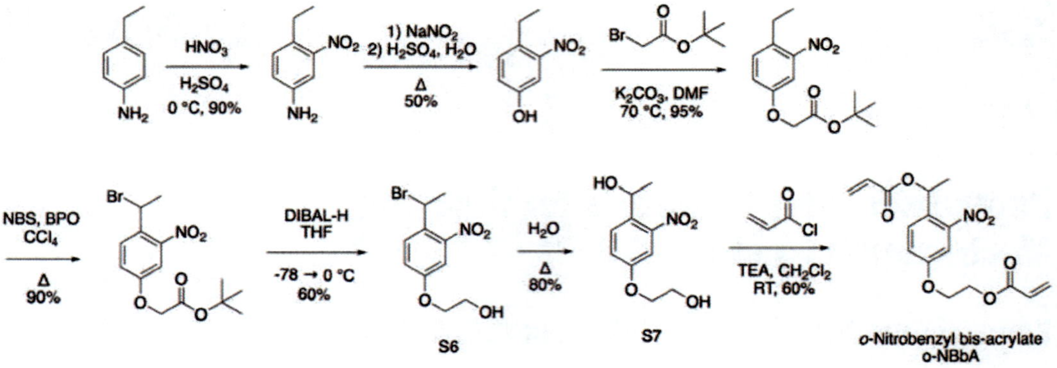

Fig. 1 Schematic representation of the synthesis of the photocleavable cross-linker o-NBbA from *p*-ethyl aniline. *DMF*: dimethylformamide; *NBS*: *N*-bromosuccinimide; *BPO*: benzoylperoxide; *THF*: tetrahydrofuran; *TEA*: triethylamine; *RT*: room temperature

patterns. However, despite the research that has been ongoing for close to two decades since its initial discovery, the mechanism of durotaxis at the molecular level is still poorly understood. This lack of progress emphasizes the importance of developing materials that more accurately mimic the brain environment. The brain is one of the softest organs in which its stiffness lies within the 1–10 kPa range, and the mechanical environment is also highly heterogeneous [8, 9]. This characteristic of the brain makes it especially difficult to replicate since soft gels are difficult to prepare and even more challenging to fabricate post-synthesis due to their structural instability [10, 11]. The high mechanical heterogeneity of the brain requires methods that allow more elaborate control of the physical properties of the hydrogel substrate at the micron scale.

In view of these challenges, we developed an *o*-nitrobenzyl (o-NB)-based photocleavable cross-linker, *o*-nitrobenzyl bis-acrylate (o-NBbA), and incorporated it into a polyacrylamide gel to render the gel photoresponsive (Fig. 1). A critical attribute of this new cross-linker is its low molecular weight which maximizes the number of cross-links obtainable from a given mass of material, allowing the synthesis of polyacrylamide with the desired stiffness range to be created. The incorporation of the o-NB cross-linker into the network structure of the gel was facile as it was copolymerized with acrylamide to yield polyacrylamide gels that did not require additional functionalization processes. The o-NB photolabile functionality was chosen for its high one-photon photolysis efficiency and high deprotection yields [12, 13]. Studies in the past have utilized o-NB-functionalized molecules in hydrogels for creating channels via photodegradation [14, 15] or modulating its mechanical properties [16–18]. While most of these studies utilized polyethylene glycol-based gels, we aimed to focus on polyacrylamide gels that are most commonly used in durotaxis studies. The resulting photoresponsive gel was soft, exhibiting a stiffness close to

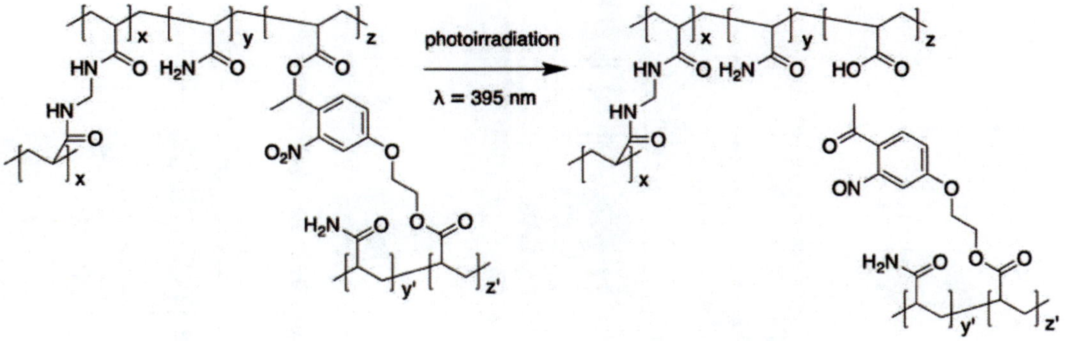

Fig. 2 Schematic representation of the chemical structure of the o-NBbA incorporated polyacrylamide gel, before and after photoirradiation. The photoresponsive gel comprises strands of polyacrylamide cross-linked by o-NBbA or bis-acrylamide cross-linker molecules. Irradiation with 395 nm light causes the o-NBbA to cleave which reduces the overall cross-linking density of the gel, leading to softening of the gel. After photocleavage, o-NBbA does not release any soluble byproducts into the gel environment

the geometric mean of the brain in the lower stiffness (~9.5 pKa) range [8]. Stiffness patterns could be spatiotemporally created in the gel with high precision by adjusting the exposure of UV light (Fig. 2). The photolysis of o-NBbA did not produce soluble byproducts that could affect cellular experiments. Additionally, the surface of the gel was coated with fibronectin patterned in a one-dimensional manner, oriented in the direction of the stiffness gradient, which was designed and fabricated to quantify and qualify cellular movement more easily (Fig. 3a). The line-patterned extracellular matrix (ECM) also allowed easier reproduction of experimental conditions in computational simulations, which was especially useful in the present study since the goal of collecting in vitro migration data was to integrate it with in silico data obtained using a motor-clutch-based computational model to ultimately examine the molecular components that dictate biased cellular movement. We direct readers that are interested in the full scope of the research to our published research article [19]. This report focuses on the details of the material science aspect, namely, the synthesis of the o-NBbA photocleavable cross-linker and its use in fabricating a photoresponsive gel suitable for durotaxis studies.

2 Materials

Purchase chemical reagents and solvents from appropriate vendors. Store chemicals based on instructions provided by the manufacturer (e.g., store flammable solvents at RT in a flammable cabinet, store acryloyl chloride in a dry and well-ventilated place at 2–8 °C, etc.).

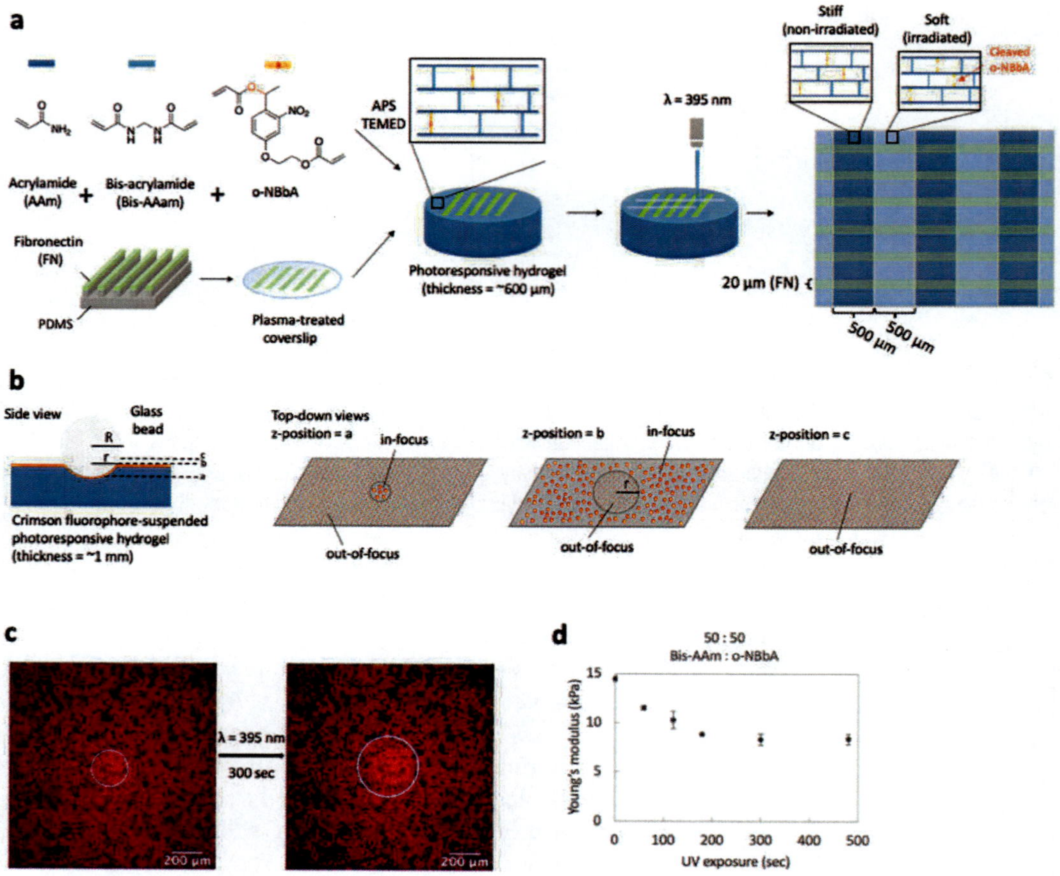

Fig. 3 Preparation and characterization of photoresponsive hydrogels. (**a**) Schematic representation of stiffness gradient preparation for migration experiments. The photocleavable carbon-oxygen bond in o-NBbA is indicated by red color. (**b**–**d**) Stiffness characterization by bead indentation. (**b**) Placing a silica bead (d = 1 mm) on a crimson fluorosphere (d = 200 nm)-coated gel surface creates an indentation. The crimson fluorospheres allow this indentation to be visualized and analyzed via microscopy. By imaging various depths by focusing on different z-positions, the crimson fluorospheres yield various 2D plane images that contain regions that are in focus and out of focus, which allows the measurement of the bead contact radius (*r*). (**c**) Representative fluorescence images focusing on the "b" z-position, wherein the area that is indented due to the bead is out of focus and the rest of the region is in focus. The out-of-focus region becomes wider as the gel softens with photoirradiation. (**d**) Plot obtained from the experiment, depicting hydrogel elasticity as a function of UV exposure. Mean ± SD of n = 3 measurements

2.1 o-Nitrobenzyl Bis-acrylate (o-NBbA) Photocleavable Cross-Linker

1. TLC plate.
2. Silica gel.
3. pH paper.
4. Filter paper.
5. Sat. NH_4Cl (aq).
6. Sat. NaCl (aq).
7. 0.1% $NaHCO_3$ (aq).

8. Dry ice.
9. Liquid nitrogen.
10. Nuclear magnetic resonance (NMR) sample tubes.

2.2 1D-Fibronectin-Coated Photoresponsive Gel: Preparation

1. Glass bottom culture dish.
2. 3-Aminopropyltrimethyoxy silane.
3. Micropipette and tips.
4. Plastic transfer pipette.
5. Desiccator.
6. 0.5% glutaraldehyde in PBS.
7. 40% (w/v) acrylamide solution.
8. 2% (w/v) bis-acrylamide solution.
9. 50 mM o-NBbA (in DMSO).
10. 1 M HEPES (pH 7.0).
11. 71.7 mM acrylic acid N-hydroxysuccinimide ester (in DMSO).
12. Ammonium persulfate (APS).
13. Tetramethylethylenediamine (TEMED).
14. Dulbecco's phosphate-buffered salt solution (DPBS or PBS).
15. PDMS stamps.
16. 100 μg/mL fibronectin.
17. 12 mm glass coverslips.
18. 25 mm glass coverslips.
19. 4% BSA.

2.3 1D-Fibronectin-Coated Photoresponsive Gel: Characterization

1. Silica bead (1 mm diameter).
2. Rabbit anti-fibronectin antibody.
3. Goat anti-rabbit IgG-rhodamine.

2.4 Instruments

1. Chemical hood.
2. Glassware.
3. Stirring hot plate.
4. Magnetic stir bar.
5. Bath sonicator.
6. Rotary evaporator.
7. Plasma cleaner.
8. Glass column for flash chromatography.
9. HPLC columns.
10. NMR spectrometer.

11. Electrospray ionization mass spectrometer (ESI-MS).
12. Gas chromatography mass spectrometer (GC-MS).
13. Epifluorescence microscope (with objective, LED source, DAPI filter set, stage, etc.).

3 Methods

Conduct all procedures at RT unless otherwise specified. Prepare all aqueous solutions using deionized water unless otherwise specified.

3.1 Abbreviations

AAm	acrylamide
APS	ammonium persulfate
BSA	bovine serum albumin
DI	deionized
DMF	dimethylformamide
DMSO	dimethyl sulfoxide
ECM	extracellular matrix
HEPES	(4-(2-hydroxyethyl)-1-piperazineethanesulfonic acid)
Hex	hexanes
HRMS	high-resolution mass spectrometry
NHS	N-hydroxysuccinimide
NMR	nuclear magnetic resonance
PBS	phosphate-buffered saline
PDMS	polydimethylsiloxane
RT	room temperature
TEA	triethylamine
TEMED	tetramethylethylenediamine
TFA	trifluoroacetic acid
THF	tetrahydrofuran
TLC	thin-layer chromatography

3.2 Synthesis of o-NBbA

This protocol assumes the reader is knowledgeable of basic procedures for organic synthesis. If unfamiliar with the art of organic chemistry required to perform the synthesis described in this protocol, we direct the reader to an Organic Chemistry Laboratory course text book that provides detailed descriptions for the standard techniques used in this work [20].

3.2.1 2-Nitro-4-ethyl Aniline (S2)

1. Add p-ethyl aniline (5 g, 41.3 mmol) dropwise to a cold solution of concentrated H_2SO_4 (30 mL) immersed in an ice water bath and stirred for 5 min (*see* **Note 1**).
2. In a separate flask, mix 5.3 mL of 70% HNO_3 (82.6 mmol) with an equal volume of H_2SO_4, and add dropwise to the reaction vessel (*see* **Notes 2** and **3**).

3. Stir 15 min at 0 °C. Thin-layer chromatography (TLC) analysis (Hex:EtOAc, 2:1, v/v) will indicate complete conversion to the product.
4. Quench the reaction by pouring the mixture into 200 mL of ice water. Filter and wash the resulting precipitate with H_2O (~200 mL) to obtain compound S2 (6.2 g, 90%).

Characterization

^1H NMR (500 MHz, $CDCl_3$) ppm 1.099 (t, J = 7.5 Hz, 3H), 2.612 (q, J = 7.0 Hz, 2H), 5.558 (s, 2H), 6.804 (dd, J = 8.0, 2.5 Hz, 1H), 7.041 (d, J = 2.5 Hz, 1H), 7.095 (d, J = 8.5 Hz, 1 H)

^{13}C NMR (100 MHz, DMSO-d6) 149.341, 147.864, 131.605, 124.059, 118.842, 107.919, 24.489, 15.309

HRMS (m/z): $[M]^+$ calculated for $[C_8H_{10}N_2O_2]^+$ 166.0737, found 166.0737.

3.2.2 4-Ethyl-3-nitrophenol (S3)

1. Dissolve compound S2 (6.2 g, 37.3 mmol) in a mixture of H_2SO_4 and H_2O (1:3, v/v, 25–50 mL) by sonication (if sonification does not yield a homogenous suspension, add a few mL of THF to dissolve the solid S2, and then add the THF mixture to the mixture of aqueous H_2SO_4).
2. Dissolve $NaNO_2$ (3.86 g, 56.0 mmol) in H_2O (2.5 mL), and slowly add the mixture to the reaction flask dropwise.
3. Stir 1.5 h at RT.
4. Add H_2SO_4:H_2O (4:3, v/v, 75 mL) in a separate flask and heat to reflux (*see* **Note 4**).
5. To the refluxing mixture, add the S2 solution dropwise, and stir for 30 min.
6. Quench the mixture with ice water (50 mL), and extract the product with EtOAc (3 × 75 mL).
7. Dry the organic layer with Mg_2SO_4, remove solvent in vacuo, and purify the crude product by silica gel flash chromatography (Hex:EtOAc, 2:1) to obtain S3 (3.11 g, 50%) as a yellow oil after solvent evaporation in vacuo.

Characterization

^1H NMR (500 MHz, $CDCl_3$) ppm 1.249 (t, J = 7.5 Hz, 3H), 2.842 (q, J = 8.5 Hz, 2H), 7.030 (dd, J = 8.5, 2.5 Hz, 1H), 7.230 (d, J = 8.5 Hz, 1H), 7.383 (d, J = 2.5 Hz, 1H)

^{13}C NMR (100 MHz, $CDCl_3$) 154.247, 149.515, 132.391, 131.301, 120.733, 111.444, 25.662, 15.199

HRMS (m/z): $[M\text{-}H]^-$ calculated for $[C_8H_8NO_3]^-$ 166.0510, found 166.0524.

3.2.3 Tert-butyl 2-(4-ethyl-3-nitrophenoxy)acetate (S4)

1. Dissolve compound S3 (3.11 g, 18.6 mmol) and tert-butyl 2-bromoacetate (4.35 g, 22.3 mmol) in DMF (25 mL).
2. Add solid K_2CO_3 (5.14 g, 37.2 mmol) to the reaction flask and stir at 70 °C for 1.5 h until TLC analysis (2:1 Hex:EtOAc, v/v) indicates complete conversion to the product.
3. Remove solvent in vacuo and redissolve the product in 100 mL EtOAc.
4. Wash the organic layer with saturated NH_4Cl (50 mL) and brine; then dry over Na_2SO_4.
5. Remove solvent in vacuo to afford S4 (4.97 g, 95%) as a yellow oil.

^{1}H NMR (500 MHz, $CDCl_3$) ppm 1.253 (t, J = 7.5 Hz, 3H), 1.5 (s, 9H), 2.857 (q, J = 7.5 Hz, 2H), 4.554 (s, 2H), 7.116 (dd, J = 8.5, 2.5 Hz, 1H), 7.277 (d, J = 8.5 Hz, 1H), 7.391 (d, J = 3 Hz, 1H)

^{13}C NMR (100 MHz, $CDCl_3$) 167.393, 156.365, 149.457, 132.315, 132.213, 120.627, 110.004, 83.095, 66.052, 28.181, 25.764, 15.122

HRMS (m/z): $[M + Na]^+$ calculated for $[C_{14}H_{19}NO_5Na]^+$ 304.1155, found 304.1160.

3.2.4 Tert-butyl 2-(4-(1-bromoethyl)-3-nitrophenoxy)acetate (S5)

1. Dissolve compound S4 (4.97 g, 17.7 mmol), N-bromosuccinimide (3.8 g, 19.5 mmol) and benzoylperoxide (0.2 g, 1 mmol) in CCl_4 (100 mL), and reflux for 4 h.
2. Cool the reaction mixture to RT and wash with 100 mL of 0.1% $NaHCO_3$ (aq) and brine; then dry over Na_2SO_4.
3. Remove solvent in vacuo, and purify the crude product by silica gel flash column chromatography (3:1 Hex:EtOAc, v/v) to afford S4 (5.7 g, 90%) as a yellow oil.

^{1}H NMR (500 MHz, $CDCl_3$) ppm 1.498 (s, 9H), 2.054 (d, J = 7 Hz, 3H), 4.571 (s, 2H), 5.787 (q, J = 7 Hz, 1H), 7.184 (dd, J = 8.5, 3 Hz, 1H), 7.299 (d, J = 2.5 Hz, 1H), 7.784 (d, J = 9 Hz, 1H)

^{13}C NMR (100 MHz, $CDCl_3$) 167.003, 157.759, 148.001, 131.149, 130.812, 120.703, 109.733, 83.372, 66.015, 42.063, 28.185, 27.372

HRMS (m/z): $[M\text{-}Br]^+$ calculated for $[C_{14}H_{18}NO_5]^+$ 280.1179, found 280.1163.

3.2.5 2-(4-(1-bromoethyl)-3-nitrophenoxy)ethan-1-ol (S6)

1. Dissolve compound S5 (5.7 g, 15.9 mmol) in 100 mL THF, and cool down to −78 °C using an acetone/dry ice bath.
2. Add DIBAL-H (39.8 mmol, 1 M in THF) to the reaction flask, and stir at −78 °C (by immersing the reaction vessel in a

container with dry ice and acetone) for 20 min, and then stir for an additional 2 h at 0 °C using a water/ice bath.

3. Once TLC analysis (3:1 Hex:EtOAc, v/v) indicates complete conversion to the product, quench the reaction by slowly adding 30 mL H_2O to the mixture, and then add 5% HCl (aq) solution until the aqueous solution becomes acidic (pH = ~4, as judged by pH paper).
4. Vigorously mix the biphasic mixture in a separatory funnel. Separate the organic layer, wash with brine (an aqueous solution saturated with NaCl, 100 mL), and dry over Na_2SO_4 (by adding the solid particles of Na_2SO_4 into the flask containing the organic layer, allowing it to absorb water, and removing the solid by filtration afterward).
5. Remove solvent in vacuo, and purify the crude product by silica gel flash column chromatography to obtain S6 (3.23 g, 60%) as a yellow oil.

^1H NMR (500 MHz, $CDCl_3$) ppm 2.056 (d, J = 5 Hz, 3H), 4.006 (dd, J = 4.5, 4.5 Hz, 2H), 4.142 (dd, J = 4, 4 Hz, 2H), 5.785 (q, J = 7 Hz, 1H), 7.201 (dd, J = 8.5, 2.5 Hz, 1H), 7.356 (d, J = 2.5 Hz, 1H), 7.783 (d, J = 8.5 Hz, 1H).

^{13}C NMR (100 MHz, $CDCl_3$) 158.563, 148.152, 131.046, 130.281, 120.514, 109.654, 70.175, 61.250, 42.129, 27.342

HRMS (m/z): $[M\text{-}Br]^+$ calculated for $[C_{14}H_{18}NO_5]^+$ 210.0761, found 210.0761.

3.2.6 1-(4-(2-hydroxyethoxy)-2-nitrophenyl)ethan-1-ol (S7)

1. Dissolve S6 (3.23 g, 11.1 mmol) in 250 mL H_2O, and reflux for 1 h until TLC analysis (1:1 Hex:EtOAc, v/v) indicates complete conversion to the product.
2. Extract the product with EtOAc (3 × 50 mL), wash the organic layer with brine, and dry over Na_2SO_4.
3. Remove solvent in vacuo to obtain S7 (2.0 g, 80%), and use it for the next step without further purification.

^1H NMR (500 MHz, $CDCl_3$) ppm 1.540 (d, J = 6.4 Hz, 3H), 3.997 (dd, J = 4.6, 4.6 Hz, 2H), 4.120 (dd, J = 7.1, 7.1 Hz, 2H), 5.341 (q, J = 6.2 Hz, 1H), 7.201 (dd, J = 8.8, 2.8 Hz, 1H), 7.410 (d, J = 2.7 Hz, 1H), 7.734 (d, J = 8.8 Hz, 1H)

3.2.7 1-(4-(2-(acryloyloxy)ethoxy)-2-nitrophenyl)ethyl Acrylate (o-NBMA)

1. Add TEA (3.5 eq) to a solution of S7 (2.0 g, 8.88 mmol) and acryloyl chloride (26.6 mmol) in CH_2Cl_2 (75 mL).
2. Stir the mixture for 24 h at RT.
3. Wash the mixture with H_2O and brine, and then dry over Na_2SO_4.

4. Remove solvent in vacuo, and purify the crude material by silica gel flash column chromatography (2.5:1 Hex:EtOAc, v/v) to obtain S8 (1.79 g, 60%) as a yellow oil.

^1H NMR (500 MHz, $CDCl_3$) ppm 1.653 (d, J = 6.5 Hz, 3H), 4.253–4.272 (m, 2H), 4.517–4.536 (m, 2H), 5.849 (dd, J = 16.5, 1.5 Hz, 1H), 5.87 (dd, J = 16.5, 1.5 Hz, 1H), 6.135 (dd, J = 33, 10.5 Hz, 1H), 6.135 (dd, J = 10.5, 1.5 Hz, 1H), 6.333 (dd, J = 6.5, 6.5 Hz, 1H), 6.425 (dd, J = 38.5, 1.5 Hz, 1H), 6.425 (dd, J = 4, 1 Hz, 1H), 7.181 (dd, J = 8.5, 2.5 Hz, 1H), 7.471 (d, J = 2.5 Hz, 1H), 7.547 (d, J = 8.5 Hz, 1H)

^{13}C NMR (100 MHz, DMSO-d6) 165.360, 164.609, 157.825, 148.572, 132.269, 132.046, 128.769, 128.284, 127.974, 127.952, 120.606, 109.422, 67.267, 66.655, 62.517, 21.219

HRMS (m/z): $[M + Na]^+$ calculated for $[C_{16}H_{17}NO_7Na]^+$ 358.0897, found 358.0888.

3.3 Silane-Glutaraldehyde Activation of Glass Bottom Culture Dishes

1. This procedure is based on a method generally accepted in the field [21].
2. Activate the surface of the 35 mm cover-glass bottom (Fisher, 50-305-810) with 0.1 N NaOH solution for 15 min.
3. Remove the NaOH solution with a pipet.
4. Add 75 μL of 3-aminopropyltrimethyoxy silane to the bottom of each dish. Using the back end of a pipette tip, smear the silane reagent to cover the entire glass surface. Incubate for 5 min on a rotator to maintain an even surface coating.
5. Wash the dishes by adding DI H_2O and incubating for 30 min on the rotator. Exchange the H_2O every 10 min. Dispose the first wash as chemical waste (*see* **Note 5**).
6. Remove H_2O and transfer the dishes to a chemical hood. Using a transfer pipet, add 0.5% glutaraldehyde in PBS to cover the entire glass surface. Incubate for 30 min (*see* **Note 6**).
7. Transfer the glutaraldehyde solution to an appropriate hazardous waste container. Wash the dishes by adding them to a H_2O-filled large beaker and carefully placing it on a rotator. Exchange the water every 10 min.
8. Let the dishes air dry. Once completely dried, store them in a desiccator (*see* **Note 7**).

3.4 Preparation of Photoresponsive Polyacrylamide Gels

1. This procedure is based on a method for preparing polyacrylamide gels generally accepted in the field [21]. The composition of the prepolymer solution components is adapted from a protocol that produces 20 kPa gels (*see* **Note 8**).
2. Prepare a 990 μL prepolymer mixture with 250 μL of 40% (w/v) acrylamide solution, 25 μL of 2% (w/v) bis-acrylamide,

32.5 μL of 50 mM o-NBbA (in DMSO), 10 μL of 1 M HEPES (pH 7.0) solution, 40 μL of 71.7 mM acrylic acid N-hydroxysuccinimide ester (in DMSO, 4% of the total volume), and 632.5 μL of DI H_2O (*see* **Notes 9–12**).

3. Degas the prepolymer mixture for 30 min by placing the Eppendorf tube containing the mixture in a vacuum desiccator with the lid open.
4. Prepare fresh 10% (w/v) APS in H_2O.
5. Initiate polymerization by adding 6 μL 10% (w/v) APS solution and 4 μL TEMED.
6. Immediately after initiation, pipette 200 μL of the gel solution onto an activated glass culture dish, and cover with a fibronectin-patterned glass coverslip face down (fabricated as described below) (*see* **Note 13**).
7. Allow polymerization for 30 min.
8. Add 1 mL PBS to the dish, and carefully remove the coverslip using forceps.
9. Rinse the gels with 1 mL PBS three times to remove unpolymerized reactants.

3.5 Preparation of 1D-Fibronectin Micropatterns on the Gel Surface

1. This procedure is based on a method generally accepted in the field [22].
2. PDMS stamps (~1 × 1 cm) containing topographical patterns (21 μm width, 40 μm spacing) were fabricated by photolithography and provided by the M. Piel laboratory (Dept. of Cell Biology, Inst. Curie).
3. Ink the patterned side of the stamp with 100 μg/mL fibronectin in PBS for 1 h.
4. During the inking process, prepare activated coverslips by rinsing them with ethanol and treating with plasma for 60 s.
5. After 1 h inking process, dry the fibronectin-inked stamps using a stream of air.
6. Stamp the fibronectin-coated stamp onto the coverslip. Place a 20 g weight on top of the stamp and wait 30 min.
7. Remove the weight and stamp from the coverslip (*see* **Note 14** and **15**).
8. Place the coverslip face down so that the fibronectin-pattern contacts the prepolymer solution described in the gel preparation procedure mentioned above, immediately upon the initiation of polymerization.
9. Allow polymerization for 60 min.
10. Add PBS to the dishes, wait 15 min, and carefully remove the coverslip using forceps.

11. Incubate the gels with 4% BSA in PBS for 45 min to passivate the remaining activated esters.
12. Rinse the gels with 1 mL PBS three times (*see* **Note 16**).

3.6 Fabrication of Stepwise Stiffness Gradients by Controlled UV Exposure

1. Place the fibronectin-patterned photoresponsive gel on the stage of an epifluorescence microscope. The one used here was a Nikon Eclipse Ti-E epifluorescence microscope equipped with a Plan Fluor 10×/0.30NA objective (Nikon). The control software was NIS-Elements (Nikon) (*see* **Note 17**).
2. Using phase-contrast imaging, select two regions of the gel such that they are "A" mm (A > 2) apart and the hypothetical line connecting the two regions runs across the fibronectin patterns perpendicularly. Record the locations based on the grids on the bottom of the dishes.
3. Adjust the field diaphragm lever so that the diameter of the illuminated area on the substrate is 500 μm (*see* **Note 18**).
4. Initiate fluorescence imaging using a 395/25 nm LED (315 mW) and DAPI filter set with LED fluorescence illumination from a SpectraX light Engine (Lumencor) (*see* **Note 19**).
5. Subsequently, capture a time lapse movie of the two regions at 0 s intervals for "15 × A" min. Leave the active shutter open during stage movement.
6. **Steps 1–5** result in a 500 μm × "A" mm region irradiated such that all the photolabile cross-linkers in the exposed region are cleaved (*see* **Notes 20–22**).
7. Repeat **steps 1–5** in regions parallel to and 500 μm apart from the first irradiated area.
8. The final product is a gel that exhibits alternating 500 μm wide stiff (~15 kPa) and soft (~8 kPa) regions.

3.7 Stiffness Characterization by Bead Indentation (Calibration of Photoirradiation time vs. photocleavage)

1. Prepare a prepolymer solution based on the protocol described above.
2. After initiating polymerization, pipette >300 μL of the prepolymer solution onto an activated glass culture dish to prepare a thick (>1 mm) photoresponsive hydrogel (*see* **Note 23**).
3. Cover the polymerizing prepolymer solution with a 25 mm no. 1.5 circular coverslip.
4. After polymerization for 30 min, add 1 mL PBS to the dish and carefully remove the coverslip using forceps, and wash the gel subsequently with 1 mL PBS three times.
5. Add 200 nm crimson fluorospheres suspended in PBS to the gel.

6. Wait 1 h for the fluorospheres to settle by gravity on the gel surface. The red fluorospheres coated on the gel now allows the researcher to determine the position of the gel surface via focusing and visualization using the microscope.
7. Place a silica bead (1 mm diameter) on the gel surface. The weight of the silica bead creates an indentation on the gel surface, which results in two different areas when imaging the surface of the gel. The region that is non-indented remains in focus, while the area that is indented becomes out of focus. In other words, the crimson fluorospheres allow the depth of indentation to be determined as they will come into focus at different depths than the overall non-indented surface.
8. Initiate photoirradiation of the gel. As the gel gradually becomes softened via photoirradiation, the silica bead starts to sink deeper into the gel. Accordingly, the indented area (area out of focus) becomes wider.
9. At each irradiation time point, measure the contact radius (r) of the silica bead to the gel surface using a microscope and its imaging software according to the following method. First, acquire z-stack fluorescent images. Then, choose the plane wherein the surface area in contact with the bead (indented) is out of focus and the rest of the surface (non-indented) is in focus. Finally, measure the radius of the area out of focus with the measurement tools of the imaging software.
10. Calculate the bead indentation depth (δ) using the contact radius (r) determined above and the bead radius (R) determined using the microscope and its imaging software (Fig. 3b–c) according to Eq. (1):

$$\delta = R - \sqrt{R^2 - r^2} \tag{1}$$

11. Using the obtained indentation depth, calculate the Young's modulus (E) using the Poisson ratio of the hydrogel (v) and buoyancy corrected bead force (f) according to the Hertz solution:

$$E = \frac{3(1 - v^2)f}{4R^{1/2}\delta^{3/2}} \tag{2}$$

wherein f is calculated according to Eq. (3):

$$f = V(B - W)g \tag{3}$$

using the volume of the glass bead (V), density of the glass bead (B), density of water (W), and the gravitational constant (g).

For polyacrylamide gels, $\nu = 0.3$–0.5 (here, $\nu = 0.3$ was used). The glass bead density was measured to be ~2600 kg/m^3 (Fig. 3d).

3.8 Characterization of Fibronectin Density by Immunofluorescence Detection

1. Prepare a fibronectin-coated photoresponsive gel following the protocol stated above.
2. Optionally introduce stiffness patterns via irradiation as described above.
3. Rinse the gel five times with PBS.
4. Incubate the gel with rabbit anti-fibronectin antibody (1:500 dilution of stock, v/v) for 60 min at 37 °C.
5. Rinse the gel five times with PBS.
6. Incubate the gel with goat anti-rabbit IgG-rhodamine (1:1000 dilution of stock, v/v) for 45 min at RT.
7. Finally, rinse the gel with PBS five times prior to epifluorescence imaging (*see* **Note 24**).

4 Notes

1. All chemical synthesis procedures should be conducted with a strong safety awareness. The researcher should be trained on the proper use or conduct of equipment, apparatus, chemical hood, lab coat, safety google, safety shower, evacuation or shutdown protocol, waste disposal, etc., before conducting chemical synthesis.
2. For synthetic procedures that involve acidic conditions such as in the synthesis of S2, we suggest neutralizing samples before performing NMR analysis since the acidity may cause shifting of proton peaks and hence misinterpretation of collected data.
3. Acidic conditions may also affect TLC analysis of the product.
4. For synthesis procedures that require high temperature conditions with an oil bath such as in the synthesis of S3, we suggest preparing a fresh oil bath and ensuring that silicone oil is used. Using other types of oil such as mineral oil that is not suitable for heating may cause flammability safety hazards.
5. The silane-glutaraldehyde activation process involves multiple rinsing steps with water. We noticed that the coverslips were prone to damage during rinsing and that the grid-500 high glass-bottom dishes were especially fragile. We suggest taking special care such as washing with low water pressure to minimize loss.
6. Procedures that involve glutaraldehyde should be conducted in a chemical fume hood, and apparatuses such as pipette tips or conical tubes that come in contact with the reagent should be properly disposed as hazardous waste.
7. The activated coverslips could be used up to a minimum of 2 weeks when stored properly in a vacuum desiccator.

8. The initial and final stiffness values of the photoresponsive gel prepared in this study were 15 and 8 kPa, which was achieved by replacing 50 mol% of the bis-acrylamide content with o-NBbA based on a prepolymer composition that yields a 20 kPa polyacrylamide gel. While 50% reduction in stiffness after complete photocleavage of the o-NBbA was achieved as expected, the o-NBbA containing gels were overall softened (initial stiffness reduced from 20 to 15 kPa and final stiffness from 10 to 8 kPa) compared to polyacrylamide. We hypothesize that the hydrophobicity of o-NBbA is responsible for this reduced stiffness. Previous reports indicated that the hydrophobicity of bis-AAm cross-linkers may lead to clustering and phase separation of the molecules at high concentrations, resulting in the reduction of the effective cross-linker concentration and hence softening of the gel [23]. Similarly, we suspect that the hydrophobicity of o-NBbA, which is much higher than bis-AAm, may cause phase separation even at low cross-linker concentrations. The visual observation of turbidity in the photoresponsive prepolymer solution increases with o-NBbA content and supports this hypothesis.
9. We stored stock solutions of o-NBbA in DMSO in a brown vial at −80 °C. Diluted aliquots were prepared as needed.
10. We suggest minimizing light exposure while conducting experiments with o-NBbA (purification of the cross-linker, preparation of the prepolymer solution, preparation of the gel, etc.).
11. NHS-ester acrylic acid is extremely water sensitive. Hydrolyzed NHS-ester acrylic acid is the major factor that can cause failure to transfer fibronectin patterns from the coverslip to the gel surface. To prevent such failure, we suggest preparing aliquoted stock solutions of NHS-ester acrylic acid in anhydrous DMSO and storing them in −80 °C in a sealed container with a desiccant. We suggest disposing of unused material from any aliquot once it has been opened to air for an experiment.
12. Aqueous solutions of acrylamide, bis-acrylamide, and HEPES buffer should be stored at 4 °C. Although we noticed these reagents can be used for months when stored properly, we suggest preparing fresh reagents at least every 3 months to ensure reproducibility. For example, acrylamide may hydrolyze to acrylic acid and ammonia and alter the mechanical properties of the polyacrylamide gel product.
13. We used high grid-500 glass-bottom dishes (Fisher, 50-305-810) which allowed locating the area that was UV exposed, i.e., where the gel was softened (Fig. 4). We suggest using other non-grid 35 mm dishes until mastering the gel-preparation steps since these dishes are particularly costly.

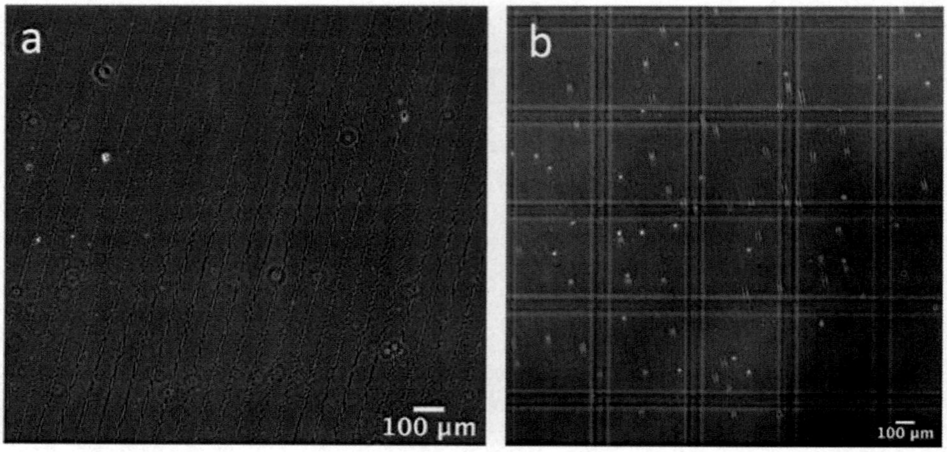

Fig. 4 Adhesion of U-251MG cells onto 1D fibronectin coated on the surface of photoresponsive gels. (**a**) Bright field microscopy confirms successful fabrication of uniform fibronectin line-patterns on the surface of the gels. (**b**) U-251MG cells only adhere to the fibronectin-coated regions of the gel, demonstrating the effect of fibronectin patterns in confining the direction of cellular migration. The checkered patterns shown in the background of the image result from the grid-patterns on the glass-bottom dishes, which assist in recording the location of individual cells and UV-irradiated regions of the gel

14. After using PDMS stamps for fibronectin printing, immerse the stamps in 70% ethanol and sonicate for 10 min. Rinse the stamps with 70% ethanol, let the stamps air dry, and store in a vacuum desiccator.
15. We noticed that with proper maintenance, the PDMS stamps are quite resistant to wear and tear even after extensive repetitive use. However, we suggest checking their condition occasionally using an optical microscope.
16. We noticed most failures in fibronectin patterning of the gel were caused by using deactivated, hydrolyzed NHS-ester acrylic acid. In case of a failed experiment, we suggest pinpointing the responsible fibronectin-transfer step by carefully observing the fibronectin-printed coverslip after each step (transfer of fibronectin from PDMS stamp to coverslip and from coverslip to gel surface). The line-patterned fibronectin is visible by optical microscopy and often even to the naked eye (Fig. 4).
17. Although we here report the use of a specific type of microscope and settings thereof to introduce stiffness gradients to the photoresponsive gel, we note that any microscope designed to perform epifluorescence imaging should be capable of conducting the procedure as described with some modification of the specific settings employed.
18. Using a 40× objective and controlling the field diaphragm under our microscope settings, the diameter of the

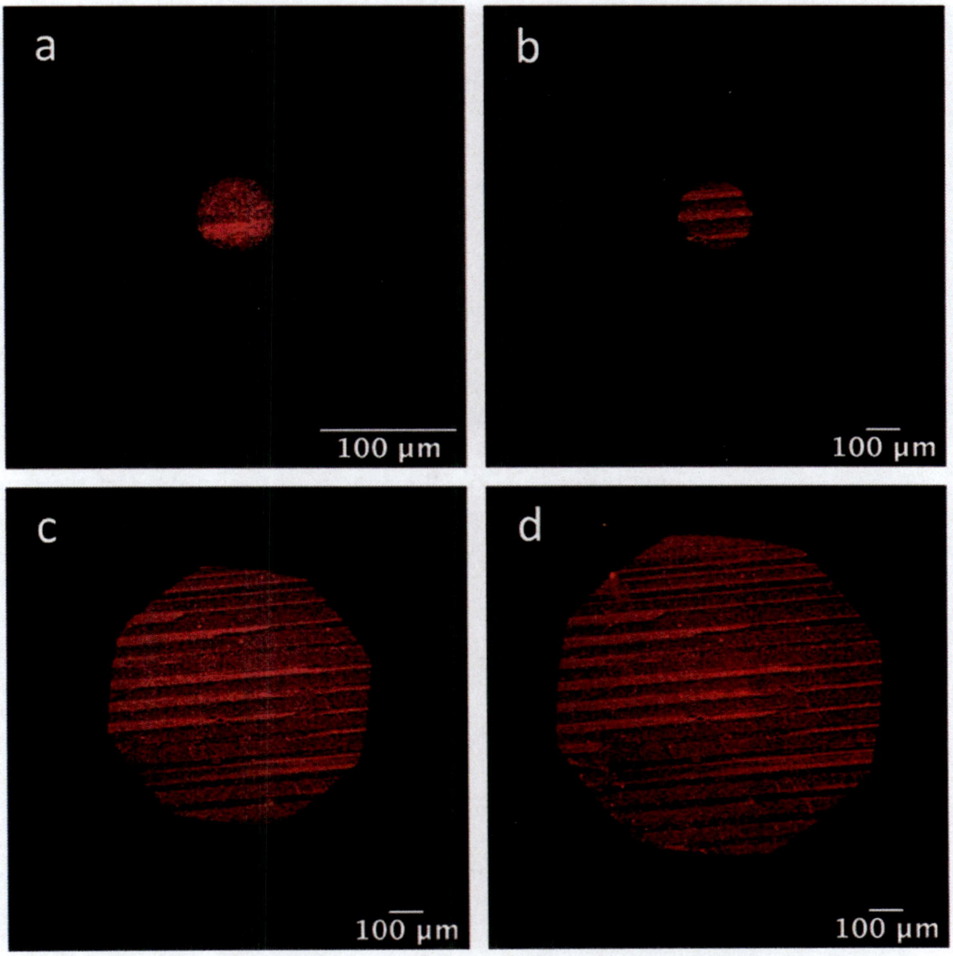

Fig. 5 Control of the UV-irradiation region by adjusting the objective and field diaphragm of the microscope. The red area represents the surface of the gel that is excited by the LED on the microscope through an mCherry filter. The scale bar in all images represents 100 μm. (**a**) The diameter of the irradiated region can be adjusted to be as low as d = 59 μm using a 40x objective with the field diaphragm nearly closed. (**b**–**d**) show that the area of the irradiated region can be controlled by opening and closing the field diaphragm. All images were taken with a 10x objective, with opening of the field diaphragm from b to d. Diameter of the irradiated regions is measured as d = 218, 774, and 978 μm, from b to d, respectively

LED-irradiated region could be controlled down to 59 μm (Fig. 5). An even smaller irradiation area would be achievable using a higher objective, which demonstrates the ability to fabricate elaborate, high-resolution stiffness patterns comprising spots, lines, planes, etc., using this methodology.

19. Photopatterning of the gel should be conducted wearing UV-protective safety goggles.

20. We noticed that as the o-NBbA in the gel is photolyzed and the gel becomes softened, the thickness of the photoirradiated

regions increase due to swelling. However, the degree of swelling was minimal, especially when only selective regions of the gel were photoirradiated. For example, the 1D patterned gels used in this study showed only a 20 μm difference in thickness between the irradiated and nonirradiated regions of the gel, which translates to a 3% increase after photoirradiation. These numbers are well below the cellular dimensions of most cells including U-251MG cells that were the focus of our studies. Therefore, we believe the impact of topography on cellular studies would be minimal.

21. One negative side effect from swelling of the gel was crease formation. The correlation between swelling and creasing was confirmed by observing deswelling and un-creasing of the gels when immersed in high salt concentration PBS or 10 v/v% methanol (Fig. 6). Previous reports also support the connection between swelling and creasing. Hayward and coworkers demonstrated that a thin layer of gel fixed to a rigid substrate experiences anisotropic swelling, which introduces equibiaxial compressive strain on the gel surface and in turn surface crease formation [10, 11]. The photoresponsive gels were particularly more susceptible to this phenomenon since the goal was to replicate the brain environment, which consists of low stiffness regions (<25 kDa) [8]. Soft gels with low cross-linking density tend to swell to a larger extent than stiff gels due to their structural instability, which leads to a higher tendency to crease. As our goal was to observe the effect of substrate stiffness on cell migration, with all other factors controlled to be consistent, it was desired to prepare flat, crease-free gel

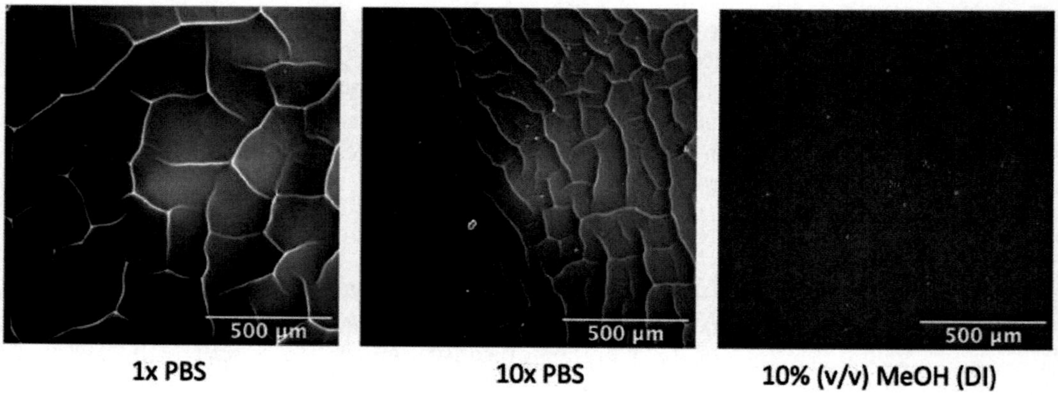

Fig. 6 Creasing of the photoresponsive gel surface under various solvent conditions. The degree of creasing of the gel immersed in 1x PBS reduces when the solvent is changed to 10x PBS, which eventually becomes crease-free when immersed in 10% (v/v) MeOH in water. Results indicate that creasing is a result of swelling of the gel. Phase-contrast images were obtained using a Nikon Eclipse Ti-E microscope equipped with an Andor Zyla 5.5 sCMOS camera (Andor Technology) and Plan Fluor 10x/0.25NA objective (Nikon)

surfaces. Therefore, multiple aspects that were explored to prevent creases included adjusting the interaction between the gel and glass substrate and adjusting the solvents of the prepolymer solution. Controlling the thickness of the gel was found to be the most simple and effective solution. Maintaining a thickness above 600 μm led to crease-free gels, which was in accordance with previous studies that showed an inverse correlation between the gel thickness and the onset point of creasing [10]. Irradiating selective regions of the gel instead of the whole gel also helped preventing creases as the degree of swelling of the irradiated regions could be offset by the surrounding nonirradiated regions.

22. A potential solution to preventing surface creases would be to render the o-NBbA more hydrophilic. One option could be functionalizing the photocleavable cross-linker with water-soluble groups such as a charged carboxylic acid or poly(ethylene glycol) group.

23. Using thick gels for stiffness characterization purposes is critical. The contact area of silica beads placed on thin gels may be affected by the underlying glass substrate, resulting in false (higher than actual stiffness values) results.

24. Comparison of the fibronectin density between the irradiated and nonirradiated region demonstrated that light does not affect the surface-coated ECM (Fig. 7).

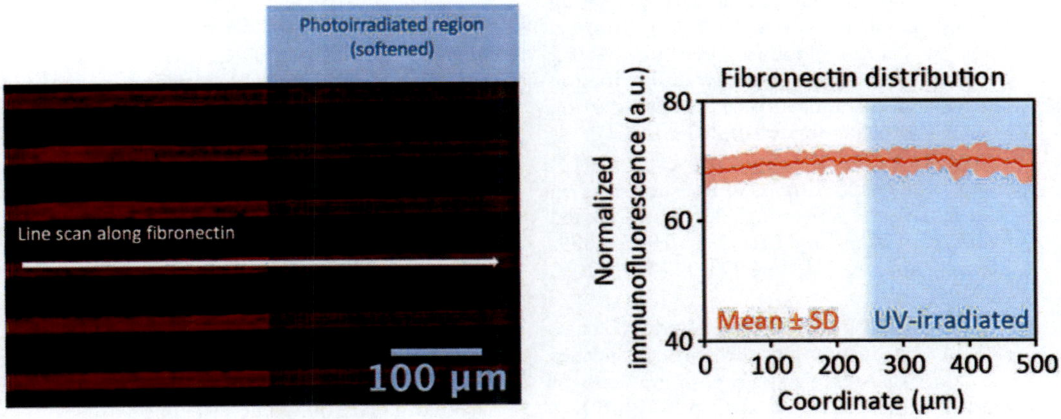

Fig. 7 Density of the microprinted fibronectin in the UV-irradiated and nonirradiated regions of the photoresponsive gel. Immunofluorescence detection of the fibronectin density by fluorescence microscopy indicates that photocleavage of the gel does not result in changes in fibronectin density

Acknowledgments

This work was supported by NIH grant R35 GM141853 to MDD and NIH grants U54 CA210190 and R01 CA172986 to DJO. We thank Aleksi Isomursu for generating the plot used here for Figure 7.

References

1. Ostrom QT, Gittleman H, Fulop J et al (2015) CBTRUS statistical report: primary brain and central nervous system Tumors diagnosed in the United States in 2008-2012. Neuro-Oncol 17:iv1–iv62. https://doi.org/10.1093/neuonc/nov189
2. Giese A, Bjerkvig R, Berens ME, Westphal M (2003) Cost of migration: invasion of malignant gliomas and implications for treatment. J Clin Oncol 21:1624–1636. https://doi.org/10.1200/JCO.2003.05.063
3. Lo CM, Wang HB, Dembo M, Wang YL (2000) Cell movement is guided by the rigidity of the substrate. Biophys J 79:144–152
4. Isenberg BC, DiMilla PA, Walker M et al (2009) Vascular smooth muscle cell Durotaxis depends on substrate stiffness gradient strength. Biophys J 97:1313–1322. https://doi.org/10.1016/j.bpj.2009.06.021
5. Hadden WJ, Young JL, Holle AW et al (2017) Stem cell migration and mechanotransduction on linear stiffness gradient hydrogels. Proc Natl Acad Sci 114:5647–5652. https://doi.org/10.1073/pnas.1618239114
6. Kim TH, An DB, Oh SH et al (2015) Creating stiffness gradient polyvinyl alcohol hydrogel using a simple gradual freezing-thawing method to investigate stem cell differentiation behaviors. Biomaterials 40:51–60. https://doi.org/10.1016/j.biomaterials.2014.11.017
7. Jiang FX, Yurke B, Schloss RS et al (2010) Effect of dynamic stiffness of the substrates on neurite outgrowth by using a DNA-crosslinked hydrogel. Tissue Eng Part A 16:1873–1889. https://doi.org/10.1089/ten.TEA.2009.0574
8. Bouchonville N, Meyer M, Gaude C et al (2016) AFM mapping of the elastic properties of brain tissue reveals kPa μm−1 gradients of rigidity. Soft Matter 12:6232–6239. https://doi.org/10.1039/C6SM00582A
9. Guimarães CF, Gasperini L, Marques AP, Reis RL (2020) The stiffness of living tissues and its implications for tissue engineering. Nat Rev Mater 5:351–370. https://doi.org/10.1038/s41578-019-0169-1
10. Trujillo V, Kim J, Hayward RC (2008) Creasing instability of surface-attached hydrogels. Soft Matter 4:564–569. https://doi.org/10.1039/B713263H
11. Yoon J, Kim J, Hayward RC (2010) Nucleation, growth, and hysteresis of surface creases on swelled polymer gels. Soft Matter 6:5807–5816. https://doi.org/10.1039/C0SM00372G
12. Mahmoodi MM, Abate-Pella D, Pundsack TJ et al (2016) Nitrodibenzofuran: a one- and two-photon sensitive protecting group that is superior to brominated Hydroxycoumarin for thiol caging in peptides. J Am Chem Soc 138:5848–5859. https://doi.org/10.1021/jacs.5b11759
13. Wieboldt R, Gee KR, Niu L et al (1994) Photolabile precursors of glutamate: synthesis, photochemical properties, and activation of glutamate receptors on a microsecond time scale. Proc Natl Acad Sci U S A 91:8752–8756
14. Kloxin AM, Kasko AM, Salinas CN, Anseth KS (2009) Photodegradable hydrogels for dynamic tuning of physical and chemical properties. Science 324:59–63. https://doi.org/10.1126/science.1169494
15. Wong DY, Griffin DR, Reed J, Kasko AM (2010) Photodegradable hydrogels to generate positive and negative features over multiple length scales. Macromolecules 43:2824–2831. https://doi.org/10.1021/ma9023679
16. Yang C, Tibbitt MW, Basta L, Anseth KS (2014) Mechanical memory and dosing influence stem cell fate. Nat Mater 13:645–652. https://doi.org/10.1038/nmat3889
17. Rosales AM, Vega SL, DelRio FW et al (2017) Hydrogels with reversible mechanics to probe dynamic cell microenvironments. Angew Chem Int Ed Engl 56:12132–12136. https://doi.org/10.1002/anie.201705684
18. Ramanan VV, Katz JS, Guvendiren M et al (2010) Photocleavable side groups to spatially alter hydrogel properties and cellular interactions. J Mater Chem 20:8920–8926. https://doi.org/10.1039/C0JM00734J

19. Isomursu A, Park K-Y, Hou J et al (2022) Directed cell migration towards softer environments. Nat Mater. https://doi.org/10.1038/s41563-022-01294-2
20. Mohrig J, Alberg D, Hofmeister G et al (2014) Laboratory techniques in organic chemistry, 4th edn. Macmillan Learning
21. Wang YL, Pelham RJ (1998) Preparation of a flexible, porous polyacrylamide substrate for mechanical studies of cultured cells. Methods Enzymol 298:489–496. https://doi.org/10.1016/s0076-6879(98)98041-7
22. Polio SR, Smith ML (2014) Patterned hydrogels for simplified measurement of cell traction forces. Methods Cell Biol 121:17–31. https://doi.org/10.1016/B978-0-12-800281-0.00002-6
23. Denisin AK, Pruitt BL (2016) Tuning the range of polyacrylamide gel stiffness for Mechanobiology applications. ACS Appl Mater Interfaces 8:21893–21902. https://doi.org/10.1021/acsami.5b09344

Chapter 10

Analysing Mechanically Evoked Currents at Cell-Substrate Junctions

Surabhi Shrestha, Jessica Richardson, and Kate Poole

Abstract

The precise study of mechanically activated ion channels requires a combination of electrophysiology to directly measure channel-mediated ionic flux and a means to apply meaningful mechanical stimuli to activate the channel. In metazoans, individual cells in vivo experience mechanical inputs at the cell-substrate interface where cells form connections to the local microenvironment. To study such processes in vitro, a technique is required where mechanical stimuli can be applied to cells via connections with an underlying substrate. Here, we outline the methodology for combining whole-cell patch-clamp electrophysiology (to monitor transmembrane currents) with elastomer pillar arrays that can be deflected (to apply stimuli to cells). This quantitative technique can be used to assess changes in sensitivity and kinetics of mechanically evoked currents when cell intrinsic or cell extrinsic factors are manipulated.

Key words Mechanically activated ion channels, Elastomeric pillar arrays, Electrophysiology, Mechanotransduction, PIEZO1

1 Introduction

Mechanically activated (MA) ion channels are pore-forming membrane proteins that are largely closed at rest. The open probability of these channels increases with increasing mechanical input, leading to a localised flux of ions across the membrane [1]. The composition of these transmembrane currents is dependent on the selectivity of the specific channel. MA ion channels display a range of selectivities, including non-selective cation channels, such as PIEZO1 [2, 3], and K^+-selective channels, such as TREK-1 [4]. A defining feature of MA channels is the direct coupling of the mechanical input to the gating mechanism, with latencies between stimulus and channel activation of less than 5 ms and activation time constants of less than 2 ms [5, 6]. These rapid kinetics and varying ion selectivities necessitate the use of patch-clamp electrophysiology to directly study current flux and to ensure

Ronen Zaidel-Bar (ed.), *Mechanobiology: Methods and Protocols*,
Methods in Molecular Biology, vol. 2600, https://doi.org/10.1007/978-1-0716-2851-5_10,

sufficient temporal resolution. There are a number of methodologies available in order to apply meaningful mechanical stimuli to cells, in combination with patch-clamp measurements. For example, high-speed pressure clamp [7] (HSPC) can be used to locally stretch a patch of membrane, and indentation [8] with a glass rod can be used to globally compress a cell. Significant insights regarding the activation of MA channels in the apical surface of cells have been generated using these techniques. The use of HSPC in particular has been instrumental in the characterisation of MA channels that mediate stretch-activated currents, such as TREK-1 and PIEZO1 [9–12]. However, in metazoans, an important functional compartment for mechanotransduction in adherent cells is the cell substrate interface [13–15]. Within this interface, there are several distinct types of mechanosensing molecules. The study of mechanosensing in the cell-substrate interface has largely focused on the action of mechanosensors such as integrins and strain gauge proteins that are found in focal adhesion structures [16–19]. Recent data have clearly demonstrated that some MA channels are also activated by mechanical inputs applied via the cell-substrate interface, including PIEZO1, PIEZO2, TRPV4 and TACAN [20–25]. In addition, the newly identified ELKIN1, a putative MA channel, supports mechanically evoked currents within the cell-substrate interface [26]. The study of activation of MA channels within this interface requires a methodological approach where stimuli are propagated to cells via their connections to the substrate [20]. Here, we outline an approach using elastomeric pillar arrays as a deformable substrate to apply stimuli while monitoring the subsequently evoked currents using whole-cell patch-clamp electrophysiology. The flexibility of this technique further allows manipulation of both cell intrinsic and cell extrinsic factors that can regulate MA channel activation, such as bulk substrate mechanics [22], the composition of the extracellular matrix [26] and intracellular accessory proteins [20, 21].

2 Materials

2.1 Fabricating Negative Masters to Cast Arrays

1. Microfabricated positive masters (*see* **Note 1**).
2. Trichloro(1H,1H,2H,2H-perfluorooctyl)silane, 97% (*see* **Note 2**).
3. Polydimethylsiloxane (PDMS) (*see* **Note 3**).
4. Vacuum desiccator.
5. Oven capable of heating to 80 °C.
6. Large weigh boats.
7. Straight razor blades.

2.2 Casting Pillar Substrates

1. Silanised negative masters (prepared in **steps 3.1** and **3.2**).
2. PDMS.
3. Vacuum desiccator.
4. Oven capable of heating to 110 °C.
5. Square glass coverslips (thickness 2).
6. Low pressure plasma cleaner.

2.3 Sample Preparation

1. Adherent cells and appropriate media (*see* **Note 4**).
2. Cell dissociation solution (*see* **Note 5**).
3. Plastic culture dish, 35 mm × 10 mm.
4. Pillar array (prepared in **step 3.3**).
5. Fast curing, two-component epoxy.
6. Low-pressure oxygen plasma system.

2.4 Whole-Cell Patch-Clamp Electrophysiology

1. Filamented, thick-walled glass to prepare recording electrodes and glass stimulators.
2. Pipette puller.
3. Microforge.
4. Intracellular buffer (IC): 10 mM NaCl, 135 mM KCl, 1 mM $MgCl_2$, 10 mM HEPES, 1 mM EGTA, adjust to 7.2–7.3 pH using KOH, 280–285 osmolarity.
5. Extracellular buffer (EC): 140 mM NaCl, 4 mM KCl, 2 mM $CaCl_2$, 1 mM $MgCl_2$, 4 mM glucose, 10 mM HEPES, adjust to 7.4 pH using NaOH, 295 osmolarity.
6. Recording electrodes; prepared from filamented glass using pipette puller to have a final resistance of 3–6 MΩ.
7. Glass stimulator: filament glass pulled to give micropipettes which have been sealed using a microforge (*see* **Note 6**).
8. Patch-clamp rig including amplifier, digitiser, appropriate software, micromanipulator for controlled movement of the patch-clamp head stage, appropriate silver chloride wires for both recording and reference electrodes.
9. Inverted light microscope, with 10 × and long distance 40 × objective (adjustable coat, bright field contrast).
10. CCD or sCMOS camera (pixel size <8.5 μm).
11. Nano-stimulator (*see* **Note 7**).
12. Optical table to affix all elements of patch-clamp set-up.
13. Faraday cage to enclose entire patch-clamp rig.

3 Methods

Carry out all methods at room temperature, unless otherwise indicated.

3.1 Design and Obtain Positive Masters

1. Positive masters should be designed with a hexagonally packed array of cylinders. Microfabricated masters can be obtained with multiple patterned regions to allow parallel casting of multiple negative masters (Fig. 1a, b) with consistent dimensions. There are some constraints on the dimensions of the cylindrical elements within the array; elements should be at least 5 μm high to allow access with stimulating probe. The-spring constant of each pilus in the final will depend on the height and diameter of each cylinder (Eq. 1), where E is the bulk elastic modulus of the material, d is the diameter and H is the height of each cylinder in the array.

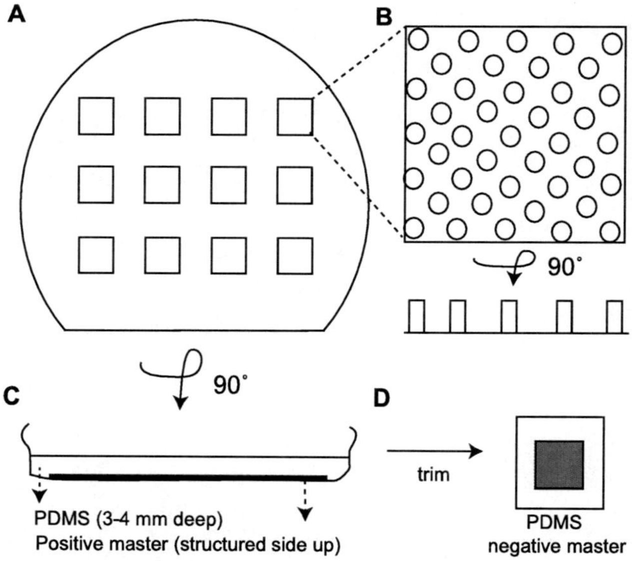

Fig. 1 Design of positive masters and creating negative masters. (**a**) When microfabricating positive masters, create a substrate with multiple repeating structured regions that will correspond to individual arrays (approx. 10 mm × 10 mm each). (**b**) Design the structured regions to contain a hexagonally packed array of cylindrical elements of defined size; note in this figure, these features are not to scale. A view from the side would reveal the cylinders of defined and precise dimensions that constitute the positive master array. (**c**) To create negative masters for ongoing casting of arrays for experiments, place silanised positive master structured-side up in a weigh boat, and carefully cover with PDMS before curing. (**d**) After curing, use a straight razor to cut out the individual negative masters to use for casting arrays (here patterned region is shaded grey)

Table 1
Validated dimensions of cylindrical elements within pillar arrays (*see* Note 8)

Diameter (μm)	Height (μm)	Centre-to-centre spacing (μm)	Spring constant *k* (nN/nm), E = 2.1 mPa
3.5	6	7	0.2
4.5	5	7.5	1.0
4.5	5	9	1.0
4.5	5	11.5	1.0
6.5	5	11.5	4.4

$$k = \frac{3\pi E d^4}{64 H^3} \quad (1)$$

Increasing the length or decreasing the diameter can lead to arrays that collapse after casting. See Table 1 for array dimensions that have been successfully employed for these experiments.

2. Microfabrication of masters can be outsourced to commercial providers but requires a deep etching capacity to ensure sufficient height of the array elements. With careful handling, positive masters can last years, being periodically employed to cast negative masters for ongoing use (*see* **Note 9**).

3.2 Preparing Negative Masters

1. Silanise the positive master by carefully transferring the master, structured-side up, to a vacuum desiccator in a fume hood.
2. Place a clean, dry coverslip on either side of the positive master and pipette 5 μl of Trichloro(1H,1H,2H,2H-perfluorooctyl) silane, 97% onto each coverslip.
3. Seal the desiccator, apply vacuum and leave overnight.
4. Take care to flush the bottle of Trichloro(1H,1H,2H,2H-perfluorooctyl)silane, 97% with nitrogen before storage.
5. After overnight incubation, remove the coverslips and dispose.
6. Test that the silanisation has created a sufficiently hydrophobic surface by pipetting 10 μl of milliQ water onto the surface of the positive master (in an unstructured region), and check the contact angle of the droplet. If the contact angle is high, the silanisation has been effective. If the contact angle is low, repeat the silanisation step (*see* **Note 10**).
7. Prepare 40 ml of PDMS from the two components provided in the Sylgard 184 kit; the ratio of bulk elastomer to curing agent is 10:1. Mix the two components extremely well to ensure even mixing, and then degas in a vacuum desiccator for at least 30 min, or until all bubbles have been extracted from the PDMS.

8. Carefully place silanised positive master into a disposable weigh boat, structured-side up, and pour degassed PDMS mix gently over the top, to a depth of 3–4 mm (Fig. 1c). Place in oven, pre-heated to 80 °C, for 15 min.
9. Remove from oven and allow to cool slightly before very gently cutting away the weigh boat. Turn the sandwich of PDMS and positive master over and gently trim excess PDMS from the underside of the positive master.
10. Turn the PDMS-positive master again to place the positive master on the underside of the PDMS. While the positive master remains sitting against a clean, flat surface, very gently deform the PDMS up and away from the positive master. This deformation should cause the PDMS to peel smoothly away. Keep gently deforming the PDMS until it is separated from the positive master.
11. Immediately return the positive master to a secure, dust-free location for storage until negative masters need to be replaced.
12. Use a straight razor to cut out individual negative masters from the block of PDMS, allowing 2–3 mm of clearance around the patterned region (Fig. 1d).
13. Before using negative masters to cast arrays, silanise with Trichloro(1H,1H,2H,2H-perfluorooctyl)silane, 97%, as described in **steps 1–5** above (*see* **Note 11**).

3.3 Casting Pillar Arrays

1. Mix 3.6 mg bulk elastomer and 0.4 mg curing agent from the Sylgard 184 kit, ensuring complete mixing of the two components. Place in a vacuum desiccator, and degas under vacuum for 30 min. Preheat the oven to 110 °C.
2. Place silanised PDMS-negative masters, pattern-side up, on glass dish. After degassing the PDMS mixture, carefully drop sufficient PDMS on the patterned side of the negative masters, and carefully spread to cover the entire patterned surface, minimising contact with the surface of the negative master. Leave to sit for 30 min.
3. Place a glass coverslip for each array in a plasma system and activate by treating with plasma oxygen for 90 s. Immediately place activated coverslips on top of the PDMS-coated-negative master. Apply gentle pressure to ensure a thin layer of PDMS between the negative master and the glass coverslip.
4. Place masters in the preheated oven for exactly 1 h to allow for the PDMS to cure (*see* **Note 12**). After curing, gently peel the coverslips with the cured array affixed off the negative masters. The pillar arrays are stable and can be used over a period of weeks if stored carefully in a dust-free environment.
5. Store negative masters in a clean, dust-free environment until the next use.

3.4 Preparing Cell Samples on Pillar Arrays

1. Mix equal parts of a two-component epoxy, and apply small amounts to each corner of the glass coverslip on which a pillar array is cast. Affix the glass coverslip on a plastic Petri dish and allow the epoxy to cure.
2. Release cells from tissue culture flasks or prepare primary cells, as per standard protocols.
3. Cells should be seeded on pillar arrays 2–36 h prior to experiment, depending on experimental design. Longer incubation times are possible; however, patching efficiency decreases with time. When using tissue culture cells, acutely (within a few hours of isolation) release cells from tissue culture plastic using an enzyme-free cell dissociation buffer. When tissue culture cells are to be assessed the following day, they can be released from plastic using trypsin solutions as per standard protocols. If tissue culture cells are to be studied, take care to never let cells overgrow during passaging, as this will result in cells where it is challenging to form a high resistance seal with the patch pipette. For primary cells, take care in isolating to minimise exposure to mechanical disruption and enzyme treatment, as harsh conditions can result in cells where it is challenging to form a high resistance seal.
4. As cells are detaching or in the final step of primary cell isolation, place culture dish with affixed pillar array in a plasma cleaner and treat with plasma oxygen for 90 s (*see* **Note 13**).
5. Immediately seed cells in a Petri dish containing the array at a low density. Allow cells to settle and adhere to array.

3.5 Measuring MA Currents

1. Prepare recording pipettes by using a pipette puller. To avoid dust and fine particles accruing on pipettes, pipettes should be pulled at most a day before or on the day of experimentation. For whole-cell patch clamp, thick-walled capillary glass can be used. Parameters for preparing glass pipettes should result in tapered pipettes with a final resistance of 3–6 MΩ (*see* **Note 14**). Fire polish pipettes before use using a microforge. Ensure there are no structural defects at the tip of the pipette, and handle carefully to avoid touching the fragile, fine end against any surface.
2. Equilibrate EC and IC to room temperature. EC can be stored at 4 °C for approximately 2–4 weeks and IC at −20 °C for a year.
3. Set up electrophysiology equipment, software and micromanipulator.
4. Remove media from culture dish and gently wash with EC. Fill dish with 2 mL EC, mount on inverted light microscope and place the reference electrode in the EC buffer at the side of the dish.

5. Locate a cell of interest that is healthy and is located where there are pili accessible by the stimulator. The cell should be clearly attached to the subjacent pili. However, be mindful of whether the cell has filopodia extended over the edge that could be damaged by the glass stimulator.
6. Attach a sealed glass stimulator to the nano-stimulator, and, using a low power objective (i.e. 10 ×), carefully manoeuvre the stimulator towards the cell of interest so that the shadow of the glass probe can be seen when the top of the pillar array is in focus.
7. Fill recording pipette with IC using a syringe with a long fine needle, ensuring to only fill enough so that the recording electrode is submerged. Make sure there are no bubbles. Attach pipette onto the head stage and lower the stage to approach the surface of the EC. Ensure that a syringe is attached via tubing to the pipette holder of the head stage, and gently apply a small amount of outward pressure before submerging the recording pipette in EC. Take note of the initial resistance when a test pulse is applied. If the resistance is outside of the 3–6 MΩ range, discard the pipette and start again. Ensure the pressure is released before removing the pipette.
8. Once the recording pipette and stimulator are coarsely positioned near the cell to be studied, change to a 40× objective, and ensure the recording pipette is positioned so that it is approaching the apex of the cell. Use the coarse movement of the micromanipulator controlling the recording pipette movement to carefully approach so that the recording pipette is just above the cell.
9. Adjust the focal plane so that the tops of the pili are in clear focus, ensuring a well-contrasted image to aid with post hoc analysis. Change the micromanipulator controlling the recording pipette to fine mode, and gently lower the recording pipette, keeping an eye on the pipette resistance in the test pulse. Continue approaching the cell with the recording pipette until a 10% increase in resistance is noted in the recording software.
10. Release outward pressure and gently apply inward pressure to form a high-resistance seal (> 1 GΩ) between the pipette and the membrane of the cell. Adjust the holding potential to an appropriate setting for the cells to be studied (depending on the expected transmembrane potential of the cells, often −80 or − 60 mV). Apply a quick pulse of inward pressure to disrupt the cell membrane within the patched region of membrane. Switch amplifier to 'whole-cell' mode, and charge the membrane capacitance using the whole-cell compensation. Once a stable whole-cell configuration is established, carefully compensate series resistance to at least 60% (*see* **Note 15**).

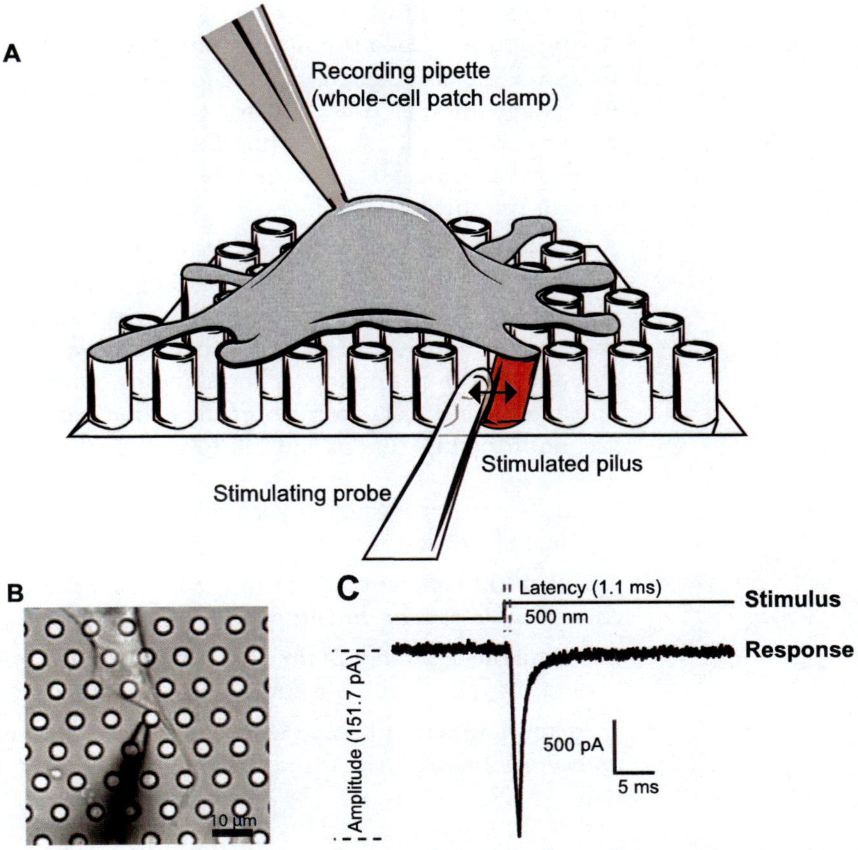

Fig. 2 Schematic of experimental setup. (**a**) Cells can be cultured on top of a pillar array. First, whole-cell patch-clamp mode is established, allowing continuity between the intracellular solution of the pipette and the inside of the cell. Mechanical stimuli are then applied by positioning a stimulating probe adjacent an individual pilus subjacent to the cell. Stimuli are applied by serially deflecting the selected pilus with the stimulating probe. Data is collected by (**b**) imaging the pilus with each deflection and (**c**) recording the resulting mechanically evoked currents via the recording pipette. This example trace was recorded from melanoma WM-266/4 cells at a holding potential of −60 mV. Latency, amplitude and stimulus size are all marked for this specific current

11. Using fine movements, approach the glass stimulator (which should already be visible in a live image from the camera) towards a pilus of interest, subjacent to the cell. Position the stimulator towards the middle of the height of the pilus (Fig. 2a). Be careful not to touch or bump the cell with the stimulator during this movement; monitoring the transmembrane current during this positioning allows an assessment of whether the cell has been touched by the glass probe during the approach.
12. Capture an image of the cell and stimulator (Fig. 2b). Start recording in voltage-clamp mode, at an appropriate holding potential. Initiate a series of pillar deflections by using the

nano-stimulator to move the pilus, ensuring to take pictures during and after each stimulus. Ensure the stimulus is applied for 0.5–1 s to allow sufficient time to capture an image so that the extent of pillar deflection for each measurement can be precisely determined in post hoc analysis. It is recommended to allow at least 10 s between successive stimuli to avoid current rundown.

13. Apply a series of stimuli ranging from 5 nm to 1000 nm. To generate stimulus response curves, these stimuli should be randomised. Alternatively, to calculate the threshold of activation, apply stimuli with increasing magnitudes. For each cell, apply at least five stimuli over the stimulus range, and, where possible, collect data from more than one pilus per cell to account for pilus-to-pilus and cell-to-cell variation.

3.6 Analysis

3.6.1 Characterising Currents

1. Open current recordings in Clampfit or other electrophysiology analysis software.
2. Calculate the difference between the baseline and peak of the current to determine amplitude.
3. Calculate the difference in time between the start of the stimulus and the start of the current to determine latency.
4. To determine activation and inactivation time constants, use a mono-exponential fit of the start and end of the curve, respectively.

3.6.2 Calculating Pillar Deflection

1. Export series of images on ImageJ or similar image editing software. Crop pilus of interest from the series and export as a series of TIFF files.
2. Use an appropriate analysis program to apply a 2D Gaussian fit of the intensity values of each pilus. This analysis will allow the centre of the pilus in each image (as x, y coordinates) to be determined, as each pilus acts as a light guide when visualised using bright field microscopy.
3. Calculate the difference between the centre of the pilus in successive images to derive the magnitude of the deflection.

4 Notes

1. Microfabricated negative masters can be purchased or obtained from various sources. We have utilised Bonda Technology Pte Ltd. Dimensions of masters are a critical factor for both handling of arrays, and for experimental interpretation, arrays should be at least 5 μm high.
2. This silane is both corrosive and toxic in liquid and vapour phases and should be handled and stored with care.

3. PDMS is often purchased as Sylgard 184.
4. A fundamental requirement for this technique is that cells should be adherent and thus form contacts with the subjacent pillar array. Beyond that requirement, any adherent cell type should be amenable to study with this technique. To date, the following cell types have been studied: primary dorsal root ganglia neurons [20, 21], primary chondrocytes [23], melanoma cells [26] and HEK-293 T cells [22, 25–27]. Cells should be cultured in appropriate media, with no specific adaptations required for successful maintenance on pillar arrays.
5. Cultured cell lines should be released from tissue culture plastic for seeding on pillar arrays using an enzyme-free cell dissociation buffer for analysis within 8 h of subculture. Cells to be assayed the following day can be released from tissue culture plastic using solutions containing trypsin.
6. Glass stimulators should be sealed at end to avoid damaging the delicate cell membrane but should remain approximately 2 μm in diameter to allow controlled deflection of the pilus.
7. To date, the Kleindiek, Germany, MM3A-LS nano-stimulator has been utilised for these experiments. Any stimulator capable of fine, controlled movements of a mounted glass probe could be utilised.
8. These values correspond to the dimensions specified in the design of master for microfabrication. After creating negative masters and casting pillar arrays, test arrays should be imaged using scanning electron microscopy to determine precise dimensions [22, 28].
9. Handle positive masters with care, these can be expensive to fabricate but can also be used for years if handled appropriately. They are easily shattered so should always be stored in a safe, dust-free environment. Avoid handling the patterned surface in any way.
10. Never proceed to casting negative masters if the silanisation of the positive master has been ineffective; this will lead to potential destruction of the microfabricated master, increasing cost. Positive masters should be silanised immediately prior to each use to ensure efficacy.
11. Negative masters do not need to be silanised with each use but will require periodic re-silanisation. In dry climates, arrays can be silanised every 4–5 uses; in more humid climates, arrays may need re-silanising every 2–3 uses. If pillar arrays become difficult to pull away from the negative master, the master is in need of re-silanisation.

12. These curing conditions will result in an elastic modulus of the PDMS of around 2.1 MPa [22, 23]. This value can be varied by altering the ratio between the bulk elastomer and the curing agent, adjusting curing time and adjusting curing temperature [22].
13. Pillar arrays can be coated with extracellular matrix (ECM) molecules. If ECM components are to be added to the array, apply an appropriate dilution of the molecule of interest to the array after plasma cleaning, and allow to adhere for 30–60 min. Wash excess ECM solution off the array and seed cells on coated arrays [26]. To restrict coating to the tops of the arrays, see protocol provided previously [20, 28, 29].
14. An excellent source of information regarding the appropriate pipette characteristics for various experiments is the Pipette Cookbook from the Sutter Instrument Company (https://www.sutter.com/MICROPIPETTE/cookbook.html).
15. An excellent resource for details on electrophysiological recordings, including whole-cell patch clamp is the Axon Guide from Molecular Devices (https://www.moleculardevices.com/en/assets/user-guide/dd/cns/axon-guide-to-electrophysiology-and-biophysics-laboratory-techniques#gref).

References

1. Martinac B, Poole K (2018) Mechanically activated ion channels. Int J Biochem Cell Biol 97. https://doi.org/10.1016/j.biocel.2018.02.011
2. Coste B, Xiao B, Santos JS et al (2012) Piezo proteins are pore-forming subunits of mechanically activated channels. Nature 483:176–181. https://doi.org/10.1038/nature10812
3. Coste B, Mathur J, Schmidt M et al (2010) Piezo1 and Piezo2 are essential components of distinct mechanically activated cation channels. Science 330:55–60. https://doi.org/10.1126/science.1193270
4. Brohawn SG, Su Z, MacKinnon R (2014) Mechanosensitivity is mediated directly by the lipid membrane in TRAAK and TREK1 K+ channels. Proc Natl Acad Sci U S A 111:3614–3619. https://doi.org/10.1073/pnas.1320768111
5. Corey DP, Hudspeth AJ (1983) Kinetics of the receptor current in bullfrog saccular hair cells. J Neurosci 3:962–976. https://doi.org/10.1523/jneurosci.03-05-00962.1983
6. Christensen AP, Corey DP (2007) TRP channels in mechanosensation: direct or indirect activation? Nat Rev Neurosci 8:510–521. https://doi.org/10.1038/nrn2149
7. Besch SR, Suchyna T, Sachs F (2002) High-speed pressure clamp. Pflugers Arch Eur J Physiol 445:161–166. https://doi.org/10.1007/s00424-002-0903-0
8. McCarter GC, Reichling DB, Levine JD (1999) Mechanical transduction by rat dorsal root ganglion neurons in vitro. Neurosci Lett 273:179–182. https://doi.org/10.1016/S0304-3940(99)00665-5
9. Lewis AH, Grandl J (2015) Mechanical sensitivity of Piezo1 ion channels can be tuned by cellular membrane tension. elife 4:e12088. https://doi.org/10.7554/eLife.12088
10. Coste B, Mathur J, Schmidt M et al (2010) Piezo1 and Piezo2 are essential components of distinct mechanically activated cation channels. Science (80-) 330:55–60. https://doi.org/10.1126/science.1193270
11. Alloui A, Zimmermann K, Mamet J et al (2006) TREK-1, a K+ channel involved in polymodal pain perception. EMBO J 25:2368–2376. https://doi.org/10.1038/sj.emboj.7601116

12. Brohawn SG, Su Z, MacKinnon R (2014) Mechanosensitivity is mediated directly by the lipid membrane in TRAAK and TREK1 K $^{+}$ channels. Proc Natl Acad Sci 111:3614–3619. https://doi.org/10.1073/pnas.1320768111
13. Richardson J, Kotevski A, Poole K (2021) From stretch to deflection: the importance of context in the activation of mammalian, mechanically activated ion channels. FEBS J 1–23. https://doi.org/10.1111/febs.16041
14. Poole K, Moroni M, Lewin GR (2014) Sensory mechanotransduction at membrane-matrix interfaces. Pflugers Arch Eur J Physiol 467: 121–132
15. Poole K (2022) The diverse physiological functions of mechanically activated ion channels in mammals. Annu Rev Physiol 84:307–329. https://doi.org/10.1146/annurev-physiol-060721-100935
16. Xia S, Kanchanawong P (2017) Nanoscale mechanobiology of cell adhesions. Semin Cell Dev Biol 71:53–67. https://doi.org/10.1016/j.semcdb.2017.07.029
17. Revach OY, Grosheva I, Geiger B (2020) Biomechanical regulation of focal adhesion and invadopodia formation. J Cell Sci 133. https://doi.org/10.1242/jcs.244848
18. Hytönen VP, Wehrle-Haller B (2016) Mechanosensing in cell-matrix adhesions - converting tension into chemical signals. Exp Cell Res 343:35–41. https://doi.org/10.1016/j.yexcr.2015.10.027
19. Wang N, Tytell JD, Ingber DE (2009) Mechanotransduction at a distance: mechanically coupling the extracellular matrix with the nucleus. Nat Rev Mol Cell Biol 10:75–82. https://doi.org/10.1038/nrm2594
20. Poole K, Herget R, Lapatsina L et al (2014) Tuning piezo ion channels to detect molecular-scale movements relevant for fine touch. Nat Commun 5:3520. https://doi.org/10.1038/ncomms4520
21. Wetzel C, Pifferi S, Picci C et al (2017) Small-molecule inhibition of STOML3 oligomerization reverses pathological mechanical hypersensitivity. Nat Neurosci 20. https://doi.org/10.1038/nn.4454
22. Bavi N, Richardson J, Heu C et al (2019) PIEZO1-mediated currents are modulated by substrate mechanics. ACS Nano 13:13545–13559. https://doi.org/10.1021/acsnano.9b07499
23. Servin-Vences MR, Moroni M, Lewin GR, Poole K (2017) Direct measurement of TRPV4 and PIEZO1 activity reveals multiple mechanotransduction pathways in chondrocytes. elife 6. https://doi.org/10.7554/eLife.21074
24. Beaulieu-Laroche L, Christin M, Donoghue A et al (2020) TACAN is an Ion Channel involved in sensing mechanical pain. Cell 0. https://doi.org/10.1016/j.cell.2020.01.033
25. Sianati S, Schroeter L, Richardson J et al (2021) Modulating the mechanical activation of TRPV4 at the cell-substrate Interface. Front Bioeng Biotechnol 8:608951. https://doi.org/10.3389/fbioe.2020.608951
26. Patkunarajah A, Stear JH, Moroni M et al (2020) TMEM87a/Elkin1, a component of a novel mechanoelectrical transduction pathway, modulates melanoma adhesion and migration. elife 9:e53308. https://doi.org/10.7554/eLife.53308
27. Tay A, Sohrabi A, Poole K et al (2018) A 3D magnetic hyaluronic acid hydrogel for Magnetomechanical Neuromodulation of primary dorsal root ganglion neurons. Adv Mater 30: 1800927. https://doi.org/10.1002/adma.201800927
28. Sianati S, Kurumlian A, Bailey E, Poole K (2019) Analysis of mechanically activated ion channels at the cell-substrate Interface: combining pillar arrays and whole-cell patch-clamp. Front Bioeng Biotechnol 7:47. https://doi.org/10.3389/fbioe.2019.00047
29. Ganz A, Lambert M, Saez A et al (2006) Traction forces exerted through N-cadherin contacts. Biol Cell 98:721–730. https://doi.org/10.1042/BC20060039

Chapter 11

Quantifying Strain-Sensing Protein Recruitment During Stress Fiber Repair

Shailaja Seetharaman, Stefano Sala, Margaret L. Gardel, and Patrick W. Oakes

Abstract

A family of proteins have been identified that recognize damaged, strained actin filaments in stress fibers. These proteins are often referred to as strain- or force-sensing and trigger downstream signaling mechanisms that can facilitate repair at these strain sites. Here we describe a method using high-resolution microscopy to screen and quantify the mechanosensitive recruitment of proteins to these stress fiber strain sites. Strain sites are induced using spatially controlled illumination of UV light to locally damage actin stress fibers. Recruitment of potential strain-sensing proteins can then either be compared to (Blanchoin, Physiol Rev 94, 235–263, 2014) a known control (e.g., zyxin-GFP) or (Hoffman, Mol Biol Cell 23, 1846–1859, 2012) the pre-damaged stress fiber protein distribution. With this method, strain-sensing proteins and their dynamic association with stress fiber strain sites can be reproducibly measured and compared.

Key words Stress fiber strain sites, Laser photoablation, LIM proteins, Zyxin, Testin, Stress fibers, Mechanosensing

1 Introduction

Actomyosin stress fibers (SFs) are load-bearing structures that form a major part of the mechanotransduction machinery in adherent cells, and are important for a wide range of cellular processes including cell migration, adhesion, and wound healing [1]. They are coupled to the cellular microenvironment via focal adhesions, enabling them to change their composition and assembly dynamics in response to extracellular cues. Increased SF tension can result in local SF damage and mechanical failure, which can either be induced experimentally or occur spontaneously in living cells. During the subsequent remodeling process, mechanosensitive proteins

Authors Shailaja Seetharaman and Stefano Sala have equally contributed to this chapter.

Ronen Zaidel-Bar (ed.), *Mechanobiology: Methods and Protocols*,
Methods in Molecular Biology, vol. 2600, https://doi.org/10.1007/978-1-0716-2851-5_11,

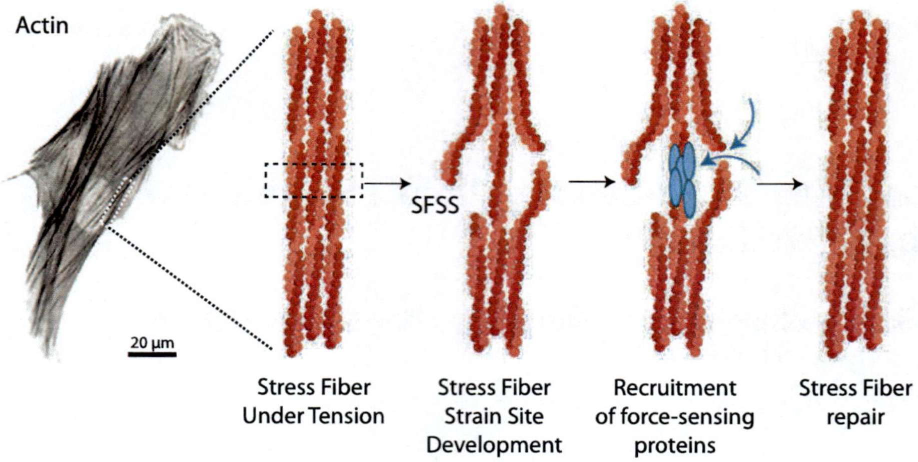

Fig. 1 Schematic of stress fiber strain site and repair. The development of a SFSS occurs either spontaneously in cells or can be induced by laser photoablation. This results in the rapid recruitment and accumulation of mechanosensitive proteins such as zyxin and testin to these SFSS. The damage sites are then repaired as zyxin recruits other proteins like VASP and α-actinin that promote actin assembly and cross-linking, respectively. (Created with BioRender.com)

such as some LIM domain proteins (e.g., zyxin, testin) localize to these stress fiber strain sites (SFSS) [2–7], and recruit actin polymerizing and crosslinking proteins, including VASP and α-actinin, to facilitate SF repair [8] (Fig. 1).

Several tools and methods have been developed to study the mechanical properties of SFs and the force-sensing mechanisms associated with SF damage and repair [9]. These include in vitro reconstitution of purified proteins and in vivo techniques such as traction force and atomic force microscopy, as well as whole-cell stretching and shearing. A powerful in vivo method to induce SF strain sites is laser photoablation. Here, we describe a method to induce SFSS by laser photoablation of individual SFs and quantify the recruitment of mechanosensitive proteins to these sites of strain [3, 5]. We describe cell culture and transfection protocols, followed by a detailed protocol for laser ablation, image acquisition, and analyses, primarily focusing on LIM domain proteins as representative mechanosensitive proteins that recognize SFSS. This method, however, can be utilized to visualize any protein of interest that might be important in SF strain-sensing or repair.

2 Materials

2.1 Cells and Culture Media

1. Zyxin$^{(-/-)}$ + EGFP-zyxin mouse embryonic fibroblasts (MEFs; cells were a gift of Mary Beckerle's laboratory, University of Utah, Salt Lake City, UT [2, 10]), or human foreskin fibroblasts (HFF) (*see* **Note 1**).

2. Fibroblast culture media: Dulbecco's Modified Eagle Medium (DMEM), 10% fetal bovine serum, 1% penicillin–streptomycin antibiotics, 1% antimycotic solution, 2 mM L-glutamine (*see* **Note 2**).

2.2 Constructs

Plasmid DNA encoding is a fluorescently tagged version of the protein(s) of interest (*see* **Note 3**). In this protocol we used the following constructs:

1. Zyxin-LCR-mCherry (LCR: LIM containing region)
2. Zyxin-FL-mCherry
3. NES-mCherry
4. GFP-Testin-FL
5. GFP-LIM 1-2-3 (the LIM domains from Testin)
6. mApple-actin

2.3 Transfection Equipment

1. TrypLE Express
2. 60 mm tissue culture dish
3. 22 × 30 mm glass coverslips
4. Neon™ transfection system (Invitrogen)
5. Neon™ transfection reagents: Resuspension Buffer R (MPK10096R) and Electrolytic Buffer E2 (MPK10096E)
6. Neon™ pipette (MPP100)
7. Neon™ tubes (MPT5)
8. Neon™ disposable tips, 100 μL (MPK10096T)
9. Cell counter

2.4 Microscope

1. Spinning Disk Confocal Unit (*see* **Note 4**)
2. Laser Launch with appropriate lasers for exciting green and red fluorophores
3. sCMOS camera
4. Spatial light modulation unit (*see* **Note 5**)

3 Methods

3.1 Transfections (see Note 6)

1. Place a sterile 22 × 30 mm coverslip into a fresh 60 mm tissue culture dish. Add 5 mL cell culture media and put the dish in the incubator at 37 °C (*see* **Note 7**).
2. Transfer 3 mL Electrolytic Buffer E2 into a Neon™ tube and place the Neon™ tube into the Neon™ pipette station.
3. Mix 110 μL Resuspension Buffer R with 1–5 μg DNA of choice. For example, we often used 2.5 μg plasmid DNA

encoding fluorescently tagged actin mixed with 2.5 μg plasmid DNA encoding, a fluorescently tagged version of the LIM protein of interest (*see* **Note 6**).

4. Detach the cells by adding 0.5 mL of TrypLE Express to an 80–90% confluent 60 mm dish. Incubate at 37 °C for 3–5 min until all the cells are completely detached (*see* **Note 8**).
5. Count the cells.
6. Transfer the appropriate volume containing 500,000 cells into a 15 mL conical and centrifuge for 3 min at 1500 rpm (*see* **Note 9**).
7. Aspirate the cell media and resuspend the cell pellet in the solution containing Resuspension Buffer R and DNA. Transfer the mixture to a 1.5 mL Eppendorf tube (*see* **Note 10**).
8. Insert a Neon™ disposable tip into the Neon® pipette.
9. Carefully aspirate the cell and DNA solution (1.5 mL Eppendorf tube) with the Neon™ tip and insert the Neon™ pipette into the Neon™ tube in the Neon™ pipette station (*see* **Note 11**).
10. Select the appropriate electroporation protocol. For fibroblasts we used 1 pulse of 1600 V applied to the cells (*see* **Note 12**).
11. Unplug the Neon™ pipette as soon as the electroporation is done and transfer the cells into the 60 mm dish containing 22 × 30 mm coverslip.
12. Place the cells in the incubator and wait 24 h before imaging to ensure sufficient protein expression (*see* **Note 13**).

3.2 Image Acquisition

1. Live cell imaging and laser photoablations are performed using confocal spinning disk microscopes that contain 405, 491, 561, and 642 nm laser lines, an sCMOS camera, motorized X, Y stage, and a spatial light modulation unit. For exact system details, *see* **Note 4**. During image acquisition, keep imaging settings constant between different cells and samples (*see* **Note 14**).
2. Transfer the imaging chamber into the microscope incubator and let the sample equilibrate for at least 5 min to avoid thermal fluctuations and hence drifting during imaging.
3. Acquire several images of the cell culture media region without cells in each channel before an imaging experiment. This will act as the background signal from camera noise, media fluorescence, etc.
4. Select cells that express the constructs of interest (in this case, the mCherry-zyxin-FL or GFP-Testin), mApple-actin, and/or GFP-zyxin using a high magnification (e.g., 60X or 63X) objective.

5. Select a SF to induce strain in, and draw a small area targeting a region of interest on the SF. A 5-μm ablation line perpendicular to the orientation of the fiber is used to induce a SFSS as depicted in Fig. 3 (*see* **Note 15**).
6. Image the "steady state" by acquiring 30 s–1 min pre-ablation, alternating between both channels (*see* **Note 16**).
7. Illuminate the ablation line with a 405 nm laser for ~1.5 s at a power density of ~350 μW/μm^2 to induce a stress fiber strain site (*see* **Note 17**).
8. Continue the timelapse for the desired duration (e.g., Supplemental Movie 1 is 3 min and 30 s; *see* **Note 18**), imaging both channels at a consistent time interval ranging from 2 to 20 s.
9. Save the ablation region for analysis.
10. Export all images (containing both the actin and LIM protein channels) as 16-bit TIFF files.

3.3 Image Analysis

Two different analyses methods to observe recruitment of proteins to SFSS are described below. In the first case, cells expressing measurements of protein recruitment (mCherry-tagged) are made in cells stably expressing GFP-zyxin. In the second case, cells do not stably express any protein, and the enrichment of the protein of interest (GFP-tagged) is measured with respect to the signal prior to the ablation. Ultimately, both analyses methods yield measurements of the recruitment of LIM proteins to the strain sites on SFs.

Intensity Relative to GFP-Zyxin The following steps will calculate the ratio of the mean fluorescence intensity of the transiently transfected mCherry tagged protein of interest to the fluorescence of the stably integrated GFP-zyxin. This provides a measurement of the enrichment of the protein of interest at the SFSS (Fig. 2).

1. *Open image files as a sequence/movie on ImageJ and identify the region of ablation in each channel.*
2. *Using the ImageJ reslice function, draw kymographs in the region of ablation in each channel* (see ***Notes 19 and 20***).
3. *Generate a line profile by drawing a line scan (shown in red, Fig. 2a, d, h) across the kymograph (through the time axis).*
4. *Plot the raw line scans as shown in Fig. 2b, e, h.*
5. *Consider the background as the region on the fiber where the future strain sites develop. Calculate intensity of the signal immediately before the SFSS is formed. Subtract this signal from both channels (mCherry-background and GFP-zyxin-background).*
6. *The background subtracted line scans are plotted in Fig. 2c, f, i.*
7. *To obtain the enrichment for each channel, determine the peak of the curves (shown as dotted circles; Fig. 2c, f, i) to get the*

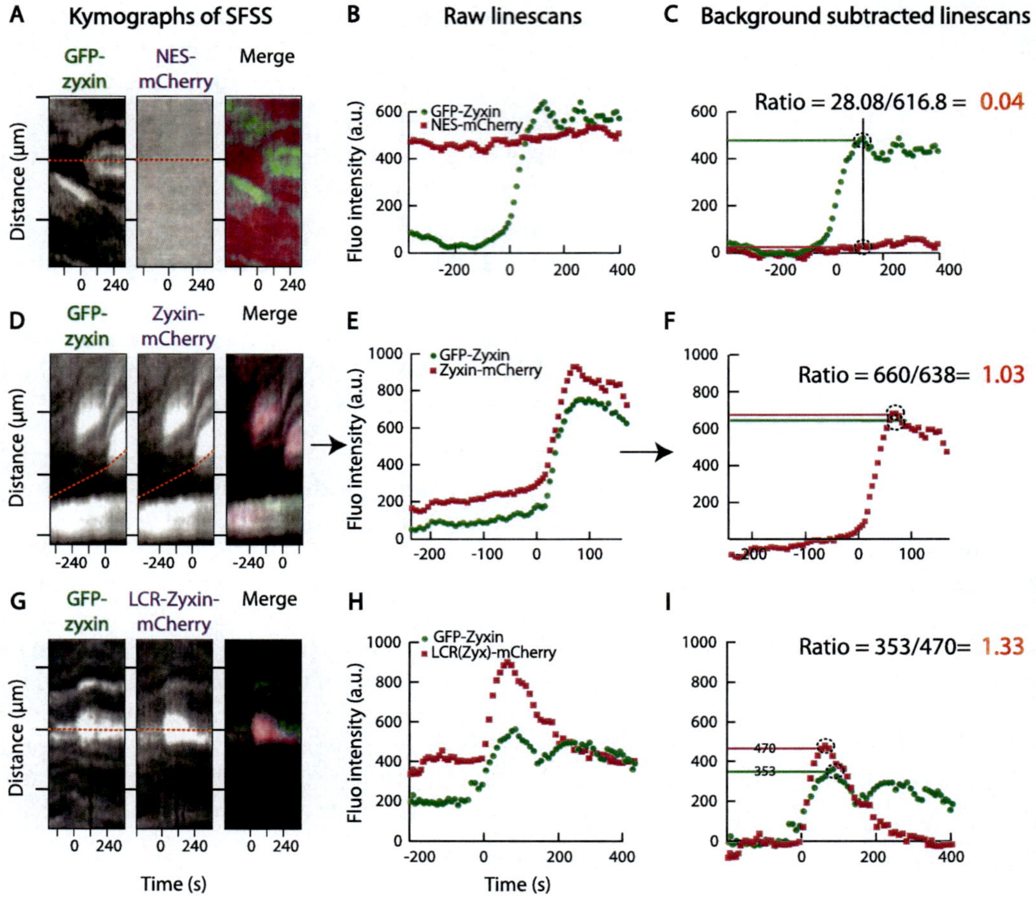

Fig. 2 Quantification of relative LIM protein recruitment to laser photoablation-induced stress fiber strain sites. (**a**, **d**, **g**) Kymographs from movies of GFP-zyxin-integrated fibroblasts that are transfected with an mCherry-construct of interest (in this case, a construct of nuclear export signal NES-mCherry (**a**) used as negative control, zyxin-mCherry as positive control (**d**), and LCR-zyxin-mCherry as the construct of interest (**g**) are transiently transfected). (**b**, **e**, **h**) Raw line scans through kymographs (marked in red) shown in (a), (d), and (g). (**c**, **f**, **i**) Background subtracted line scans where the signal before the development of a SFSS is subtracted. The peak of each curve (shown as dotted circles) is determined and averaged across two adjacent data points. This provides the enrichment of the construct at the site. This enrichment is then divided by the enrichment of GFP-zyxin to calculate the main mCherry:GFP fluorescence ratio (depicted in red) as described in the Image Analysis Subheading 3.3. Scale bar: 2 μm. (Figure adapted from Winkelman et al., 2020)

maximum fluorescence intensity, and average this with the two adjacent points (see ***Note 21***; *denoted as mCherry-construct-SFSS or GFP-zyxin-SFSS*).

8. *Divide the enrichment of mCherry-construct by the enrichment of GFP-zyxin to obtain the main metric fluorescence ratio as follows:*

$$\text{mCherry : GFP ratio} = \left(\frac{(\text{mCherry} - \text{construct} - \text{SFSS}) - (\text{mCherry} - \text{background})}{(\text{GFP} - \text{zyxin} - \text{SFSS}) - (\text{GFP} - \text{zyxin} - \text{background})}\right.$$

Intensity Relative to Pre-ablation For the enrichment of a GFP-tagged, LIM protein of interest being recruited to SFSS in fibroblasts. Simultaneous monitoring of mApple-actin signal in the region of the SFSS can confirm that the SF is damaged during ablation and that repair is occurring over the timelapse. All movies are analyzed in Python as outlined below and described in Sala et al., 2021, and Fig. 3. The code, along with example data, is available at https://github.com/OakesLab/Sala_MBoC_2021.

1. *Break down each movie into the individual stacks representing the actin and LIM protein channels.*
2. *For each stack, perform a flat field and dark field correction. Flat field corrections are performed by dividing the image by a normalized image of uniform intensity. Dark field corrections are done by subtracting an image of the camera taken without opening the shutter* (see **Note 22**).
3. *Correct each stack for photobleaching by fitting the intensity sum of each frame to an exponential decay of the form* $I = A\,e^{-B * t}$, *where* A *and* B *are fit parameters. To create a correction curve, we divide the value at* $t = 0$ *by the values at all other time points. Correct each frame by multiplying the intensities in each frame by the corrected value at that time point* (see **Note 23**).
4. *Register the corrected image stack of the LIM protein to the first frame by calculating the registration shifts using any registration routine* (see **Note 24**).
5. *Register the actin channel based on the calculated shifts of the channel representing the LIM protein.*
6. *Crop both registered stacks to a region of 121 × 121 pixels (approximately 21 × 21 μm) centered on the midpoint of the ablation line (Fig. 3).*
7. *For the cropped LIM protein stack, create a circular "ablation mask" with a radius of 20 pixels around the midpoint of the ablation line (Fig. 3).*
8. *Expand the ablation mask three times (radius of 60 pixels) to create a "ROI mask" (Fig. 3).*
9. *Make a mask of the entire cell. Multiply both the ablation mask and ROI mask by the cell mask to ensure that all downstream calculations are done only on pixels that are inside the cell.*
10. *Create a reference intensity image* (I_{ref}) *for both the actin and LIM protein stacks by calculating the average intensity of each pixel from the frames before the photoablation in each channel.*
11. *For every frame in both channels, subtract the reference image* (I_{ref}) *from the given frame* (I_t) *and divide by the reference image. This will create a relative difference intensity image* ($I_{diff} = (I_t - I_{ref})/I_{ref}$) *for each time point* (see **Note 25**).

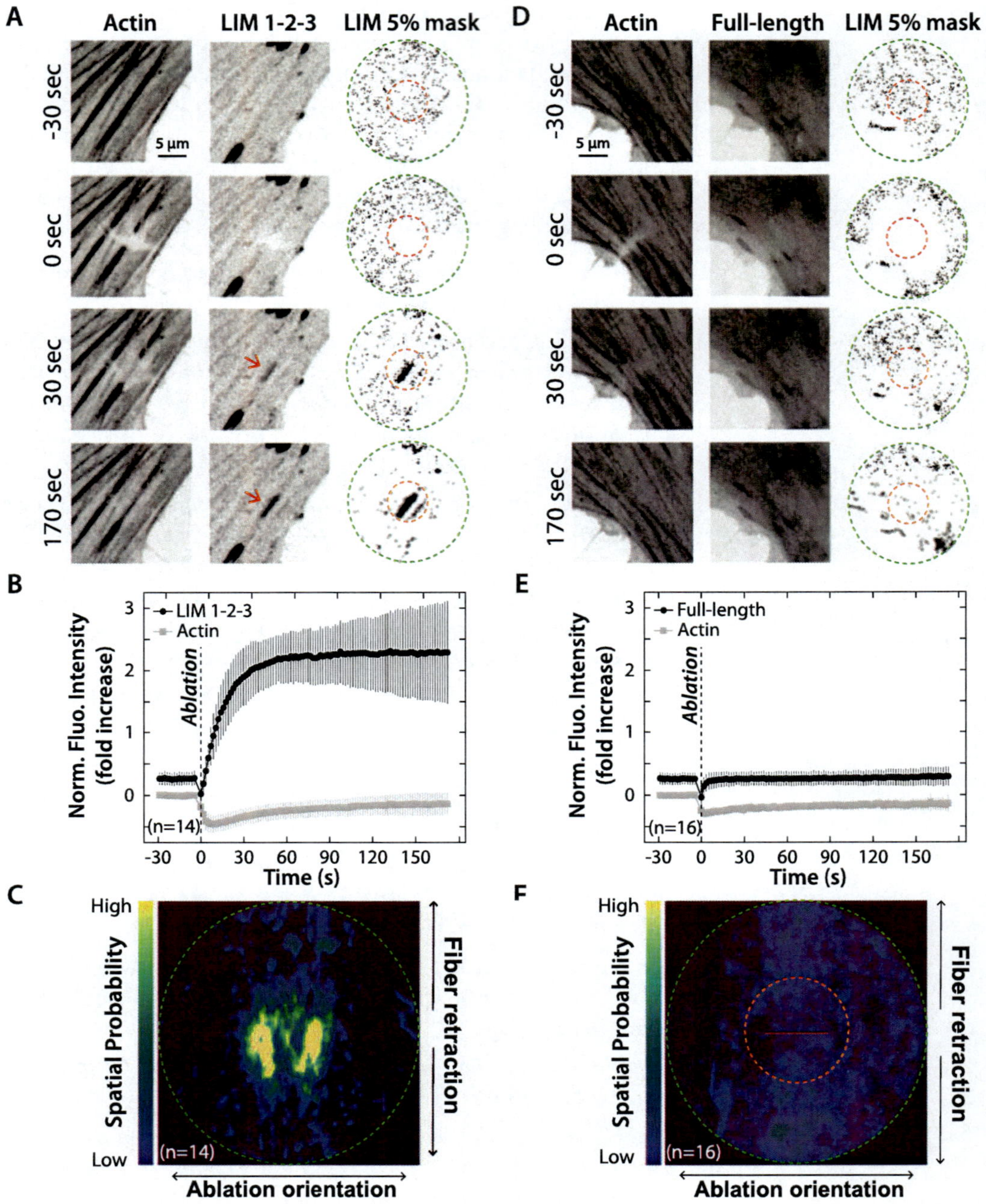

Fig. 3 Quantification of LIM protein recruitment to photoablation-induced stress fiber strain sites. (**a**, **d**) HFFs co-expressing mApple-actin and GFP-coupled versions of either a mechanosensitive LIM construct (LIM domains of testin (LIM 1-2-3) (**a**), or a non-mechanosensitive LIM construct (full-length testin (**d**). The images show the localization of the actin and LIM constructs during a laser photoablation experiment ("0 sec" indicates the photoablation timepoint, *see* Supplemental Movie S1). The LIM 5% masks show the pixels in the ROI (green dashed circle) that changed the most in intensity in the LIM channel compared to their pre-ablation state. The red arrows indicate relocation of the mechanosensitive LIM domains of testin to laser-induced stress fiber strain sites as the brightest pixels in the LIM 5% mask are located inside the ablation region (red

12. *Make a mask of the 5% brightest pixels in the LIM protein channel (LIM 5% mask) for both the ROI mask and ablation mask zones (Fig. 3a, d).*
13. *For both the LIM and actin channels, calculate the average intensity value of the pixels that match the position of the 5% brightest pixels in the ablation region (red dashed circle in Fig. 3a, d) of the LIM 5% mask. Plot these average values over time for both channels to create a trace of the normalized fluorescence intensities* (see **Note 26**).
14. ***Steps 1–13** should be repeated for every movie that is made. Based on the total number of movies, an average fluorescence intensity trace (**step 15**) and spatial probability map of recruitment (**step 16**) can be created.*
15. *To create an average trace of all analyzed movies, calculate and plot the averaged normalized fluorescence intensity value ± standard deviation of each time point for both the actin and LIM protein channels (Fig. 3b, e).*
16. *To create a spatial probability map of the LIM protein, sum all the LIM 5% masks across the entire stack for each movie (summed LIM mask). Rotate the resulting summed LIM masks so that the direction of the ablation is aligned horizontally and average them across all movies (Fig. 3c, f).*

4 Notes

1. Cells should be cultured at 37 °C with 5% CO_2 and passaged when a confluency of 80–90% is reached.
2. Imaging media can be used during acquisition: Leibovitz's L-15 medium without phenol red, 10% fetal bovine serum, 1% antibiotic, antimycotic solution.

Fig. 3 (continued) dashed circle). This relocation is not observed for the non-mechanosensitive full-length testin construct as confirmed by a random distribution of the pixels in the LIM 5% mask. (**b**, **e**) Average fluorescence traces and standard deviations of the actin intensity and intensities of either LIM 1-2-3 of testin (**b**) or full-length testin (**e**) based on a total of *n* photoablation experiments. Following the induction of a stress fiber strain site, a rapid increase of the LIM 1-2-3 signal occurs, which coincides with a loss of the actin signal as it is damaged (**b**). The LIM signal plateaus as the damaged actin gets repaired (**b**). This increase is not observed for the full-length construct, indicating it is not recruited to stress fiber strain sites (**e**). (**c**, **f**) Spatial probability recruitment maps of the LIM domains of testin (**c**) and full-length protein (**f**) based on a total of *n* photoablation experiments. The brightest LIM 1-2-3 pixels are located in the ablation region (red dashed circle), which is indicative of strain recognition. The brightest pixels of the full-length construct are distributed randomly within the ROI (green dashed circle). The red line represents the 5-μm ablation line. (Figure adapted from Sala et al., 2021)

3. In the experiments shown here, cells are transiently transfected with fluorescently tagged constructs. As protein overexpression often disrupts their natural stoichiometry and causes off target effects (e.g., mis-localization, pathway (in)activation [11]), the creation of a stable cell line using a lentiviral expression system or CRISPR technology might be beneficial to obtain similar protein expression levels and minimize heterogeneity. In addition to the type of fluorescent protein used for tagging, it is important to consider where to add the fluorescent tag. The tag can be fused either at the N- or C-terminal, depending on the structure and folding of the protein of interest, and its subsequent localization within the cell. Note that the functionality and localization of some proteins can be maintained only if the tag is placed on the appropriate terminus. Finally, the protocol described in this chapter involves time-lapse imaging. To prevent photobleaching, it is recommended that a fluorophore with high photostability, such as EGFP or mEmerald, be used for tagging.
4. The exact makeup of the microscope for these experiments is flexible. The following microscopes and setups were used for image acquisition:

   (A) Inverted Nikon Ti-E microscope (Nikon, Melville, NY) with a Yokogawa CSU-X confocal scanhead (Yokogawa Electric, Tokyo, Japan) and laser merge module containing 491, 561, and 642 nm laser lines (Spectral Applied Research, Ontario, Canada). Images were collected on Zyla 4.2 sCMOS Camera (Andor, Belfast, UK). A 405 nm laser coupled to a Mosaic digital micromirror device (Andor) was used to locally damage SF. Images were collected using a 60X 1.49 NA Apo-TIRF oil immersion objective (Nikon). All hardware was controlled using MetaMorph Automation and Image Analysis Software (Molecular Devices, Sunnyvale, CA). (B) Marianas Imaging System (Intelligent Imaging Innovations) consisting of an Axio Observer 7 inverted microscope (Zeiss) attached to a W1 Confocal Spinning Disk (Yokogawa) with Mesa field flattening (Intelligent Imaging Innovations), a Phasor photomanipulation unit (Intelligent Imaging Innovations), a motorized X, Y stage (ASI), and a Prime 95B sCMOS (Photometrics) camera. Illumination is provided by a TTL triggered multifiber laser launch (Intelligent Imaging Innovations) consisting of 405, 488, 561, and 637 lasers, using a 63 × 1.4 NA Plan-Apochromat objective (Zeiss). Temperature and humidity are maintained using a Bold Line full enclosure incubator (Oko Labs). The microscope is controlled using Slidebook 6 Software (Intelligent Imaging Innovations).

5. Like microscopes, the spatial light illumination module is flexible, as long as sufficient power can be concentrated in small region. For these experiments an Mosaic digital multimirror (DMM) device (Andor) or a Phasor holographic photomanipulation unit (Intelligent Imaging Innovations) were used.
6. For transient transfection, electroporations are preferred since they usually yield high transfection efficiencies and protein expression levels. For certain cell types, however, a low viability is observed post electroporation in which case we recommend using other transfection methods and optimizing transfection conditions to improve cell survival. The quantity of transfected DNA can vary. High concentrations of transfected DNA can lead to low cell viability post electroporation. In the experiments described in Fig. 2, cells expressing high levels of transfected zyxin display an increased accumulation of zyxin at SFSS, and is inversely correlated with paxillin recruitment to the SFSS (*see* Fig. 2G-I and S2A in Winkelman et al., 2020 [3]). Thus, it is important to note that high expression levels of proteins can lead to competition in force-sensing.
7. Any size coverslip and imaging chamber can be used. For laser photoablation experiments, however, cells should be plated as single cells following transfection. We therefore recommend using a coverslip and tissue culture dish size that match these requirements (*see* **Note 9**).
8. Limit the time that the cells are in the TrypLE Express solution to increase cell viability. Gently tapping the tissue culture dish can speed up their detachment.
9. The number of cells chosen for transfection should match the optimal plating density for imaging (*see* **Note 7**).
10. Avoid creating air bubbles when the cells are resuspended in Resuspension Buffer R as it can cause arcing during electroporation which results in increased cell death.
11. Avoid creating air bubbles when the cells and DNA are aspirated with the Neon™ tip (*see* **Note 10**).
12. Voltage and number of pulses are cell type-dependent and crucial for optimal cell viability and transfection efficiency. We refer to the ***cell line data and transfection parameters*** listed on the manufacturer's website (www.thermofisher.com).
13. If a significant amount of cell death is observed 2–3 h post transfection, the cell media should be changed.
14. Although this method can be used to locally induce SFSS in cell monolayers, the experiments depicted here were carried out in single cells. Parameters such as cell density and cell–cell contacts must be taken into consideration and kept similar across the various samples. In addition, cells with abnormal nuclei,

blebbing, retraction, or experiencing phototoxic effects must be excluded during data acquisition and analysis. We found that using ~2.5 mW of power (measured at the objective) is sufficient to produce high signal-to-noise intensity measurements with minimal phototoxic effects with both our 488 and 561 nm lasers.

15. It is very important to draw the ablation line perpendicular to the orientation of the stress fiber to ensure that all the individual actin filament that build up the selected fiber are illuminated. In this way, a subset of the actin filaments will be stochastically damaged, which causes the remaining undamaged fibers to be stretched under the increased load. Successful strain induction is obtained when recoil and thinning of the ablated stress fiber are observed.
16. The successful induction of a stress fiber strain site strongly depends on the illumination duration and intensity. If the illumination intensity is too low and/or the duration too short, insufficient damaged is caused to the stress fiber and hence no strain will be induced (no recoil or thinning are observed – though photobleaching will likely still occur). In contrast, if the illumination intensity is too high and/or the duration too long, actin filaments can be fully severed, resulting in a severed stress fiber that is no longer under tension. We therefore recommend optimizing the illumination intensity and duration when performing laser photoablations on a different sample.
17. A few frames (at least 10–15 images) should be acquired before the ablation event as these pre-ablation frames are used to create a reference intensity image to quantify potential LIM protein recruitment to stress fiber strain sites.
18. The representative movie is acquired for a total duration of 3 min and 30 s. In this case, the duration is precisely chosen because on the one hand, it allows enough time for the damaged actin to be repaired and for the LIM proteins that are recruited to these damaged actin sites to reach a plateau (*see* Fig. 3). On the other hand, it limits the amount of time the cells are exposed to light-reducing phototoxic effects. However, depending on the nature and purpose of the experiment, the duration of the movie can vary.
19. If cells or the stage moved significantly, align stacks before analysis. If SFs move significantly, use a python script in ImageJ to build kymographs from user-drawn line segments (as in [3]).
20. Always draw kymographs of the same length and angle with respect to the SF orientation.

21. If there was no obvious signal for the construct of interest being tested, then the time point at which GFP-zyxin signal was maximal was used in measuring the mCherry-construct signal.

22. To create flat fields, 30–50 fields of a soluble dye solution are taken (e.g., 100 mg/mL of FITC or Rose Bengal for green and red dyes). A flat field can then be calculated by taking the median of the stack of images (the median is preferred to the mean to avoid aggregates in any given image). The darkfield image is then subtracted from the median image, and the subsequent product is normalized by dividing by the maximum value. To correct an image, the darkfield image is first subtracted, and then the result is divided by the flat field image.

23. We often observe a two-step photobleaching response with mApple and mCherry constructs. This can be seen as a sudden dip in fluorescence intensity within the first few frames of a time-lapse experiment, followed by a slower and steadier photobleaching response. To avoid artifacts from this initial rapid photobleaching, we recommend disregarding the first 5 frames (i.e., the first few seconds of the steady state imaging, well before the ablation event) of each movie for analysis.

24. There are several plug-ins in FIJI that can accomplish image registration. We used a sub-pixel registration developed by Guizar-Sicairos et al [12]. This implementation can be downloaded in our GitHub repository (https://github.com/OakesLab/Sala_MBoC_2021).

25. The creation of a relative difference image allows the quantification of the highest intensity pixels due to enrichment (stress fiber strain site recruitment) of a LIM protein compared to its localization in the pre-ablation state. This avoids the detection of the brightest pixels before photoablation that do not change in intensity after the photoablation and thus avoids including LIM protein intensities in cellular compartments such as focal adhesions or the nucleus.

26. Inspection of the signal in the actin channel can provide insight into whether proteins are being recruited to an actual strain site. A damaged fiber should show a decrease in intensity in the actin channel that then increases over time, indicating that repair is taking place. If the actin channel instead increases concomitant with the LIM protein, it can suggest that the protein is simply recognizing all actin filaments and not strained filaments in particular. Repair is completed when actin intensity in these regions returns to its baseline value prior to ablation.

Acknowledgments

This work was supported by NIH R01 GM104032 to M.L.G. and NSF CAREER Award #2000554 to P.W.O.

References

1. Blanchoin L, Boujemaa-Paterski R, Sykes C, Plastino J (2014) Actin dynamics, architecture, and mechanics in cell motility. Physiol Rev 94: 235–263
2. Hoffman LM, Jensen CC, Chaturvedi A, Yoshigi M, Beckerle MC (2012) Stretch-induced actin remodeling requires targeting of zyxin to stress fibers and recruitment of actin regulators. Mol Biol Cell 23:1846–1859
3. Winkelman JD, Anderson CA, Suarez C, Kovar DR, Gardel ML (2020) Evolutionarily diverse LIM domain-containing proteins bind stressed actin filaments through a conserved mechanism. Proc Natl Acad Sci 117:25532
4. Oakes PW et al (2017) Optogenetic control of RhoA reveals zyxin-mediated elasticity of stress fibres. Nat Commun 8
5. Sala S, Oakes PW (2021) Stress fiber strain recognition by the LIM protein testin is cryptic and mediated by RhoA. Mol Biol Cell 32: 1758–1771
6. Yoshigi M, Hoffman LM, Jensen CC, Yost HJ, Beckerle MC (2005) Mechanical force mobilizes zyxin from focal adhesions to actin filaments and regulates cytoskeletal reinforcement. J Cell Biol 171:209–215
7. Sun X et al (2020) Mechanosensing through direct binding of tensed F-actin by LIM domains. Dev. Cell 55:468–482.e7
8. Smith MA et al (2010) A zyxin-mediated mechanism for actin stress fiber maintenance and repair. Dev Cell 19:365–376
9. Kassianidou E, Kumar S (2015) A biomechanical perspective on stress fiber structure and function. Biochim Biophys Acta 1853:3065–3074
10. Hoffman LM et al (2003) Targeted disruption of the murine zyxin gene. Mol Cell Biol 23:70–79
11. Schwinn MK, Steffen LS, Zimmerman K, Wood KV, Machleidt T (2020) A simple and scalable strategy for analysis of endogenous protein dynamics. Sci Rep 10:8953
12. Guizar-Sicairos M, Thurman ST, Fienup JR (2008) Efficient subpixel image registration algorithms. Opt Lett 33:156–158

Part IV

Measuring Forces Applied by Cells on the Extracellular Matrix

Chapter 12

Quantification of Invadopodia Formation and Matrix Degradation Activity

Jubina Balan Venghateri and Benjamin Geiger

Abstract

Cancer cells possess a remarkable capacity to dissociate from a primary tumor, invade the surrounding tissues and vasculature, and eventually form metastases in distant organs. This complex and multistep process remains one of the major causes of mortality in cancer patients worldwide. Multiple studies have highlighted the role of actin-rich structures called invadopodia ("invasive feet"), which adhere to the matrix, contain and secrete matrix-degrading proteinases, and apply protrusive forces generated by the actin cytoskeleton, which drive the invasive process. Here, we describe a fluorescent microscopy-based protocol for imaging and quantifying both invadopodia formation and matrix degradation.

Key words Invadopodia, Extracellular matrix (ECM), Fluorescently labeled gelatin, Gelatin degradation, Metastasis, Cancer cell lines, Actin cytoskeleton, Invasion, Fluorescent microscopy, Image processing

1 Introduction

Globally, metastasis continues to be the primary cause of cancer-related deaths [1]. Cancer research has shed much light on the complex molecular mechanisms involved in cancer initiation, progression, and eventually metastatic dissemination [2–4]. In recent years, significant efforts have been directed toward understanding the cellular machinery that drives the metastatic cascade [4–6]. It is broadly accepted that comprehensive characterization of the invasive process will help develop effective antimetastatic cancer therapy. The spread of tumor cells to distant organs, where they form secondary, metastatic lesions, involves a series of steps whereby cells are dissociated from the primary tumor and actively penetrate through tissue barriers (e.g., the basement membrane) into the surrounding connective tissues. These invading tumor cells can, individually or collectively, undergo intravasation through the vascular endothelium and disseminate throughout the body. A fraction

Ronen Zaidel-Bar (ed.), *Mechanobiology: Methods and Protocols*,
Methods in Molecular Biology, vol. 2600, https://doi.org/10.1007/978-1-0716-2851-5_12,

of these circulating cancer cells may subsequently "extravasate" and colonize distant organs, forming metastases [2, 7]. Both intravasation and extravasation are critical steps in the metastatic invasion, demanding a high spatial and temporal coordination between the adhesion to the extracellular matrix (ECM) or to the vessel wall, the proteolytic degradation of the extracellular matrix, and the generation of mechanical protrusive forces that are applied to the surrounding environment [8].

This complex process is accomplished by specialized actin-rich protrusive subcellular structures referred to as "invadopodia," which are commonly formed by cancer cells and contain the multi-molecular machinery needed to mediate the adhesive–proteolytic–protrusive processes. They possess an adhesive integrin-rich ring that connects them to the matrix, they also contain different matrix metalloproteinases (MMPs) on their surface and secrete them, and they develop a mechanically robust actin core that polymerizes and produces the protrusive force [9]. Bearing similarity to invadopodia are podosomes, observed in various normal cells such as osteoclasts, macrophages, dendritic cells, etc. Invadopodia are, though, unique in their capacity to invade solid tissues [10, 11] and differ from other actin-based protrusive structures like lamellipodia and filopodia [8].

Degradation of extracellular matrix by invadopodia in cancer cells can be demonstrated using a variety of ex vivo approaches, including a widely used 2D technique of preparing a fluorescently labeled ECM, seeding invasive cancer cells onto the matrix, and visualizing the "dark lesions" microscopically. This image-based approach, originally introduced using fluorescently tagged fibronectin by Wen-Tien Chen [12], is still highly useful for detecting invadopodia and quantifying matrix degradation. Here, we describe a modified, stepwise protocol for accurately quantifying both invadopodia formation and the consequent degradation of the underlying gelatin matrix.

2 Materials

1. **Cell culture:** Biosafety cabinet, tissue culture CO_2 incubator, cell counter or hemocytometer, water bath at 37 °C, sterile Dulbecco's phosphate-buffered saline (DPBS) without calcium and magnesium, penicillin–streptomycin solution, Trypsin EDTA solution with phenol red (0.25% Trypsin and 0.05% EDTA in Pucks Saline A).
2. **Cancer cell lines shown here for demonstration invadopodia formation and matrix degradation:** A375, CSK-A375 (knockdown of C-terminal *Src kinase* using shRNA *in* A375 cell line), MDA-MB-231, and IGR-1 cell lines.

3. **Tissue culture media for growing the cancer cell lines:** DMEM (435 mL), 10% fetal bovine serum (50 mL), 2 mM glutamine (5 mL), 2 mM sodium pyruvate (5 mL), and 100 U/mL penicillin–streptomycin (5 mL). Store the medium at 4 °C.
4. **Poly-L-lysine solution for coating plates:** Prepare a 1:1 mixture of a sterile-filtered solution of (10 μg/mL) poly-L-lysine and DPBS. Always prepare fresh solution while starting to prepare the fluorescently labeled cross-linked gelatin matrix plate.
5. **Gelatin solution (2 mg/mL):** Weigh 0.2 g of gelatin and dissolve it in 100 mL sterile DPBS. Dissolve the gelatin by intermittent stirring and keep the solution at 37 °C for 1 h. After the gelatin is completely dissolved, use a 0.22 μm Steritop bottle-top filter. Store the gelatin solution at 4 °C for further use.
6. **Fluorescent labeling of gelatin:** Gelatin (2 mg/mL) can be fluorescently labeled using a wide variety of commercially available fluorophores that are used for covalent protein modification. Labeling the gelatin with these fluorophores was performed according to the dye manufacturer's instructions. We have used here a commercial labeling kit from Thermo Fisher Scientific (Alexa Fluor 488). Each vial of the kit contains a reactive dye that can label 1 mg of protein. Prepare 1 M sodium bicarbonate stock solution by dissolving 8.4 g of sodium bicarbonate solution in 100 mL distilled water. Vortex or stir the solution until the sodium bicarbonate is completely dissolved. Add 50 μL of the 1 M sodium bicarbonate stock solution to 600 μL of 2 mg/mL gelatin solution (*see* **item 5**). Thaw the fluorophore vial with a magnetic stir bar at room temperature. Add the sodium bicarbonate-buffered gelatin mixture to the vial containing the dye and mix at room temperature for 1 h. During the incubation time, load a gel filtration column with the purification resin provided by the fluorophore manufacturer or use other gel filtration resins that are used for separating proteins from the low molecular weight dye. Load the gelatin + dye reaction mixture onto the column and recover the labeled gelatin (first colored peak). Store the labeled gelatin at 4 °C in a dark bottle for up to 1 week.
7. **Cross-linker solutions:** 10% solution of *N*-hydroxysuccinimide (NHS) in distilled water. 10% solution of *N*-(3-dimethylaminopropyl)-*N′*-ethylcarbodiimide hydrochloride (EDC hydrochloride) in distilled water.
8. **Fixation solution:** Prepare 3% paraformaldehyde solution in DPBS.

9. **Staining solutions:** Dissolve tetramethylrhodamine B isothiocyanate in 300 μL of 100% methanol. Aliquot and store the solution in the dark at −20 °C. Dissolve 20 mg of 4′,6-diamidino-2-phenylindole dihydrochloride in 1 mL of sterile water. Store the solution in the dark at −20 °C.
10. **Selection of plates for the assay:** For optimal results, particularly when using immersion-oil objectives with a relatively short working distance, it is recommended to use glass-bottom plates.

3 Methods

The steps involved in the invadopodia formation and gelatin degradation experiment are shown diagrammatically in Fig. 1. The detailed protocol used for coating plates with the cross-linked gelatin matrix is described below.

3.1 Preparation of Alexa Fluor 488-Gelatin Coated Plates

The gelatin matrix can be fluorescently labeled or non-labeled, depending on the nature of the experiment. Perform all procedures at room temperature except when specified otherwise.

1. Plan the experiment according to the number of samples required to be tested.

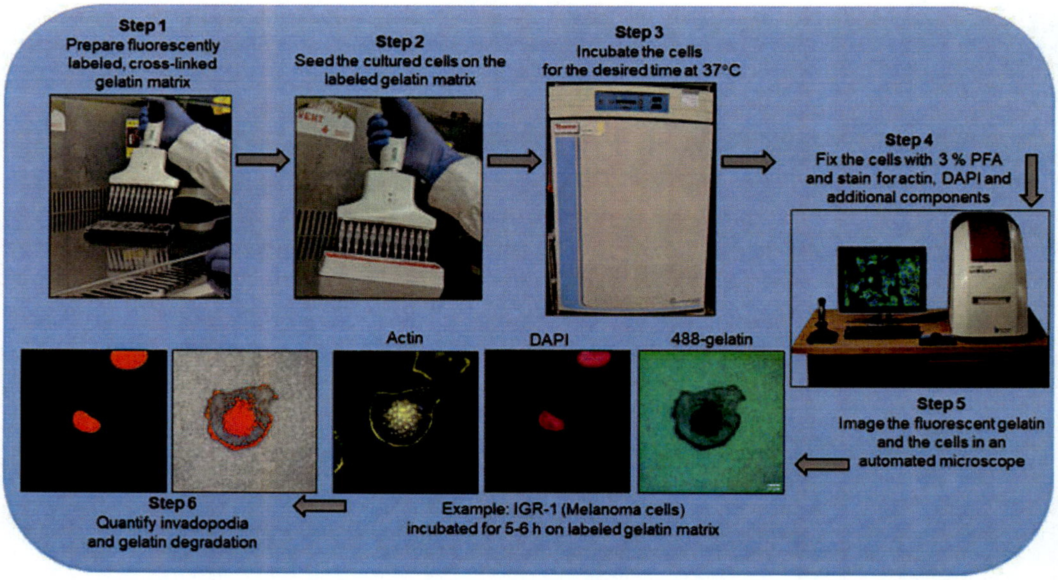

Fig. 1 Graphic description of the workflow for testing invadopodia formation and gelatin degradation in cultured cancer cell lines

2. Choose the type of plates desired for the experiment. The volume of solution required for each template plate is given as follows:

Template of plate	Volume/well
96 well	40 μL
24 well	100 μL
6 well	300 μL
35 mm MatTek	100 μL

1. We present here results obtained with a 96-well glass-bottom plates. Coat the 96-well glass-bottomed plate with a 1:1 mixture of sterile poly-L-lysine solution and sterile DPBS. Incubate the plates for 20 min at room temperature in a biosafety cabinet.
2. Then, wash the plates (3×) with DPBS.
3. To prepare 100 μL cross-linked gelatin, mix 82.5 μL gelatin, 12.5 μL 10% NHS, and 5 μL 10% EDC.
4. Mix the non-labeled gelatin and labeled gelatin in a 1:10 ratio.
5. Pipette 40 μL of the prepared cross-linked fluorescently labeled gelatin mixture for each well in the 96-well glass-bottom plate (*see* **Note 1**). Cover the plate with aluminum foil and incubate for 1 h at room temperature in the biosafety cabinet.
6. After 1 h, wash the plate 3× with 100 μL of DPBS (*see* **Notes 2** and **3**).
7. After three washes with DPBS, add 100 μL of DPBS to each well and sterilize the plate using 30 min UV under the biosafety cabinet. The prepared cross-linked fluorescently labeled gelatin plates can be stored at 4 °C for up to 1 week (*see* **Notes 3** and **4**).

3.2 Plating Cancer Cells on the Labeled Gelatin Matrix

Below are a few examples of specific experiments that are useful for studying invadopodia formation using gelatin degradation assay.

3.2.1 Experimental Approach for Screening Different Cell Lines for Invadopodia Formation Using the Gelatin Degradation Assay

1. Continue from **step 3.1**. Seed 1 × 10^4 cells/well in a 96-well plate, freshly prepared on fluorescently labeled cross-linked gelatin. Incubate the cells at 37 °C in a tissue culture incubator under a humid atmosphere of 5% CO_2 for 5–6 h (*see* **Note 5**).
2. After the incubation stage, remove the plate and aspirate the media. Perform the permeabilization–fixation step in a chemical hood.
3. Add 100 μL of 0.5% Triton-X-100 in 3% PFA to each well and keep it at room temperature.

4. After 3 min, aspirate the PFA-Triton-X-100 solution and add 100 μL of 3% PFA only.
5. Incubate for an additional 20 min at room temperature.
6. Aspirate the PFA and wash the wells (3×) with DPBS.
7. The cells are now permeabilized and fixed and can be stored at 4 °C for a few days or proceed to the next step for labeling the cells.
8. Aspirate the DPBS and add TRITC-phalloidin and DAPI (1: 500) prepared in DPBS and incubate for 30 min. For labeling of the cells with invadopodia-specific markers (e.g., Tks5 or cortactin), use specific antibodies, as follows:
9. Wash the wells (3×) with DPBS.
10. After the final wash, add 100 μL DPBS to the wells to prevent drying of the gelatin matrix.
11. For best results, immediately proceed with imaging. F-actin, nuclei along with an invadopodia-specific marker (e.g., proteins such as Tks5 or cortactin) labeling is the best way for accurately counting the number of invadopodia per cell and calculating the percentage of cells with invadopodia.
12. For the labeling of the cells, proceed from **step 10** above.
13. In case of co-labeling with specific invadopodia marker, use primary antibody and incubate for 30 min. Wash the cells with DPBS (3×).
14. Then add the appropriate secondary antibody and further incubate for 30 min at room temperature. Wash the wells with DPBS (3×) (*see* **Note 6**).
15. Next, add TRITC-phalloidin and DAPI (1:500) in DPBS. Wash the wells with DPBS (3×).
16. After the final wash, add 100 μL DPBS to the wells to prevent drying of the gelatin matrix. For best results, proceed immediately with imaging and quantification of the invadopodia and the underlying gelatin matrix degradation. An example of invadopodia formation and gelatin degradation by different cancer cell lines (MDA-MB-231 breast carcinoma and IGR-1 melanoma) using actin and DAPI is shown in Fig. 2. Further, labeling for the invadopodia-specific marker Tks5 along with actin and DAPI is shown in Fig. 3.

3.2.2 Experimental Approach for Testing the Effect of Specific Gene Knockdown on Invadopodia Formation Using Gelatin Degradation Assay

1. Following Subheading 3.1: For experiments involving knockdown of the expression of a specific gene of interest using either siRNA or shRNA, perform the knockdown and incubate the cells for the time required for an efficient knockdown.
2. Then, seed the knocked-down cells on labeled gelatin matrix and incubate the cells at 37 °C in a tissue culture incubator under a humid atmosphere of 5% CO_2 for 5–6 h.

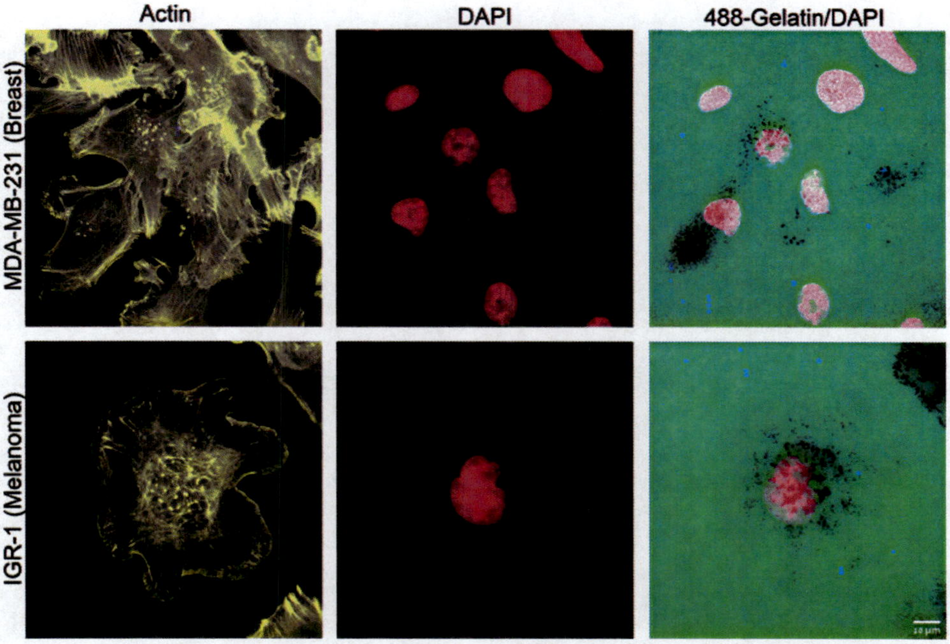

Fig. 2 Different modes of invadopodia and matrix degradation were displayed by the cancer cell lines. MDA-MB-231 and IGR-1 cells were plated on an Alexa Fluor 488-gelatin matrix for 5–6 h. The cells shown here were stained for actin and DAPI. Invadopodia (visualized by actin-rich yellow dots), nuclei (magenta), and local gelatin matrix degradation (dark spots on the green Alexa488-gelatin) can be clearly visualized. Images were acquired with 60X (1.42) oil objective in a DeltaVision Elite microscopy system (Applied Precision, Issaquah, WA). The scale bar is 10 μm

3. Then continue with the processing of the sample as described in Subheading 3.2.1, **steps 1–15**.
4. As an example, the effect of stable knockdown of C-terminal Src kinase (CSK) in A375 melanoma cells, in comparison with the parental A375 cells, is shown in Fig. 4.

3.2.3 Experimental Approach for Screening Inhibitors of Invadopodia Formation Using Gelatin Degradation Assay

1. In case of experiments involving screening of inhibitors that affect invadopodia activity. Continue from subheading 3.1 above.
2. Plate the cells on a labeled gelatin plate and incubate them on the matrix at 37 ° C in a humidified atmosphere of 5% CO_2 incubator for 1.5 h.
3. After the initial adhesion of the cells to the gelatin substrate, add the desired concentration of the invadopodia modulators.
4. Incubate for an additional 4.5 h (total of 6 h) (*see* **Note 7**).
5. Then, continue with Subheading 3.2.1, **steps 1–15** above. For example, we show here the effect of Marimastat, a broad-spectrum matrix metalloproteinase inhibitor, on gelatin degradation (Fig. 5).

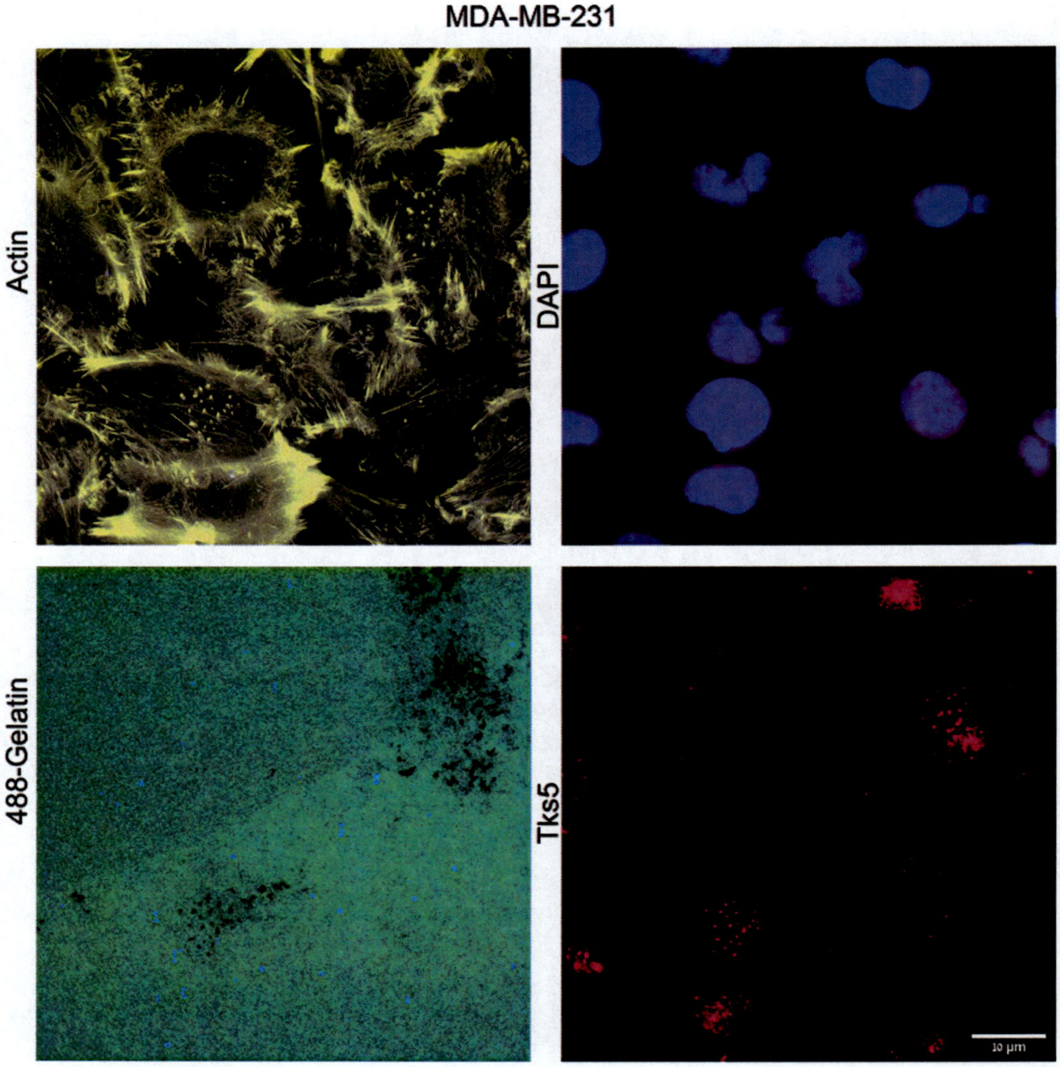

Fig. 3 Co-localization of actin and Tks5 (an invadopodia-specific marker) in MDA-MB231 cells. MDA-MB-231 and IGR-1 cells were plated on an Alexa Fluor 488-gelatin matrix for 5–6 h and stained for actin (TritC-phalloidin, yellow dots), DAPI (blue), and Tks5 (magenta dots). Gelatin matrix degradation is visualized as dark spots on the green matrix layer. Images were acquired using 60X (1.42) oil objective in a DeltaVision Elite system. The scale bar is 10 μm

3.2.4 Application of Microscopy-Based Image Processing for Quantification of Invadopodia Formation and Gelatin Degradation (in ImageJ)

1. Open all the captured TIFF images (typically acquired using 40X or 60X objectives) in ImageJ software (https://imagej.nih.gov/ij/download.html).
2. Select the DAPI images from the folder. Choose the command Image *tab→Adjust→Threshold* in ImageJ software.
3. Adjust threshold for DAPI image so that all the nuclei are covered, as shown in Fig. 6.

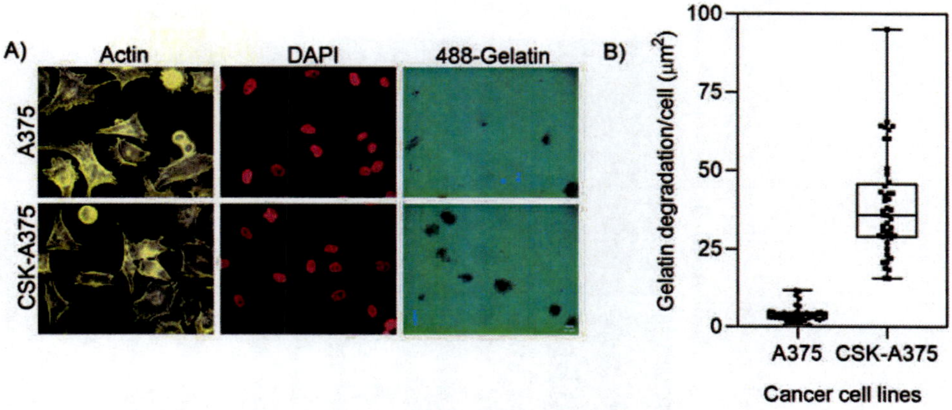

Fig. 4 Quantification of gelatin matrix degradation in A375 (melanoma) and CSK-A375 (stable knockdown of c-Src in A375 cell line). (**a**). Cells were seeded on an Alexa Fluor 488-gelatin matrix and cultured for 5–6 h, then fixed and stained for actin (TritC-phalloidin) and DAPI. Images were acquired using 40X/0.75 air objective in a WiScan® Hermes Automated High Content Imaging System (IDEA Bio-Medical Ltd.). A total of 36 fields (6×6) were imaged in each well, and the gelatin degradation/cell (μm^2) was quantitated using ImageJ software. (**b**) A box and whiskers plot showing the calculated gelatin degradation/cell (μm^2) for each cell line, based on three independent experiments $\pm$ SEM is shown

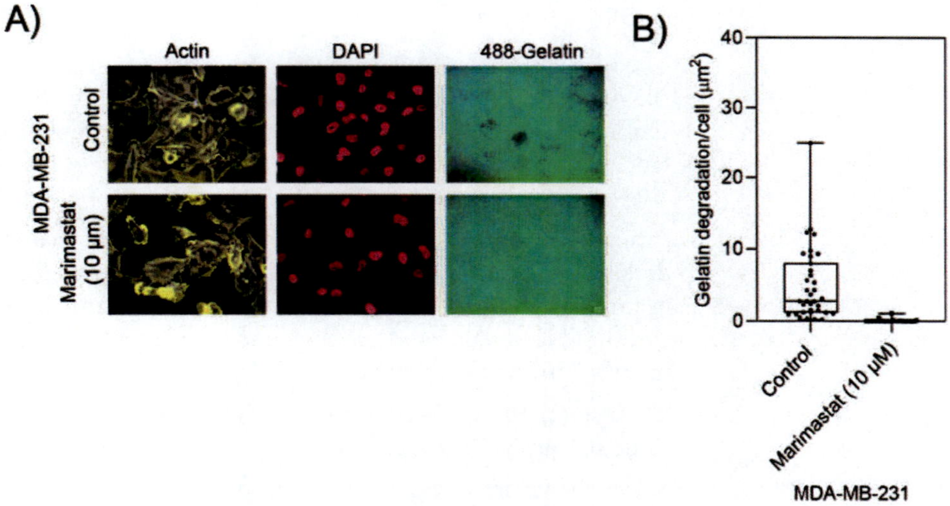

Fig. 5 The effect of the broad-spectrum metalloproteinase inhibitor, Marimastat, on invadopodia formation and matrix degradation. MDA-MB-231 and IGR-1 cells were plated on an Alexa Fluor 488-gelatin matrix for 1.5 h. Subsequently, Marimastat (10 μM, final concentration) was added, and the cells were incubated for an additional period of 4 h. The cells were then fixed and stained for actin (TritC-phalloidin) and DAPI. Images were acquired using 40X/0.75 air objective in a WiScan® Hermes Imaging System (IDEA Bio-Medical Ltd.), and matrix degradation was calculated as described above

4. Then select the command *Analyze tab→Particle analyze*. Set the value in the size tab (μm^2) as 50. Now select the obtained values and copy in an Excel sheet. This denotes the number of the cell count.

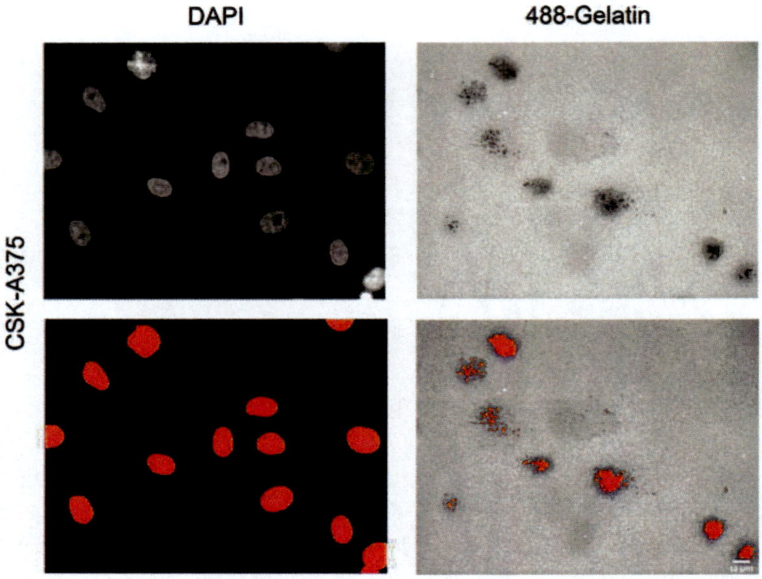

Fig. 6 Segmentation of gelatin degradation areas and actin in invadopodia (upper panel), using the ImageJ software. The segmented areas are marked in red (lower panel). The scale bar is 10 μm

5. The next step is to calculate the gelatin degradation observed in the 488-gelatin matrix: Open all TIFF images from the Alexa Fluor 488-gelatin matrix image. In case of the threshold for the labeled gelatin matrix, choose the command: *Image tab→Adjust→Threshold* again (*see* **Note 8**).
6. Adjust the lower threshold intensity in such a way that all the dark spots visualized in degraded gelatin are covered. For example, *see* Fig. 6. Using the *Set* button, confirm the lower threshold level value and use the same intensity for all the fields included in the experiment.
7. Then select the command *Analyze tab→Particle analyze*. Put value 0.1 μm^2 in the size tab.
8. Select the values and copy in Excel spreadsheet.
9. Finally, in the Excel sheet, calculate the sum of all the values obtained from the images in different fields, both cell count and the gelatin-degraded column values individually, and calculate using the formula: *Total gelatin degraded area/sum of the values of cell count*.
10. The results obtained by this calculation is the relative gelatin degraded area (in μm^2) per cell. The gelatin degradation values for each captured field of images for control or treated cells can be represented in *box and whiskers plot* with their respective ± SEM (*see* Figs. 4 and 5).

4 Notes

1. Always be careful and gentle while adding the cross-linked fluorescently labeled gelatin mixture to the plates. Add the solution mixture slowly without introducing any bubbles into the well.
2. During the washing step, after introducing the cross-linked, fluorescently labeled gelatin mixture on the plates, be careful not to disrupt the gelatin matrix with the pipette mechanically.
3. Do not let the fluorescently labeled cross-linked gelatin matrix dry in the test well. Always keep the cross-linked fluorescently labeled gelatin matrix hydrated with DPBS while storing it at 4 °C.
4. The quality of fluorescence of the cross-linked gelatin matrix deteriorates upon storage for more than a week. Thus, for precise and reproducible imaging, it is recommended to perform the experiment on freshly prepared, fluorescently labeled, and cross-linked gelatin.
5. As a part of the experimental setup, for each control and test condition, prepare the fluorescently labeled cross-linked gelatin matrix in duplicate wells or plates.
6. If using Alexa Fluor 488-labeled gelatin, always remember to avoid using Phalloidin or secondary antibodies labeled with a green, overlapping fluorophore.
7. Optimal incubation time may vary for different cell lines and needs to be determined empirically. The typical incubation time for invadopodia formation by cancer cell lines can be in the range of several hours. It is noteworthy that invadopodia often display a short life cycle (in the order of minutes). Further, the gelatin degradation is essentially irreversible; the degree of co-localization of invadopodia and focal gelatin degradation sites is expected to decrease upon long incubation times.
8. While quantifying the gelatin matrix degradation at the sites of invadopodia, using ImageJ software, always decide on an appropriate threshold by considering all the image fields of the well. Maintain the threshold value preferably in a range to mark gelatin degradation areas and not include areas displaying artifacts or scratches. This will help to reduce errors in the degradation values and provide an accurate estimation of the level of gelatin matrix degradation.

Acknowledgments

The studies described herein are funded by the Israel Science Foundation. The authors are grateful to Dr. Or-Yam Revach for her contribution to the development of the assay in our laboratory.

References

1. Dillekas H, Rogers MS, Straume O (2019) Are 90% of deaths from cancer caused by metastases? Cancer Med 8(12):5574–5576
2. Riggi N, Aguet M, Stamenkovic I (2018) Cancer metastasis: a reappraisal of its underlying mechanisms and their relevance to treatment. Annu Rev Pathol 13:117–140
3. Lambert AW, Pattabiraman DR, Weinberg RA (2017) Emerging biological principles of metastasis. Cell 168(4):670–691
4. Pachmayr E, Treese C, Stein U (2017) Underlying mechanisms for distant metastasis - molecular biology. Visceral medicine 33(1): 11–20
5. Eccles SA, Welch DR (2007) Metastasis: recent discoveries and novel treatment strategies. Lancet 369(9574):1742–1757
6. Sun Y, Ma L (2015) The emerging molecular machinery and therapeutic targets of metastasis. Trends Pharmacol Sci 36(6):349–359
7. Welch DR, Hurst DR (2019) Defining the hallmarks of metastasis. Cancer Res 79(12): 3011–3027
8. Revach OY, Grosheva I, Geiger B (2020) Biomechanical regulation of focal adhesion and invadopodia formation. J Cell Sci 133(20)
9. Paz H, Pathak N, Yang J (2014) Invading one step at a time: the role of invadopodia in tumor metastasis. Oncogene 33(33):4193–4202
10. Ayala I, Baldassarre M, Caldieri G, Buccione R (2006) Invadopodia: a guided tour. Eur J Cell Biol 85(3–4):159–164
11. Murphy DA, Courtneidge SA (2011) The 'ins' and 'outs' of podosomes and invadopodia: characteristics, formation and function. Nat Rev Mol Cell Biol 12(7):413–426
12. Chen WT, Olden K, Bernard BA, Chu FF (1984) Expression of transformation-associated protease(s) that degrade fibronectin at cell contact sites. J Cell Biol 98(4): 1546–1555

Chapter 13

Measuring Cellular Traction Forces with Micropillar Arrays

Malak Amer and Haguy Wolfenson

Abstract

Accurately evaluating cellular forces is critical for studying mechanosensing and mechanotransduction processes, and it necessitates sensitive measurements on the piconewton scale. Here we describe a specialized method that employs elastic polydimethylsiloxane (PDMS) micropillar arrays, which cells can adhere to and bend. The flexibility of the pillars correlates with their heights; the longer they are, the easier they are to bend. Thus, an array of taller pillars mimics a relatively soft substrate that readily yields in response to cellular forces. Tracking cell movements and pillar displacements using live-cell microscopy enables the calculation of cellular forces and the tracking of their dynamic features throughout early and late stages of cell spreading on the pillars. This technique offers the advantage of high spatial and temporal resolution analyses and constitutes a method to investigate the effect of substrate rigidities on cellular functions.

Key words Cellular forces, PDMS, Micropillar arrays, Live-cell imaging, Displacement, time curves

1 Introduction

Mechanosensing is the process by which cells test the mechanical properties of their nearby environment [1]. Mechanosensing of the extracellular matrix (ECM) involves the application of actomyosin-based forces to the matrix via integrin-based adhesions. ECM resistance to the cellular forces triggers the recruitment of intracellular structural and signaling proteins to the adhesions. Thus, a physical link is forged between the ECM and the cytoskeleton, which is pivotal in transmitting forces through the adhesions and in conveying mechanical signals from inside and outside the cell [2, 3].

Cellular forces can be measured by using polydimethylsiloxane (PDMS) micropillar arrays (coated with ECM proteins of choice) as substrates for the cells to adhere to. The initial position of the pillars is recorded prior to cellular contact, and once cells settle on them, the pillar displacements are tracked periodically throughout cellular attachment and spreading. The displacements of individual pillars

Ronen Zaidel-Bar (ed.), *Mechanobiology: Methods and Protocols*,
Methods in Molecular Biology, vol. 2600, https://doi.org/10.1007/978-1-0716-2851-5_13,

are used to calculate the cellular forces required to displace them by multiplying the pillar displacement distance by the pillar spring constant [4].

Importantly, the pillar arrays can be fabricated in several heights, with taller pillars being more flexible, thus simulating a "soft" substrate that offers relatively low resistance. This enables the investigation of mechanical stiffness on cellular behavior, without altering the chemical composition of the PDMS substrate.

The pillar diameters too have a biological significance; pillars with diameters of 0.5 μm or lower allow for the analysis of contractile unit (CU) forces associated with early mechanosensing [4] and nuclear actin cap fibers [5], whereas 2 μm-diameter pillars offer an insight into later and higher cellular forces involving actin retrograde flow and cell spreading/motility [6, 7].

Herein, we describe the fabrication of pillar arrays, along with two possible methods of coating the pillars with ECM proteins. The first method involves immersion of the entire array in the coating solution and the second method entails stamping the proteins directly on the top surfaces of the pillars [8]. The stamping method allows cells to form adhesions only on the tops of the pillars, which is advantageous when using cells that tend to sink along the lengths of the pillars.

A wide range of analyses may be extracted from one live-cell imaging experiment on micropillar arrays. Some relevant measures are shown in the pillar displacement–time graph (Fig. 1). Other measurements include the rate of pillar displacement and the average cellular forces of specific cell types. The cells can also be fixed and immunostained on the pillar arrays upon completion of the live experiment, for further investigations. Moreover, simultaneously tracking pillar displacements and fluorescently labeled proteins of interest may provide insights into protein localization or expression levels during specific stages of cellular force generation [4 , 7].

2 Materials

2.1 Equipment

1. Analytical weighing balance.
2. Oven.
3. Plasma surface treatment machine or corona treater.
4. Live-cell brightfield microscope, equipped with a 37 °C chamber and a CO_2 supply (in the absence of which, neutral pH level can be maintained by adding 20 mM Hepes to the imaging media described in Subheading 2.2 below).
5. 35 mm glass-bottom dishes, #0 coverslip
6. Micropillar molds (*see* **Note 1**).
7. Scalpel, forceps.

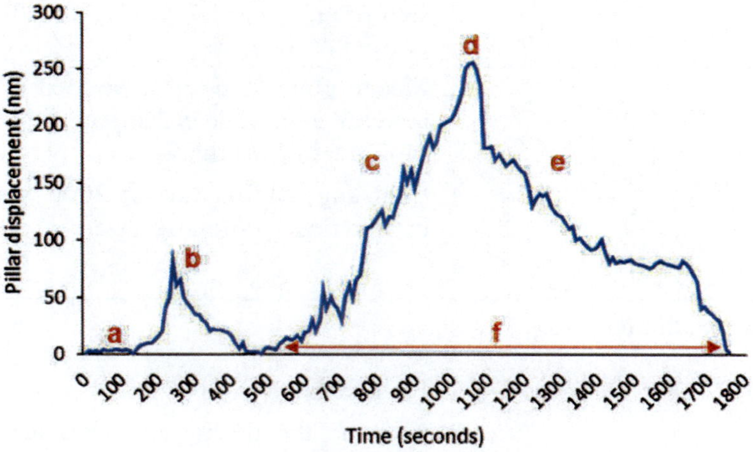

Fig. 1 A graph illustration of the displacement of a single 2 μm-diameter pillar as a function of time. (**a**) Minor fluctuations represent background signal, prior to cellular contact; (**b**) the cell catches and releases the pillar; (**c**) the cell pulls the pillar, causing an increase in pillar displacement; (**d**) the pillar reaches maximal displacement; (**e**) the cell gradually releases the pillar, allowing it to return to its original position (displacement 0); (**f**) the pillar holding time

2.2 Solutions and Media

1. Solution suitable for live-cell imaging, e.g., cell culture media or Hanks' Balanced Salt Solution (HBSS).
2. Cell culture media, e.g., Dulbecco's Modified Eagle Medium (DMEM), supplemented with 10% fetal bovine serum (FBS), 1% penicillin–streptomycin.
3. ECM proteins for coating the micropillar arrays, for example:

   for coating by immersion: 10 μg/mL fibronectin in PBS × 1;

   for coating by stamping: 50 μg/mL fibronectin, 5 μg/mL fluorescently labeled fibronectin in sterile, deionized water.
4. 0.2% (w/v) Pluronic solution in sterile, deionized water (for stamping).
5. 100% ethanol (EtOH).
6. Phosphate-buffered saline (PBS) × 1.

2.3 PDMS Mixture

Final composition for pillars: 10 parts elastomer base, 1 part elastomer curing agent, e.g., 1 g base, 0.1 g curing agent.

Final composition for stamps: 20 parts elastomer base, 1 part elastomer curing agent. The PDMS ratio suitable for making stamps may vary depending on the pillar heights.

1. Weigh the PDMS base and curing agent respectively into a falcon tube. The base is a viscous material; handle with care and cut the 1 mL pipette tips when weighing the PDMS.

2. Centrifuge briefly to collect the PDMS components at the bottom of the tube.
3. Mix thoroughly using a wooden or metal stick for approximately 3 min, while making sure to collect and mix the material from the sides of the tube.
4. Centrifuge the mixture at 2000 × *g*, for >1 min, until it is bubble-free and homogenized.

3 Methods

3.1 Pillar Fabrication

1. Preheat the oven to 60 °C.
2. Examine the quality and cleanliness of the pillar mold by inspecting it under direct light. Disturbances in the light diffraction pattern indicate that the molds are either damaged or unclean. The molds may be cleaned by a procedure involving Piranha solution [8].
3. Plasma-treat the 35 mm dishes (Fig. 2). This should be calibrated for each machine model and application to account for variance between different machines.
4. Prepare a 10:1 PDMS mixture, as described in Subheading 2.3.
5. Use a 10 μL tip to deposit a small drop of the 10:1 PDMS mixture onto the center of the mold (Fig. 3).
6. Invert the mold onto the center of the glass-bottom square section of the 35 mm dish.
7. Apply gentle pressure to the mold and place a small weight on it (e.g., a small metal nut) to prevent it from sliding.
8. Incubate the dish at 60 °C for 12 h.

3.2 Experiment Setup

3.2.1 ECM Coating by Immersion

When coating the pillars by immersion, the coating solution is applied onto the entire micropillar array. The pillars are submerged, so their lengths are also coated, and cell adhesion formation may not be restricted to the pillar tops. When using relatively dense pillar arrays (e.g., 0.5 μm-diameter pillars with 1 μm center-to-center distance), most cell types tend to stay on the pillar tops even when coating by immersion; however, some cell types, such as cancer cells, might not be restricted to the pillars tops when coating by immersion, in which case stamping (*see* Subheading 3.2.2 below) is recommended.

1. Following the PDMS curing, allow the dish to cool down at room temperature (RT) for approximately 10 min.
2. Remove the weight from the mold and perform the following procedure in a biological hood.

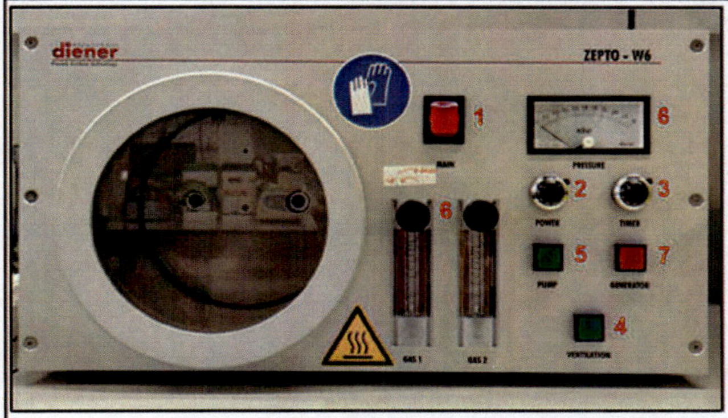

1. Turn on the **main** switch.
2. Set the **power** to 100%.
3. Set the **timer** to 1 min.
4. Switch on the **ventilation** and open the door.
5. Place the plates inside, close the door and switch on the **pump**; the vacuum seals the door shut.
6. When the **pressure** drops to 0.3 mBarr, open the valves until it increases to 0.5 mBrarr.
7. Turn on the **generator**.
8. When the treatment ends, turn off the pump, Set the pressure valves to 0, switch on the ventilation and open the door.
9. Take out the plates.
10. Close the door as in step 5 and turn off the main switch.

Fig. 2 A plasma surface treatment machine; the operation steps are shown on the right, and the machine parts corresponding to **steps 1–7** are labeled in red. Equivalent operation is to be used with plasma treaters from other manufacturers

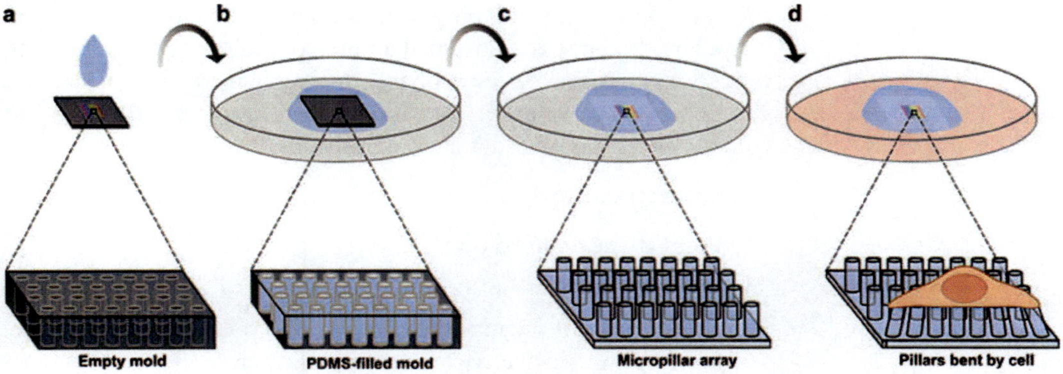

Fig. 3 A schematic of the pillar fabrication process, with magnifications shown in the bottom half. (**a**) A small drop of the PDMS mix is added to the hollow mold; (**b**) the PDMS-filled mold is flipped onto a glass-bottom dish; (**c**) after curing for 12 h at 60 °C, the mold is peeled off from the dish, leaving PDMS micropillar structures on the dish; (**d**) the micropillars are bent under the forces of seeded cells

3. Add 2 mL 100% EtOH to the dish. Once the pillars are exposed to liquid, they should not be dried, as this can cause them to collapse. They may be sealed with Parafilm and stored for up to 1 week at 4 °C before an experiment.
4. Carefully clean excess PDMS from the top of and around the mold using a scalpel and forceps. Loosen the PDMS from the edges of the mold with a scalpel, then carefully peel off the mold from the dish.
5. Wash the mold with 100% EtOH and dry it with a stream of air, without touching the center of the mold.
6. Replace the EtOH solution with PBS × 1, without allowing the pillars to dry, using repeated dilutions of the EtOH:

(i) Add 7 mL fresh PBS × 1 to the dish, so that the total volume of the solution in the dish is 9 mL.

(ii) Aspirate and discard 7 mL of the solution in the dish.

(iii) Repeat for a total of 4 times, leaving 2 mL PBS × 1 in the dish.

7. Add 20 μL fibronectin (1 mg/mL) to the solution in the dish, for a final concentration of 10 μg/mL, and gently swirl it.
8. Incubate the dish at 37 °C for 1 h.
9. Wash the pillars and replace the PBS solution with cell culture media or HBSS, as described in 3.2.1, **step 6**.

3.2.2 ECM Coating by Stamping

In this coating method, a PDMS slab is cut into cubes, one side of which is coated with the ECM coating mixture. The coated side is used to stamp the ECM coating onto the tops of the pillars. Unlike ECM coating by immersion, the stamping method avoids coating the pillar posts. The stamping method is applicable to 2 μm-diameter pillars, but not to 0.5 μm-diameter pillars, since the stamping procedure requires a dry phase during which the pillars are not immersed in solution, and the smaller diameter pillars tend to collapse more easily under such conditions.

Stamp preparation

1. Preheat the oven to 65 °C.
2. Prepare the pillars as in 3.1.
3. Gently peel off the molds while dry. Dry pillars and stamps may be stored in a dry environment indefinitely.
4. Prepare the stamps by making a 20:1 PDMS mixture sufficient to fill a 100 mm Petri dish to a height of 0.75 cm.
5. Pour the PDMS mixture into a 100 mm Petri dish.
6. Incubate the dish for 5–10 min at RT to allow any bubbles to rise to the surface and dissipate. Alternatively, place the dish in a vacuum desiccator until the bubbles dissipate.
7. Cure the PDMS at 65 °C for 2 h.
8. Remove the dish from the oven and allow it to cool at RT.
9. Use a scalpel to cut the slab of PDMS into stamps that correspond to the pillar array area. The PDMS at the sides of the dish is uneven, so it should not be used to cut stamps. One 100 mm dish can produce at least 20 stamps.
10. Immerse the stamps in ethanol for 5 min to sterilize them and clean off any particulates.
11. Dry the stamps with a stream of air.

12. Dispense the fluorescently labeled ECM coating solution onto the top of each stamp, and spread the solution along the surface using a micropipette tip. Depending on the stamp size, ~30 μL is sufficient to cover 1 stamp surface.

    Optional: cut a small corner of the side not used for stamping to indicate the coating orientation.
13. Allow the coating solution to adsorb onto the stamp for 1 h.
14. Wash the stamps twice by submerging them in a Petri dish filled with deionized water.
15. Dry the stamps with a stream of air and transfer them to a new Petri dish.

Stamping the pillars

16. Plasma-treat the pillars.
17. Use forceps to gently lower the coated side of the stamp onto the pillars, while keeping it perfectly horizontal, to avoid bending the pillars. Place a weight on top of the stamp and incubate for 3–5 min at RT.
18. Submerge the pillars in ethanol and carefully lift up the stamp.
19. Wash the pillars 3 times with deionized water, as in 3.2.1, **step 6**.
20. Add 2 mL 0.2% (w/v) Pluronic solution to the pillars, for blocking the pillar sides.
21. Incubate for **30–60 min** at RT, protected from light.
22. Wash the pillars 3 times with deionized water, then add 2 mL HBSS as in 3.2.1, **step 6**.

3.3 Visualizing the Cells

1. Place the 35 mm dish on the microscope stage and find the focus of the pillar tops using a 63X or 100X oil objective. Incubate the plate on the stage for approximately 15–20 min.
2. Trypsinize the cells, and resuspend them in cell culture media or HBSS.
3. Adjust the focus on the pillars, and seed a small volume of the cell solution onto the pillars. First seed a conservative amount, to gauge the seeding density, then add more if necessary. Ideally, there should be a few cells in the field of view, without aggregating or overlapping cells.
4. Wait 5–10 min for the cells to settle undisturbed, then select a region with good quality pillars and a reasonable seeding density, and begin the time-lapse imaging (*see* **Note 2**).

3.4 Pillar Displacement Analysis Using ImageJ/FIJI (See Note 3)

3.4.1 Prepare the File for Analysis Using ImageJ [9]

1. Open the image stack using ImageJ/FIJI. Where possible, it is recommended to use ImageJ shortcuts to save time.
2. Image → Type → 8-bit.
3. Process → Filters → Gaussian blur → Sigma (radius): 1.
4. File → save as TIFF.

3.4.2 Define the "Noise" and "Reference" Pillars and Record their Positions

1. Draw a square selection around a pillar that is not in contact with other cells, and is in sharp focus across all the frames. This pillar will represent the background noise of pillars that are not displaced.
2. Edit → Selection → Specify → width: **x**, height: **x** (enter a value for x which is large enough to fit a single pillar within the selected square, i.e., approximately equal to the pillar diameter size) → enter → move the square selection so that the pillar is at its center.
3. Plugins → Analyze → Nano Tracking → Grab → make sure that the pillar is at the center of the selection area.
4. Edit → Selection → Specify → width: [**x** + **y**], height: [**x** + **y**] (select a value for y, which will produce a square that covers the space around the selected pillar and only minutely includes the edges of neighboring pillars; typically a value of [x + 10] pixels will suffice) → enter → move the square y/2 steps upwards and to the left so that the pillar is at the center of the new selection area → set ROI (region of interest) → Run.
5. When the ROI square runs through the frames, check that it consistently tracks the selected pillar.
6. Cut the values from the log dialog → paste into a new Excel file → delete the two left columns, leaving the x and y positions of the center of the pillar, in each frame → save the Excel file as "noise.xlsx".
7. Repeat **steps 2–6** for five more pillars that will be used as references, and save the data in a seperate Excel file to be saved as "ref.xlsx".

3.4.3 Analyze the Noise and Reference Pillar Displacements Using MATLAB

1. Analyze the total displacements of the pillars in each time frame using a designated MATLAB program (*see* **Note 4**).
2. There should not be a general trend of displacements or deviations in the noise peaks, as this is indicative of a technical issue with the movie. If there are consistent high peaks in a particular frame, delete the frame in question: Image → Stacks → Delete slice → save and replace the existing file, and repeat the analysis.

3.4.4 Analyze the Pillars That Were Displaced by Cells

1. View all the frames and select the pillars that were displaced by cells throughout the experiment. Select pillars that initially were not contacted by the cell to ensure the proper identification of their zero displacement position.
2. Analyze their displacements over time, as described in 3.4.1, and save the extracted data in a separate Excel file titled "pillar.xlsx".

4 Notes

1. The mold is produced by etching a uniform array of holes (the impression of the micropillars) into a wafer [10]. Determine the appropriate pillar diameters and heights for the purposes of the experiment, and thus the necessary mold dimensions. Handle the mold gently with forceps, and avoid touching the array at the center at all times.
2. If possible, add a 600 nm long-pass filter in the illumination path to minimize photodamage to the cells. Use bright-field illumination; do not use phase contrast of differential interference contrast (DIC) due to potential distortion of the pillar top shapes. The settings to be used for the time-lapse imaging depend on the process to be tracked. Early formation of adhesions and CU can be tracked using 0.5 μm-diameter pillars, imaging the cells at 1–2 s intervals between each frame. When using 2 μm-diameter pillars, 10–20 s intervals are preferable.
3. The analysis plugin used for monitoring the pillar positions over time in a time-lapse video is "Nano_Tracking" (available upon request: haguyw@technion.ac.il). Alternatively, the "PillarTracker" plugin can be used (https://imagej.net/plugins/pillartracker).
4. The MATLAB code can be provided upon request: haguyw@technion.ac.il.

Acknowledgments

H.W. acknowledges support from the Israel Science Foundation (1738/17) and from the Rappaport Family Foundation. H.W. is an incumbent of the David and Inez Myers Career Advancement Chair in Life Sciences.

References

1. Trichet L et al (2012) Evidence of a large-scale mechanosensing mechanism for cellular adaptation to substrate stiffness. Proc Natl Acad Sci U S A 109:6933–6938
2. Geiger B, Bershadsky A (2002) Exploring the neighborhood: adhesion-coupled cell mechanosensors. Cell 110:139–142
3. Iskratsch T, Wolfenson H, Sheetz MP (2014) Appreciating force and shape-the rise of mechanotransduction in cell biology. Nat Rev Mol Cell Biol 15:825–833
4. Wolfenson H et al (2015) Tropomyosin controls sarcomere-like contractions for rigidity sensing and suppressing growth on soft matrices. Nat Cell Biol 18(1):33–42
5. Shiu J-Y, Aires L, Lin Z, Vogel V (2018) Nanopillar force measurements reveal actin-cap-mediated YAP mechanotransduction. Nat Cell Biol 20:262–271
6. Li Q, Makhija E, Hameed FM, Shivashankar GV (2015) Micropillar displacements by cell traction forces are mechanically correlated with nuclear dynamics. Biochem Biophys Res Commun 461:372–377
7. Feld L, Kellerman L, Mukherjee A, Livne A, Bouchbinder E, Wolfenson H (2020) Cellular contractile forces are nonmechanosensitive. Sci Adv 6(17). https://doi.org/10.1126/sciadv.aaz6997
8. Yang MT, Fu J, Wang YK, Desai RA, Chen CS (2011) Assaying stem cell mechanobiology on microfabricated elastomeric substrates with geometrically modulated rigidity. Nat Protocol 6(2):187–213
9. Schneider CA, Rasband WS, Eliceiri KW (2012) NIH Image to ImageJ: 25 years of image analysis. Nat Method 9(7):671–675
10. Ghassemi S et al (2008) Fabrication of elastomer pillar arrays with modulated stiffness for cellular force measurements. J Vac Sci Technol B Microelectron Nanometer Struct Process Meas Phenom 26:2549

Check for updates

Chapter 14

Imaging Cell Adhesive Force at the Single Molecule Level

Ying Tu and Xuefeng Wang

Abstract

Cell adhesive force transmitted by integrins or other mechanosensitive receptors is critical for many cellular functions and biological development. Visualization and quantification of such force have been long desired and practiced in cell mechanobiology. Here we describe integrative tension sensor (ITS), a dsDNA-based tension sensor that coverts invisible force signal to fluorescence and enables cell adhesive force imaging with ultra-sensitivity. ITS can be selectively implemented at two imaging modes: a cumulative mode that maps cell adhesive force at a high signal-to-noise ratio even with a low-end fluorescence microscope, and a real-time mode that images the force at the single molecular tension level.

Key words Integrin, Tension sensor, Mechanobiology, dsDNA, Single molecule imaging

1 Introduction

There is an increasing interest in visualizing and quantifying cell adhesive forces in the field of cell mechanobiology. In the past, cell traction force microscopies have been the mainstream tools mapping and quantifying the stress exerted by cells on substrates. By converting cell contractile force to visible substrate strain, these techniques pioneered the study of cell adhesive forces. However, as research in cell mechanobiology advances from bulk cells to some ultrafine cellular structures [1], the local force intensity becomes weaker and force distribution approaches the submicron level, therefore demanding force-imaging techniques with higher sensitivity and resolution than before. In recent years (2010–present), the development of molecular tension sensors has greatly improved the resolution and sensitivity of force visualization and quantification, hence revamping the methods for cell adhesive force study. More details can be found in this review [2]. Among these tension sensors, integrative tension sensor (ITS) is a dsDNA (double-stranded DNA) construct that converts cellular force to fluorescence at the molecular level [1, 3]. ITS evolves from another dsDNA-based technique, named tension gauge tether (TGT)

Ronen Zaidel-Bar (ed.), *Mechanobiology: Methods and Protocols*,
Methods in Molecular Biology, vol. 2600, https://doi.org/10.1007/978-1-0716-2851-5_14,

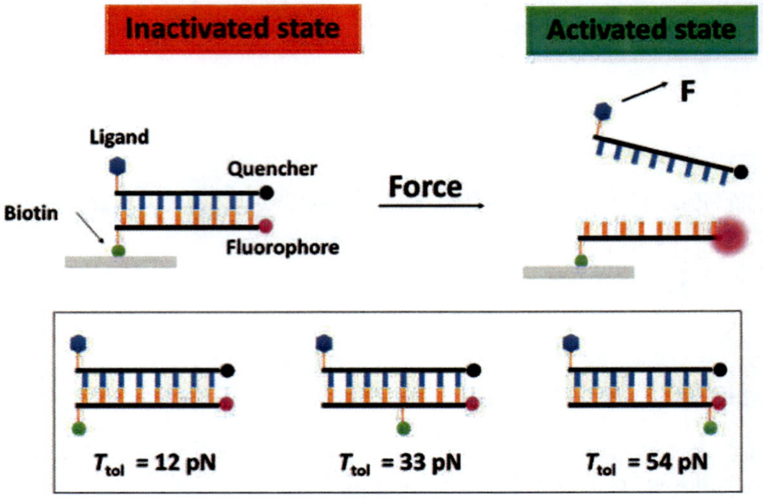

Fig. 1 A schematic of integrative tension sensor (ITS). ITS is an 18-bp dsDNA labeled with an integrin ligand, a biotin tag for surface immobilization, a quencher (blackhole quencher 2), and a fluorophore (Cy3 or Cy5). ITS is initially in an inactivated state without fluorescence due to the quencher. It becomes fluorescent after dsDNA is mechanically disassociated by the integrin–ligand transmitted force. Tension tolerance (T_{tol}) of ITS can be tuned by adjusting the anchorage site (the location of the biotin tag) at the lower stand

[4]. ITS and TGT share structural similarity but with different applications. While TGT is designed to modulate cell adhesive force by knocking down integrin-transmitted tensions to a designed molecular level, ITS is specialized in force imaging by fluorescence. This chapter introduces the synthesis of ITS and demonstrates its applications in cell adhesive force imaging. We hope that researchers would appreciate the simplicity and strength of ITS and adopt it in their research of cell mechanobiology.

ITS is typically an 18 base-paired (bp) dsDNA decorated with a fluorophore–quencher pair, a ligand targeting integrin (or other receptors) and a biotin tag for surface immobilization (Fig. 1). Immobilized on a surface, ITS is initially nonfluorescent due to the presence of the quencher. During cell adhesion, some ITS will be engaged by cells through the integrin–ligand interaction. The dsDNA can be mechanically dissociated because of the receptor–ligand transmitted force, thus freeing the fluorophore from the quencher. As a result, the ITS is activated and becomes fluorescent. ITS also has a designed force detection threshold. The dsDNA construct has tension tolerance (T_{tol}), which is the minimum molecular force required to disassociate the dsDNA. ITS with T_{tol} value is typically activated by integrin molecular tension higher than T_{tol}, therefore selectively reporting integrin tension in a designed range. T_{tol} value can be tuned by the position of ITS anchorage site (biotin location on the DNA) from DNA unzipping configuration to shearing configuration, resulting in T_{tol} transition from 12 to 54 pN correspondingly [3].

ITS can be applied in two different imaging modes: a cumulative mode and a real-time mode. Because ITS activation by cellular force is irreversible, it preserves historic force signals and enables the cumulative mode that accumulates force signals by time. The force signal remains on the surface even after cell fixation or detachment. For this reason, the force imaging by ITS is compatible with cell immunostaining, which is commonly adopted to visualize cell structures. This cumulative mode of force imaging amplifies cell adhesive force signal and can be performed even with low-end fluorescence microscopes. The signal-to-noise ratio of force imaging is comparable to that of cell structural imaging [3]. In contrast, with a strong illumination light and a sensitive camera, ITS can also operate at a real-time mode, during which force signal is imaged with single molecule imaging [1]. The molecular tension signals are constantly erased by photobleaching, so that only recent tension signals are imaged to avoid signal overcrowding. These two imaging modes can be selectively adopted by researchers, depending on the availability of the imaging system and the goals of experiments.

In this chapter, we describe the synthesis and application of ITS designed for integrin-transmitted force imaging. The ligand is a cyclic RGD (arginine–glycine–aspartic acid) peptide that binds to a variety of integrin types including integrin $\alpha_V\beta_3$, $\alpha_{IIb}\beta_3$, and integrin $\alpha_5\beta_1$. ITS can be applied to study other receptor-transmitted forces by attaching the corresponding ligands on the DNA construct.

2 Materials

This section provides the list of items required for ITS synthesis and its demonstration in cell adhesive force imaging. We strived to develop a highly accessible protocol that can be adopted in most biology labs. All chemical reactions take place in phosphate buffered saline (PBS). The requirement for biochemical equipment is minimal. In a nutshell, ITS synthesis only requires a pipette, a few centrifuge tubes, and a centrifuge.

ITS is constructed from dsDNA. We customized and ordered single-stranded DNAs with modifications from Integrated DNA Technologies. Inc. The upper DNA strand is shared by all five ITS constructs with T_{tol} values in the range of 12, 23, 33, 43, and 54 pN. Five different lower strands can be purchased to synthesize the five ITS constructs with different T_{tol} values. However, if it is only for general force imaging without concerning integrin tension level, we suggest only synthesizing the ITS with T_{tol} = 12 pN, which is the most sensitive one. ITS constructs with other T_{tol} values are optional.

Above dsDNA-based ITS constructs are applicable in most cell studies. However, some types of cells may recruit substantial DNase (deoxyribonuclease) to the cell membrane and degrade dsDNA-based ITS. Such phenomena have been observed in a few cancer cells (e.g., MDA-MB-231 human breast cancer) [5] and macrophages [6]. If force imaging in these cells is desired, PNA (peptide nucleic acid) is recommended to replace the lower stand to synthesize the ITS based on DNA/PNA duplex, which has strong DNase resistance [7]. PNA can be customized and ordered from PNA Bio. Inc.

2.1 DNA Strands and an Optional PNA Strand

Customized DNAs are listed below:

1. Upper strand: 5′-/5ThioMC6-D/CCC ACC AGC GAC GCC CGG/BHQ_2/-3′.
2. Lower strand (T_{tol} = 12 pN): 5′-/5Cy5/CCG GGC GTC GCT GGT GGG/3Bio/-3.
3. Lower strand (T_{tol} = 23 pN): 5′-/5Cy5/CCG GGC GTC GCT GG/iBiodT/GGG-3′ (Optional).
4. Lower strand (T_{tol} = 33 pN): 5′-/5Cy5/CCG GGC GTC GC/iBiodT/GGT GGG-3′ (Optional).
5. Lower strand (T_{tol} = 43 pN): 5′-/5Cy5/CCG GGC G/iBiodT/C GCT GGT GGG-3′ (Optional).
6. Lower strand (T_{tol} = 54 pN): 5′-/5Biosg//iCy5/CCG GGC GTC GCT GGT GGG-3′ (Optional).
7. Lower strand (T_{tol} = 12 pN): Cy5-O-CCG GGC GTC GCT GGT GGG-KKKK (biotin) (PNA) (Optional).

Here, ThioMC6-D is a thiol modification of C6 disulfide (S–S bond) at the 5′ end. This modification will be used for conjugating RGD ligand to the DNA. /BHQ_2/ represents a blackhole quencher 2 modified at the 3′ end. /5Cy5/ represents a Cy5 tagged at DNA 5′ end, and /iCy5/ represents a Cy5 inserted to the internal backbone of ssDNA. /3Bio/ and /5Biosg/ represent a biotin tag modified at −3′ and − 5′ end, respectively. For PNA modifications, O in PNA is an ethylene glycol linker improving PNA solubility in water. The four lysine (K) also increase water solubility of PNA. Moreover, they facilitate ITS surface immobilization by biotin–neutravidin interaction as a molecular overhang.

2.2 Chemicals and Materials

2.2.1 Chemicals Required for ITS Synthesis

1. Bond-breaker™ TCEP–HCl (77,720, Thermo Fisher Scientific).
2. Ethylenediaminetetraacetic acid (EDTA).
3. Arginylglycylaspartic acid-amine (RGD-NH_2) (PCI-3696-PI, Vivitide).
4. Sulfosuccinimidyl-4-(*N*-maleimidomethyl) cyclohexane-1-carboxylate (sulfo-SMCC) (A39268, Thermo Fisher Scientific).

2.2.2 Consumables Required for ITS Surface-Immobilization and Cell Culture

1. 35 mm glass bottom dish with 14 mm micro-well (D35-14-1.5-N, Cellvis)
2. Biotinylated bovine serum albumin (BSA-biotin).
3. PLL-PEG-biotin (optional) (customized and purchased from Nanosoft Polymers).
4. Neutravidin.
5. Dulbecco's Modified Eagle Medium (DMEM).
6. Fetal bovine serum.
7. Penicillin–streptomycin.
8. Imaging buffer.
9. Hanks' Balanced Salt Solution (HBSS).
10. 4-(2-Hydroxyethyl)-1-piperazineethanesulfonic acid (HEPES)
11. Sodium bicarbonate.
12. Cell mild detaching solution (100 mL of 10 × HBSS, 10 mL of 1 M HEPES, 10 mL of 7.5% sodium bicarbonate, 2.4 mL of 500 mM EDTA, 880 mL H_2O, and final pH 7.6).

2.2.3 Equipment and Devices

1. Centrifuge.
2. Ultrasonic bath.
3. DNA electrophoresis kit.
4. Fluorescent microscope (Nikon Ti-E microscope).
5. Fluorescent microscope (Nikon Ti-E2 TIRF (total internal reflection fluorescence) microscope).

3 Methods

3.1 Conjugation of RGD Peptide Ligand to the Upper DNA Strand (See Note 1)

1. Dissolve the upper-strand DNA (ThioMC6D-DNA-BHQ2) at a final concentration of 2 mM in PBS. Add 10 μL × 500 mM TCEP bond-breaker and 10 μL × 500 mM EDTA to 100 μL × 2 mM DNA in PBS. Let it reacts for 20 min at room temperature.
2. Dissolve 10 mg RGD-NH_2 in 200 μL ultrapure water (*see* **Note 2**) to obtain 56 mM RGD-NH_2 solution.
3. Transfer the 200 μL RGD-NH_2 solution to a vial of sulfo-SMCC (2 mg) with its bottom dipped in an ultrasonic bath (optional) which greatly facilitates sulfo-SMCC dissolution. Pipette the vial vigorously but steadily without producing bubbles. It helps to gently smash the SMCC solids in the vial with the pipette tip to assist the dissolution.

4. After sulfo-SMCC is completely dissolved (usually less than 30 s), immediately add 22 μL 10× PBS to the solution in **step 3**. Let it reacts for 10 min at room temperature.
5. Mix the two solutions obtained in **steps 1** and **4**. Let the mixture reacts for 2 h at room temperature, and place the mixture at 4 °C overnight. After this step, RGD is linked to ssDNA through maleimide-thiol reaction.
6. Purify the RGD-ssDNA from the unreacted sulfo-SMCC, RGD-NH_2 and TCEP by ethanol precipitation. Mix RGD-ssDNA solution obtained from **step 5** with 3 M NaCl solution (in water) and chilled (−20 °C) 100% ethanol at a volume ratio of 10:1:25. Place the mixed solution in a − 20 °C freezer for 30 min. RGD-ssDNA precipitates at this step.
7. Centrifuge the mixed solution at 10,000 × *g* for 30 min and discard the supernatant.
8. Add proper amount 1× PBS to DNA pellet and measure the DNA concentration using a spectrometer. The typical stocking concentration of RGD-ssDNA is 100 μM. Store in a − 20 °C freezer, the RGD-ssDNA has no apparent degradation for a few years according to our experience. This amount of RGD-ssDNA is sufficient for thousands of force assays.
9. (Optional). The above protocol usually produces RGD-ssDNA with a purity greater than 90%. The purity can be checked by DNA electrophoresis as shown in Fig. 2. The unreacted ssDNA has insignificant impact on force assays as it is quencher-labeled DNA and produces no fluorescence. It does compete with RGD-labeled ITS during surface immobilization though, and somewhat reduces the ITS surface density. If this is a concern, electrophoresis through 7.5% polyacrylamide gel can be performed to further purify RGD-ssDNA (a lagging DNA band, as shown in Fig. 2) by removing unreacted ssDNA (a leading gel band). The RGD-ssDNA gel band is diced and soaked in PBS overnight at room temperature. RGD-ssDNA diffuses to PBS and is extracted by spinning the diced gel solution through a column with a filter.

3.2 ITS Assembly

1. ITS is assembled by hybridizing the upper-strand RGD-ssDNA with the lower-strand DNA at a molar ratio of 1.2:1 in PBS (*see* **Note 3**). Place the mixture at 4 °C overnight.
2. DNA annealing at 90 °C is optional. Aliquot and store the ITS solution at −20 °C for long-term use. A typical ITS stocking concentration is 10 μM.

3.3 ITS Immobilization on Glass Surfaces

The glass surface coatings are performed in a humidity box (e.g., a pipette tip box filled with water under the tip rack) and all the reagents are chilled to around 0 °C on ice with an ice bucket.

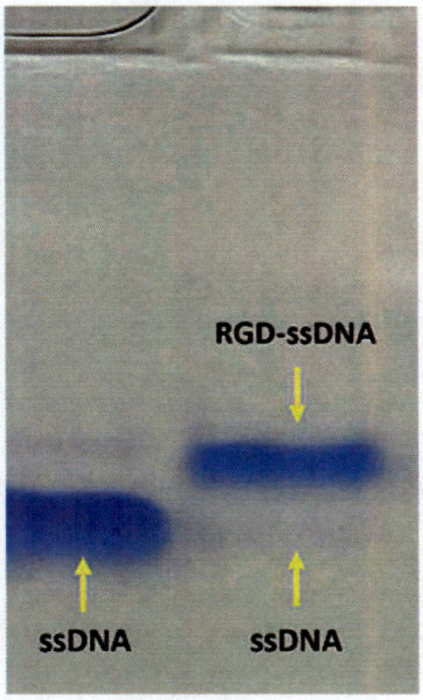

Fig. 2 Evaluating RGD-ssDNA yield by electrophoresis. The well at the left was loaded with the upper-strand DNA without RGD conjugation. The well at the right was loaded with the upper-strand DNA after RGD conjugation. RGD conjugated on the ssDNA slowed down the DNA band in the gel, separating RGD-ssDNA from unreacted ssDNA. It is clear that the yield of RGD-ssDNA is very high as the leading DNA band at the right column is extremely dim. The dark blue color of the DNA bands is from blackhole quencher 2

1. Add 200 μL × 0.1 mg/mL BSA-biotin in PBS into the well of a glass bottom Petri-dish. 2 μg/mL fibronectin (final concentration) is spiked in the BSA-biotin solution to facilitate cell adhesion.
2. Put the Petri-dish in a humidity box and incubate it for 30 min at 4 °C in a refrigerator. BSA-biotin is physically adsorbed on the glass surface.
3. Rinse the glass surface with PBS three times.
4. (Alternative). If a cleaner fluorescence background is desired, PLL-PEG-biotin can be used instead of BSA-biotin for the surface coating. PLL-PEG-biotin is a composite polymer with a PLL (poly-L-lysine) backbone decorated with numerous PEG (poly-ethylene-glycol) branches (20% of the PEG was biotinylated). The PLL facilitates cell adhesion by electrostatic attraction, and the PEG branches repel organic junks. Add 100 μg/mL PLL-PEG-biotin in water to a glass bottom Petri-dish and incubate it for 20 min at 4 °C. PLL-PEG-biotin is physically adsorbed to the surface. Rinse the glass surface with water for three times.
5. Following **steps 1–3** or **step 4**, add 200 μL × 0.05 mg/mL neutravidin in PBS into the Petri-dish well and incubate it for 30 min at 4 °C in a refrigerator.

6. Wash the surface with chilled PBS thrice. Neutravidin binds to biotin and is immobilized to the glass surface. Each neutravidin has four biotin-binding sites. Note: neutravidin can be replaced by streptavidin.
7. Incubate 200 μL × 0.1 μM ITS into the well for 30 min at 4 °C in a refrigerator. For single molecular tension imaging in the real-time mode, 200 μL × 0.01 μM ITS is used instead. Rinse the surface with PBS three times.
8. Leave PBS in the well until cell plating. The biotin tagged ITS is now immobilized on the glass surface by the biotin–neutravidin interaction.

3.4 Cell Plating on the ITS Surface

3.4.1 Cell Culture

HeLa cells are used as an example to demonstrate the cell adhesive force imaging. HeLa cells are cultured in DMEM medium (supplemented with 10% fetal bovine serum, 100 U/mL of penicillin, and 100 μg/mL of streptomycin) recommended by ATCC. Cells are passaged every 3 days. Cells are harvested when confluency reaches above 90%.

3.4.2 Cell Plating

1. To detach HeLa cells, rinse cell culture flask with a mild detaching solution (*see* **Note 4**). Keep 2 mL cell detaching solution in the culture flask and incubate it at 37 °C for 5–7 min.
2. Pipette and centrifuge the detached cells at 100–300 × *g*, and resuspend the cells in serum-free DMEM medium or imaging buffer. The final cell concentration is in the range of 10^5–10^6/mL.
3. Plate 200 μL cell solution on the ITS-coated Petri-dish and incubate it for 1–2 h at 37 °C in an incubator supplied with CO_2.

3.5 Imaging Cell Adhesive Force in a Cumulative Mode

A common epi-fluorescence microscope (Nikon Ti-E) is used for the force imaging in the cumulative mode. Figure 3 is an example image of integrin-transmitted adhesive force generated by HeLa cells. Cell fixation is optional for immunostaining.

1. Take the ITS-coated Petri-dish (with cells plated) from the incubator and mount it onto the fluorescence microscope stage.
2. Choose an objective lens with 40× or 100× magnification. Choose the corresponding filter set for the specific dye on the ITS (Cy5 here) imaging.
3. Perform the imaging with a typical setting such as an exposure time of 1 s and 30% illumination strength.

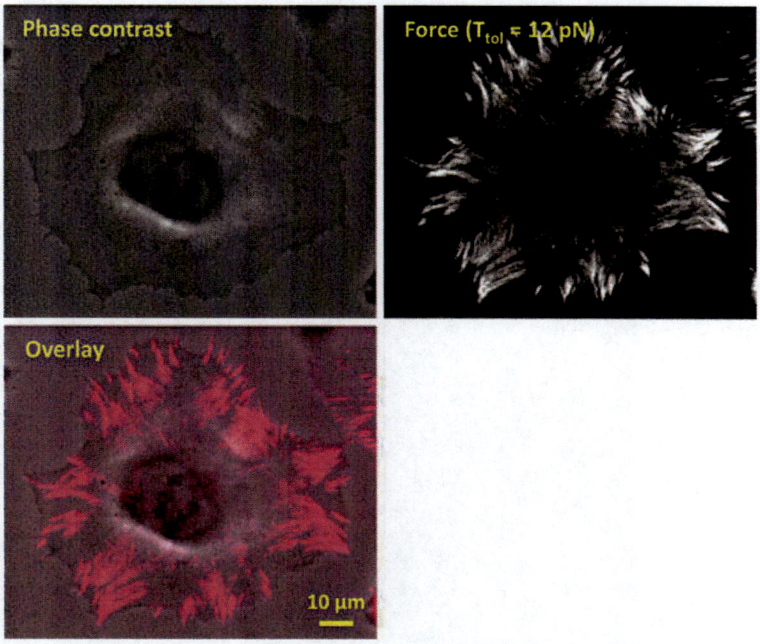

Fig. 3 Cell adhesive force imaging of HeLa cells on ITS ($T_{tol} = 12$ pN) surface in the cumulative mode. The surface is co-coated with 2 μg/mL fibronectin. HeLa cells were incubated on the ITS-coated surface for 1.5 h in a 37 °C incubator with 5% CO_2. Cells were fixed with 4% paraformaldehyde for 15 min at room temperature. The force pattern clearly consists of individual clusters (produced by focal adhesions)

3.6 Imaging Cell Adhesive Force in a Real-Time and Single Molecule Mode

Single molecule cellular force imaging in a real-time mode is performed using a Nikon Ti-E2 TIRF (total internal reflection fluorescence) microscope. Figure 4 is a single molecule force imaging of a HeLa cell in the real-time mode.

1. Prepare an ITS-coated Petri-dish (*see* **Note 5**) and mount it on the microscope stage. Keep 200 μL PBS in the well. Heat the 100× lens to 37 °C with a lens heater (optional).
2. Pre-bleach the surface using 100% laser power until the background approaches dark. Note that the Cy5 dyes in nonactivated ITS are protected from photobleaching by the quenchers.
3. Remove the PBS out from the well and add the cell solution from **step 2** of cell plating immediately. Start recording the video after finding a target cell by navigating the stage.
4. The frame interval of imaging is set to be 0.01–10 s depending on the dynamics of force events of interest, laser power of Cy5 channel is 70% (*see* **Note 6**), and exposure time is 300 ms.

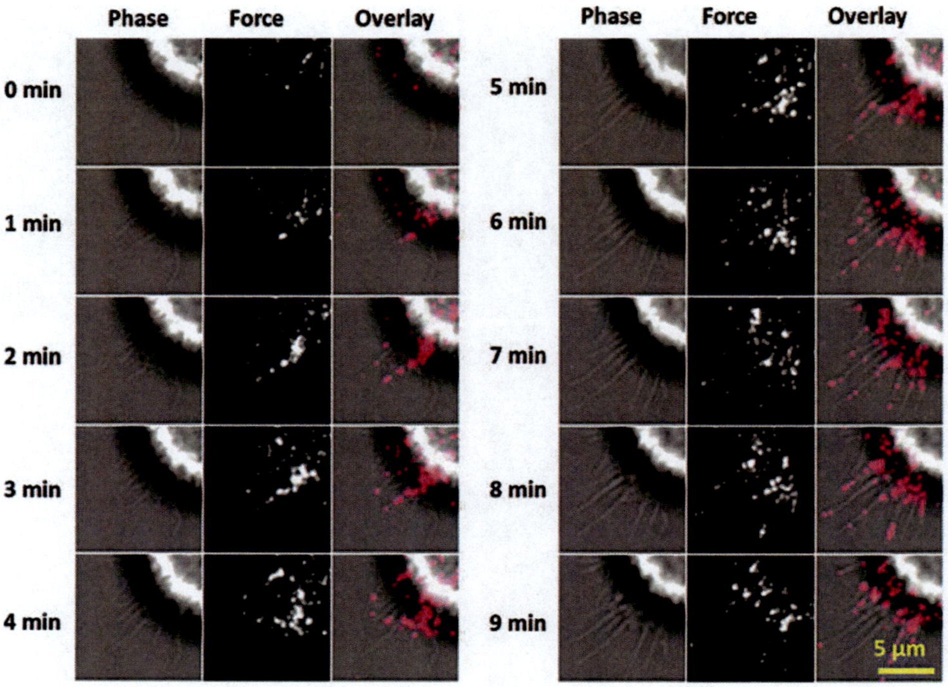

Fig. 4 Imaging single integrin tensions of HeLa cells on an ITS ($T_{tol} = 12$ pN) surface in the real-time mode. Frame interval is 10 s. During the first 10 min after cell plating, HeLa cells form numerous filopodia that elongate and generate adhesive force at the single molecule level. The imaging is recorded using TIRF, laser power is 70%, and exposure time is 300 ms. The magnification of the objective lens is 100×

4 Notes

1. Conjugating the upper DNA strand with an RGD peptide is the only chemical conjugation required for ITS synthesis. RGD is linked to DNA by a heterobifunctional linker (sulfo-SMCC) through amine–NHS ester reaction (NH_2:NHS ester) and Maleimide–thiol reaction (Maleimide:SH) as shown here: RGD-NH_2:NHS ester–Maleimide:SH-ssDNA. The thiol group (–SH) on the DNA needs to be exposed by cleaving ThioMC6-D on the DNA strand using the reducing agent TCEP.
2. **Steps 2–4** in Subheading 3.1 critically determines the purity of yielded RGD-ssDNA. The NHS ester in sulfo-SMCC is highly prone to hydrolysis. Adding RGD-NH_2 solution directly to the sulfo-SMCC vial allows NHS ester to react with RGD-NH_2 in the solution immediately. Sulfo-SMCC has poor solubility in PBS. This is why RGD-NH_2 is dissolved in water instead of PBS.

3. For procedure 3.2, excessive upper strand quencher-ssDNA-RGD is used to ensure that all the dye-ssDNA-biotin is hybridized, so that all dyes are quenched until activated by cell adhesive force.
4. For cell adhesive force assays, always use the mild cell detaching reagent introduced in cell plating. Trypsin should not be used as cell detaching reagent as it damages integrins and cells cannot regenerate lost integrins in a short term.
5. For single molecule force imaging in the real-time mode, PLL-PEG-biotin is preferred over BSA-biotin to immobilize ITS on the glass bottom Petri-dish, as PLL-PEG provides cleaner fluorescence background during imaging.
6. Strong laser power is used in single molecule mode to burst photons from Cy5 of the activated ITS, and meanwhile, to photobleach the dyes shortly. In this manner, the latest single molecular tension events are recorded without fluorescence signals overcrowding on the surface.

Acknowledgments

This work was supported by National Institute of General Medical Sciences (1R35GM128747).

References

1. Zhao Y, Pal K, Tu Y, Wang X (2020) Cellular force nanoscopy with 50 nm resolution based on integrin molecular tension imaging and localization. J Am Chem Soc 142(15):6930–6934
2. Tu Y, Wang XF (2020) Recent advances in cell adhesive force microscopy. Sensors-Basel 20(24)
3. Wang Y, Levince D, Gannon M et al (2018) Force-activatable biosensor enables single platelet force mapping directly by fluorescence imaging. Biosens Bioelectron 100:192–200
4. Wang XF, Ha T (2013) Defining single molecular forces required to activate integrin and notch signaling. Science 340(6135):991–994
5. Wang YL, Zhao YC, Sarkar A, Wang XF (2019) Optical sensor revealed abnormal nuclease spatial activity on cancer cell membrane. J Biophotonics 12(5)
6. Pal K, Zhao YC, Wang YL, Wang XF (2021) Ubiquitous membrane-bound DNase activity in podosomes and invadopodia. J Cell Biol 220(7)
7. Zhao Y, Sarkar A, Wang X (2020) Peptide nucleic acid based tension sensor for cellular force imaging with strong DNase resistance. Biosens Bioelectron 150:111959

Part V

Force Measurement Inside Cells

Check for updates

Chapter 15

Multiplexed Molecular Tension Sensor Measurements Using PIE-FLIM

Lukas Windgasse and Carsten Grashoff

Abstract

Genetically encoded Förster Resonance Energy Transfer (FRET)-based tension sensors were developed to enable the quantification of piconewton (pN)-scale forces that act across distinct proteins in living cells and organisms. An important extension of this technology is the multiplexing of tension sensors to monitor several independent FRET probes in parallel. Here we describe how pulsed interleaved excitation (PIE)-fluorescence lifetime imaging microscopy (FLIM) can be implemented to enable the analysis of two co-expressed tension sensor constructs. Our protocol covers all essential steps from biosensor expression and live cell PIE image acquisition to lifetime calculations.

Key words Förster Resonance Energy Transfer (FRET), Pulsed interleaved excitation fluorescence lifetime imaging microscopy (PIE-FLIM), Tension sensor, Molecular forces, Mechanotransduction

1 Introduction

The ability of cells to sense and respond to mechanical signals is central to many biological processes and regulated by a wide range of macromolecular structures [1–3]. A number of methods have been established to elucidate the underlying mechanobiological processes on different time and length scales [4, 5]. Förster Resonance Energy Transfer (FRET)-based tension sensors were developed to determine whether distinct proteins experience mechanical forces, to visualize where and when such molecular forces occur within cells, to quantify the magnitude of force per molecule, and to determine how many molecules of a given protein population are exposed to mechanical tension [5–9]. The technology has been utilized to study the molecular mechanics of an increasing number of mechanosensitive proteins including cell adhesion molecules like talin [7, 10, 11], vinculin [6, 12–14], metavinculin [9], and des-

Ronen Zaidel-Bar (ed.), *Mechanobiology: Methods and Protocols*,
Methods in Molecular Biology, vol. 2600, https://doi.org/10.1007/978-1-0716-2851-5_15,

moplakin [15], transmembrane receptors such as cadherins [16, 17] and integrin receptors [18], but also cytoskeletal elements like spectrin at the plasma membrane [19] or nesprin at the nuclear membrane [20, 21]. As tension sensors are genetically encoded, they can be applied to investigate processes within living cells and even in intact model organisms such as flies [22], worms [19], frogs [23], fish [24], and mice [25]. The technology has thus become an important tool to study mechanobiological processes on molecular scales.

The tension sensor method is based upon the so-called tension sensor module, a FRET pair of fluorophores that are connected by a mechanosensitive linker peptide. This linker element is designed to unfold or extend at a distinct force range leading to the separation of the donor and acceptor fluorophore and the reduction or loss of FRET. The currently available, single-molecule calibrated tension sensor modules – using different linker elements with distinct force response thresholds – allow force measurements at about 1–11 pN [5]. Importantly, the tension sensor module has to meet a number of biophysical properties including folding reversibility, loading rate insensitivity, and lack of hysteresis to allow an unambiguous interpretation of the FRET data. The relevance of these different parameters has been reviewed in detail before and we would like to refer the interested reader to the existing literature [5, 26–28].

1.1 Tension Sensor Multiplexing

Mechanosensitive proteins execute their biological function in the context of a specific macromolecular environment. Therefore, it is important to develop multiplexing technologies that enable the investigation of more than just one protein at a time. The first tension sensor multiplexing study applied two previously developed, orthogonal FRET probes (mTFP1–ShadowG and LSSmOrange–mKate2), which can be excited with the same wavelength at 440 nm [29]. Owing to the dark acceptor fluorophore (ShadowG) in the first and a long stokes-shift donor fluorophore (LSSmOrange) in the second FRET pair, emission spectra can be separated, allowing a FRET analysis by fluorescence lifetime imaging microscopy (FLIM). This strategy enabled the multiplexed investigation of two different talin tension sensors and revealed an intramolecular force gradient along the talin molecule [11]. The limitations of the technique are the suboptimal transfer efficiencies of the LSSmOrange-based FRET probe and the presence of minimal but still detectable photon contamination in the red-shifted donor channel. We here introduce a tension sensor multiplexing strategy that is based on pulsed interleaved excitation (PIE)-FLIM. This approach allows the use of efficient FRET pairs and overcomes the problem of photon contamination in the various detection channels.

1.2 Pulsed Interleaved Excitation (PIE)

The underlying principle of PIE is to excite a sample expressing, for example, two fluorescent constructs with two pulsed lasers that are synchronized on the nanosecond time scale [30, 31]. The second laser pulse is delayed to the first resulting in a sequence of interleaved excitations, which are coupled to equally synchronized detectors that record the fluorescence emission signals in separate time and emission windows. By adjusting the laser repetition rate so that the first fluorophore emission signal is completely decayed before the second fluorophore is excited, imaging artefacts by fluorescence crosstalk can be abolished (Fig. 1). The combination with FLIM allows the detection of distinct fluorescence lifetime decays and thus quantitative FRET imaging of co-expressed probes. The PIE-FLIM technique can therefore be utilized to investigate different biological activities at once, for example the activation of distinct signaling cascades within the same cell and at the same time [32]. In the following, we will detail how PIE-FLIM can be employed to multiplex two co-expressed tension sensors and enable parallelized molecular force measurements in living cells (Fig. 2).

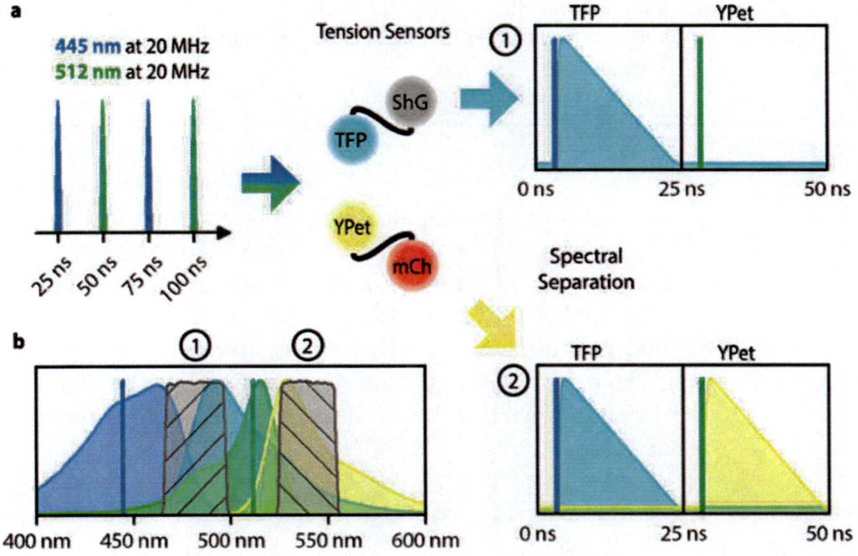

Fig. 1 Schematical depiction of a tension sensor multiplexing experiment using PIE-FLIM. (**a**) A train of interleaved excitation pulses at 445 nm (blue) and 512 nm (green) is used to excite a sample containing two tension sensor constructs, mTFP1–ShadowG (TFP-ShG) and YPet–mCherry (YPet–mCh). Each laser is pulsed at 20 MHz resulting in an overall repetition rate of 40 MHz. Right: The emission signals are recorded at distinct time and emission windows. The first detector [1] records the emission signal of mTFP1, using a 480/30 bandpass filter, 0–25 ns after the first laser pulse. The second detector records the YPet signal, isolated with a 540/30 bandpass filter, at 25–50 ns. Owing to the broad emission signal of mTFP1, detector 2 may still collect photons emitted by mTFP1. However, only the YPet signal at 25–50 ns will be considered for analysis. (**b**) Excitation and emission spectra of mTFP1 and YPet shown with excitation laser lines and employed filter sets. Blue line: 445 nm laser excitation, green line: 512 nm laser excitation, blue spectrum: mTFP1 excitation, cyan spectrum: mTFP1 emission, green spectrum: YPet excitation, yellow spectrum: YPet emission, filter 1: 480/30 ET bandpass filter, filter 2: 540/30 ET bandpass filter

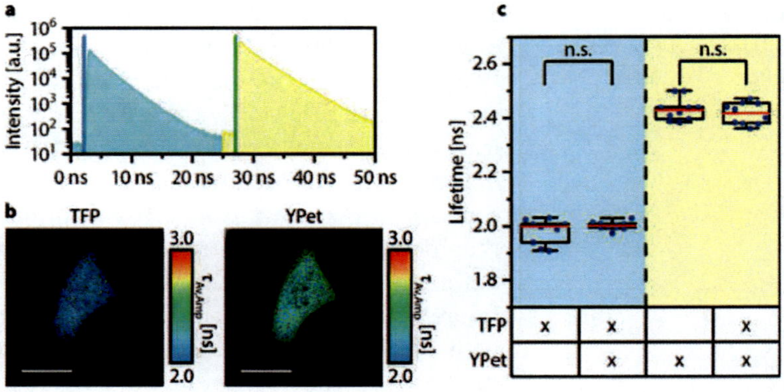

Fig. 2 Exemplary PIE-FLIM tension sensor data set. (**a**) Distribution of photon arrival times from cells co-expressing the two cytosolic tension sensor constructs, mTFP1–ShadowG and YPet–mCherry. Signals were plotted after setting time gates, thresholding, and drawing ROIs. (**b**) Representative donor lifetime images of a cell co-expressing a cytosolic mTFP1–ShadowG (TFP) tension sensor and a cytosolic YPet–mCherry sensor (YPet). Left: Cell fit to the mTFP1 lifetime. Right: Same cell fit to the YPet lifetime. Scale bar indicates 20 μm. (**c**) Calculated, amplitude-weighted average lifetimes from cells expressing either one tension sensor construct or both ($n = 10$). Red lines indicate the median value; n.s. indicates a significance value of $p > 0.05$ in the Kolmogorov-Smirnov test

2 Materials

All solutions should be kept sterile and handled in a biological safety cabinet. Waste products must be decontaminated or autoclaved before disposal to inactivate biological activities. Please note that the following protocol is optimized for work in mouse epidermal keratinocytes (MEK) and the use of mTFP1–ShadowG and YPet–mCherry-based tension sensors. Thus, adjustments may be necessary when other cell types or fluorophore combinations are used.

2.1 Transient Expression of Tension Sensor Constructs

1. Transfection Reagent Lipofectamine 3000 and P3000 Reagent.
2. Opti-MEM Reduced Serum Medium with GlutaMAX.
3. Keratinocyte Growth Medium (FAD medium): DMEM/Ham's F12 (ratio 3.5:1.1) without L-glutamine and sodium pyruvate, supplemented with 50 μM Ca^{2+}, 3.096 g/L $NaHCO_3$, 10% fetal bovine serum (FBS) (Ca^{2+}-depleted by Chelex treatment), 2 mM GlutaMAX, 1 mM sodium pyruvate, 0.18 mM adenine, 0.5 μg/mL hydrocortisone, 5 μg/mL insulin, 10 ng/mL EGF, 100 pM cholera toxin, 0.5% penicillin/streptomycin (P/S).

4. Purified cDNA of the respective expression plasmids at a total amount of 1 μg.

2.2 Stable Expression of Tension Sensor Constructs

1. Gryphon packaging cell line.
2. Growth medium: DMEM containing 4500 mg/L glucose, GlutaMAX, and pyruvate supplemented with 10% FBS and 1% P/S.
3. FAD medium (see above).
4. 2× HBS buffer, pH 7.0: 280 mM $CaCl_2$, 50 mM HEPES, 1.5 mM Na_2HPO_4; sterile filtered.
5. 2 M $CaCl_2$ in ultrapure water; sterile filtered.
6. Purified cDNA of tension sensor expression plasmids at a total amount of 20 μg.
7. Chloroquine diphosphate salt: 50 mM in PBS; sterile filtered, aliquots stored at −20 °C.
8. Polybrene Infection Reagent; aliquots stored at −20 °C.
9. Syringes (10 mL) and sterile filters (0.22 μm pore size; PES membrane).
10. Antibiotics for selection of transduced cells (depends on the utilized expression vector).

2.3 Live Cell Microscopy and Image Acquisition with PIE-FLIM

1. FAD medium without Phenol Red.
2. Immersol G.
3. Imaging dishes: μ-Dish, 35 mm.
4. Zeiss LSM 880 confocal laser scanning microscope equipped with a suitable live cell imaging objective (e.g., LD LCI Plan-Apochromat 63×, 1.2 Imm Korr DIC M27) and integrated PicoQuant FLIM module (PIE compatible): Laser Drivers: Sepia PDL 828-S, LED heads: LDH-D-C-440 and LDH-D-C-510, Photon Counting Detector: PMA Hybrid 40, TCSPC and MCS Electronics: MultiHarp 150 4 N, TimeHarp 260 PICO Dual.
5. Laser-beam splitter HC BS R514, bandpass filter for the mTFP1–ShadowG construct: 480/30 ET, and a bandpass filter for the YPet–mCherry construct: 540/30 ET.
6. Software to operate laser scanning microscope: ZEISS ZEN imaging software (black edition).
7. Software to operate FLIM module (SymPhoTime 64); hand-shake plug-in for ZEN software.

2.4 Data Analysis and Evaluation

1. SymPhoTime 64 software.
2. Microsoft Excel software.

3 Methods

The following protocols are optimized for MEK cell culture. For cell expansion and single cell imaging, cell cultures are maintained in medium with low Ca^{2+} levels. Increasing the Ca^{2+}-concentration induces keratinocyte differentiation and, for example, formation of cell–cell junctions. MEKs are grown on collagen type I coated dishes at 37 °C and 5% CO_2. Here, we describe two protocols for tension sensor expression (transient transfection and transduction), which can be combined as needed to express the FRET probes of choice.

3.1 Transient Transfection of Tension Sensor Constructs

The following procedure describes transfection in a 35 mm μ-dish. Before starting, FAD medium and Opti-MEM should be warmed to 37 °C.

1. Seed cells 16–24 h prior to transfection in a 35 mm μ-dish at a density of 3–3.5 × 10^4 cells/cm^2.
2. Prepare two 1 mL reaction vials, each containing 100 μL Opti-MEM. Add 2 μL of LF3000 to the first tube (*see* **Note 1**); add 1 μg of each plasmid and 4 μL of P3000 reagent to the second tube. Vortex both tubes well. Next, add the LF3000 mix from the first tube to the DNA mixture in the second tube and vortex again. Incubate the mixture 15–20 min at room temperature for liposomes to form. Do not shake tubes afterward.
3. Change medium in the μ-dish and add the transfection mix dropwise to the cells. Gently rock the dish to achieve an even distribution of transfected cells. Incubate cells 4–6 h, then renew medium with fresh FAD medium (*see* **Note 2**). Cells can be analyzed 24–48 h after transfection (*see* **Note 3**).

3.2 Transduction of Tension Sensor Constructs

3.2.1 Virus Production

Transfection efficiencies are often cell type-dependent and sensitive to plasmid size (with less efficient transfection rates for large cDNA constructs). In such cases, transduction of expression plasmids is recommended. The following protocol uses Gryphon cells for the production of retroviral particles. If not stated otherwise, Gryphon cells are cultivated in DMEM-based medium at 37 °C and 5% CO_2. For calcium phosphate coprecipitation, ultrapure water, 2× HBS, and 2 M $CaCl_2$ can be used and stored at room temperature.

1. Seed approximately 4–5 × 10^6 Gryphon cells onto a 10 cm dish and allow cells to adhere by overnight incubation. To ensure a high transfection efficiency, it is recommended that cells are 80% confluent at the time of transfection.
2. Before transfection, aspirate old medium from Gryphon cells and add 8 mL of fresh medium (DMEM supplemented with

10% FBS and 1% P/S, warmed to 37 °C) containing 25 μM chloroquine.

3. In a 15 mL conical tube, dilute 20 μg of plasmid DNA in ultrapure water to a total amount of 439 μL, then add 61 μL of 2 M $CaCl_2$ and vortex.
4. While vortexing, add 500 μL of 2× HBS dropwise to the DNA/$CaCl_2$ mixture.
5. Add the transfection mixture dropwise onto Gryphon cells. Gently rock the dish to achieve a homogenous distribution of the transfection mixture. If the procedure is successful, small precipitates are visible in the medium at 20–40× magnification. Transfer cells to an incubator.
6. After 4–6 h, replace medium with fresh DMEM-based medium and incubate cells at 37 °C and 5% CO_2 overnight.
7. The next evening, replace medium on Gryphon cells with 8 mL FAD medium (*see* **Note 4**).
8. 48 h after transfection, collect virus-containing supernatant in a 15 mL conical tube and add 8 mL of fresh FAD medium to the cells. Incubate the cells at 37 °C and 5% CO_2.
9. Add 8 μL polybrene to the collected supernatant and sterile filter the solution. This virus–polybrene mixture can be directly used for transduction (see below). Alternatively, virus-containing supernatant can be stored after filtering but without polybrene at −80 °C (*see* **Note 5**).
10. Harvest the supernatant every 12 h (morning/evening) up to 4 days post transfection.

3.2.2 Transduction

1. Seed MEKs at a density of about 1.5–2 × 10^4 cells/cm^2 onto a 6 cm or a 10 cm dish one day before the first transduction (*see* **Note 6**).
2. On the day of transduction, aspirate medium and add the virus polybrene mixture to the target cells; for a 6 cm dish add 2–3 mL; for a 10 cm dish add 7–8 mL. Incubate cells at 37 °C and 5% CO_2.
3. After 8–12 h, aspirate medium and add fresh FAD medium (*see* **Note 7**).
4. Fluorescence should be visible 24 h after transduction and selection of transduced cells can be started. For selection, we routinely use puromycin at a concentration of 1 μg/mL, hygromycin at 20 μg/mL, and G418 at 30 μg/mL.
5. For the transduction of the second sensor construct, repeat **steps 1–4** on selected cells (*see* **Note 8**).

3.3 Live Cell Microscopy and Image Acquisition with PIE-FLIM

Prior to force measurements, tension sensors need to be thoroughly evaluated, for example to ensure that protein functionality is not impaired. Such experiments may involve rescue experiments in knockout cells, immunostainings, or FRAP analyses; be also sure to include FRET control measurements as described previously [28]. Record data of cells expressing donor-only controls, no-force FRET controls, and tension sensors at the same experimental day and aim for 3–5 independent experimental days. To study molecular force responses, always compare tension-sensor expressing cells to identically treated no-force FRET controls. Note again that the settings detailed here are suitable for multiplexing mTFP1–ShadowG and YPet–mCherry-based biosensors.

3.3.1 Microscope and Software Setup

1. Aspirate medium from the cell culture imaging dish (μ-dish) containing transduced/transfected cells and add 1 mL of imaging medium.
2. Mount the μ-dish onto the microscopy stage. To prevent focus drift during the measurement, allow the system to equilibrate for 10–15 min.
3. Start the ZEN software (black edition) and make the following adjustments in the program.
4. Go to the "Acquisition" tab. Tick the "Show all Tools" option in the upper-left corner of the software window in the "Experiment Manager" field.
5. Use the smart setup option to set the laser and detection parameters in the ZEN software. In the pop-up window, choose "ECFP" for mTFP1 and "EYFP" for YPet; select the acquisition style "Fastest" (*see* **Note 9**). This setting leads to the simultaneous excitation with a 458 and 514 nm laser line. The emission is collected in the range of 463–509 nm for mTFP1 and in the range of 519–620 nm for YPet–mCherry. For the excitation, an MBS 458/514 filter is used in our setup.
6. Tick the "Show all" option in the "Imaging Setup" tab.
7. Add a new track for the FLIM measurement by clicking on the "+" button. Ensure that "Switch track every" stays on the option "Frame." Subsequently, perform the following changes in "Track 2": set "Mirror" to "Plate" in the "Fiberout" path, this will direct the emission signal to the external FLIM detectors. Change the filter in the "Visible Light" pathway to "Plate" and add a suitable excitation filter in the "Invisible Light" pathway. For the constructs used here, the MBS 440/510/640 filter module is an appropriate choice.
8. Set the parameters in "Acquisition Mode" as follows: Scan Mode: "Frame," Frame Size: "512" "X*Y" "512," Speed: "4" (resulting in a pixel dwell time of 16.38 μs) (*see*

Note 10), Averaging: Number "1," Bit Depth: "8 bit," Zoom: "2.2" (leading to an image size of 61.3 μm × 61.3 μm and a pixel size of 0.12 μm) (*see* **Note 11**).

9. Open the "Channels" tab. There should be two tracks visible: "Track 1" showing channels labeled as "ECFP" and "EYFP." The excitation lasers "458" and "512" should be pre-ticked. Adjust the laser intensities for fluorescence excitation (*see* **Note 12**). "Track 2" does not need to be configured, the settings will be overridden by the SymPhoTime software during the FLIM recording.
10. Tick the "Time Series" option in the "Experiment Manager" field. A "Time Serie" window will pop up. Adjust the number of cycles to set how many consecutive images will be acquired for one FLIM recording (*see* **Note 13**). In "Interval," choose "0.0" to allow rapid image acquisition.
11. Start the SymPhoTime software and make the following adjustments in the program.
12. Set a TCSPC Resolution of "80.0 ps" and use the TCSPC Mode "T3." For "Sync," use a "Trigger Level" of "−75 mV," set "Trigger Edge" to "Falling," "Offset" to "74,000 ps," and "Sync divider" to "8." For "Channel," set "Trigger Level" to "−100 mV," "Trigger Edge" to "Falling," and "Offset" to "0 ps."
13. Change the parameters for the "Laser Settings" as follows: switch "Standard" to "PIE," set "Pulse Rate" to "40 MHz" (*see* **Note 14**), activate wavelengths "445.0 nm" and "512.0 nm," and set their intensities to "67%" (*see* **Note 15**).
14. Go to "File" and create a "New Workspace."
15. Start "ZEN PicoQuant Application."
16. In the new pop-up window, the option to name files which are saved in the workspace is given. Choose a favored group name for experiments, the measured data will be listed to a folder according to that name. "File Name" gives you the option to name files separately.

3.3.2 FLIM Measurement Routine

1. Choose a cell for measurement. Use the "Locate" option in the ZEN software and find a suitable cell. Go to "Acquisition" and only enable "Track 1" in the "Channels" tab. Press the "Live" button to confirm by live cell imaging that both tension sensor constructs are expressed at suitable levels (*see* **Note 16**). Adjust the focus but avoid unnecessary exposure of the sample to laser light to minimize photo-bleaching.
2. To acquire a FLIM image, disable "Track 1" and enable "Track 2" in the "Channels" tab. Next, click on "Track 2" in the "Imaging Setup" tab. Note that the ZEN software will

automatically change the "Fiberout" from "Plate" to "Mirror" when tracks are changed; it is necessary to adjust "Mirror" to "Plate" before each FLIM measurement.

3. Press the "Measurement" button in the "PicoQuant FLIM + FCS tab" (this window was opened by starting the ZEN PicoQuant Application) to start a FLIM acquisition. The parameters set in the "Acquisition Mode" tab of the ZEN software will be now used by the SymPhoTime software.
4. A FLIM image together with the lifetime decay curves should start to build up in SymPhoTime software. After acquisition, the data will be automatically saved as a .ptu file.
5. Repeat **steps 1–4** until a sufficient number of images is recorded (*see* **Note 17**).

3.4 Data Analysis and Evaluation

Data can be directly analyzed using the SymPhoTime 64 software. In short, the channel with the signal of interest is chosen followed by thresholding of the image to remove background noise. The histogram of accumulated lifetimes is then fit by a mono- or multi-exponential tail fit. Start the analysis with the donor-only constructs since the median of its mono-exponential lifetime is used as a parameter for the multi-exponential lifetime fits of the no-force FRET control and the tension sensor sample.

3.4.1 Fitting Fluorescence Lifetimes Using the SymPhoTime Software

1. Open the SymPhoTime software. Click on "File" and choose "Open Workspace" to load the folder containing the relevant FLIM recording. The subfolders with chosen group names are listed. Click the "+" in front of the folder name to open the .ptu file. Press the "+" in front of the .ptu file to open its substructure. The files containing the signal from each detector will end on "OFLIM.pqres" (in our setup the signal from the mTFP1–ShadowG construct) and "OFLIM_1.pqres" (YPet–mCherry construct) (*see* **Note 18**).
2. To start the analysis, double click the "OFLIM.pqres" file to load the recorded data.
3. First, double click into the field showing two decay curves in the "FLIM" table. A small window pops up to set the time gate. Type either "313 Chnls" or "25.04 ns" and choose the left part of the decay curve by clicking on it. Confirm changes by pressing "OK." Now, click "Calculate FastFLIM." This updates your "Analysis" window, and the signal collected in the first channel (mTFP1–ShadowG) will be shown.
4. Click on the "Region of Interest" tab and tick "Use ROI," then press "ROI from threshold." A new window pops up where thresholding can be adjusted. By choosing "Min" "0" and "Max" "1" (*see* **Note 19**) below the "Events[Cnts]" field, a binary image is generated. In this view, the pixels with an

intensity value below the threshold value are depicted in gray. Start increasing the "Lower Threshold (Int)" by increments of 0.1% until the desired signal (e.g., signal of the subcellular structure of interest with minimal inclusion of background noise) is achieved. Confirm by pressing "OK."

5. In case additional signal needs to be excluded from the ROI, for example to evaluate potential, subcellular differences, right click into the FLIM image and choose one of the ROI tools. For example, pixels in a hand-drawn area are excluded when using "free ROI," as long as <Ctrl> on the keyboard is pressed. If more than one ROI is needed, click on "New" in the "Region of Interest" tab to generate another ROI. The thresholding and ROI isolation procedure needs to be completed for every image.
6. In the "Decay Fitting" tab, choose "n-Exponential Tailfit" as "Fitting Model" and set "Model Parameters": "n" to "1" for mono-exponential donor-only controls and "2" for all other constructs, which are expected to follow a bi-exponential decay. In the donor-only case, press "Initial fit," which initializes a global optimization of parameters, and then press "Fit All" for a local optimization. For data of the no-force FRET control and the tension sensor sample, untick τ_1 to set the median donor-only lifetime as a fixed parameter, then proceed as before. Note that the χ^2-value as a parameter for the goodness of the fit should ideally be around 1. Make sure that all data are fit with sufficiently high quality; however, do not use more model parameters than physically justified to improve this value. Data with poor fit quality should be discarded.
7. For further calculations, either create a .dat file using "Export as ASCII" or copy the fitting results with "Export to Clipboard" and transfer it to a spreadsheet, for example in Microsoft Excel.
8. Repeat **steps 3–7** for data of the YPet–mCherry construct ("OFLIM_1.pqres"). Make sure that the respective lifetime histogram (25.04–50 ns) is chosen in **step 3** for the subsequent data analysis.

3.4.2 Calculation of FRET Efficiencies

1. Open the spreadsheet containing the fitting parameters of fluorescence lifetime decays, generated above. The central variables are τ_1 (corresponding to the donor-only lifetime, τ_D) for the donor-only control and $\tau_{Av\ Amp}$ (corresponding to the donor lifetime in the presence of the acceptor, τ_{DA}) for the no-force FRET control as well as the tension sensor sample. $\tau_{Av\ Amp}$ is defined as the averaged lifetime (τ) weighted by the contributing amplitudes (A) (*see* Eq. 1).

$$\tau_{Av\ \mathrm{Amp}} = \frac{1}{A_{\mathrm{Sum}}} \sum_{k=0}^{n-1} \mathrm{A}_k \tau_k \tag{1}$$

2. Determine the median donor-only lifetime ($\tilde{\tau}_{\mathrm{D}}$) (*see* **Note 20**).
3. Calculate the FRET efficiency E_{FRET} according to Eq. 2.

$$E_{\mathrm{FRET}} = 1 - \frac{\tau_{DA}}{\tilde{\tau}_D} \tag{2}$$

4. To visualize the data, plot FRET efficiencies for the no-force FRET control and tension sensor measurements in boxplots, using suitable programs such as Origin.
5. Determine whether potential differences between both data sets are statistically significant, for example by using the Kolmogorov-Smirnov test (*see* **Note 21**).

3.4.3 Data Interpretation

1. The FRET efficiency of the no-force FRET control is expected to be higher or equal to the FRET efficiency of the tension sensor (*see* **Note 22**). A decrease in the FRET efficiency of the tension sensor sample would constitute a first indication for mechanical tension being born by the protein of interest. However, further experiments are required to ensure that a potential decrease in FRET efficiency is indeed force-specific [28]. Consider the following control experiments and always compare cells expressing the tension sensor construct with cell expressing the no-force FRET control. Note that changes in the no-force FRET control are indicative of non-specific effects.
2. If the protein of interest is known to undergo significant conformational changes, test whether inducing distinct conformational states will impact the observed FRET efficiencies in tension sensor and control samples.
3. Use chemical inhibitor, knockout, or overexpression approaches to modulate molecular tension across the protein of interest.
4. Apply mechanical stress externally, for example by using microneedles or cell stretching. Evaluate effects on tension sensor and control samples.
5. Consider point mutations that could be inserted into the protein of interest to manipulate molecular forces. To control such experiments, insert mutations into the tension sensor and control constructs and evaluate effects on both samples.

6. Compare tension sensor modules with different force sensitivities.
7. Consider the information from all control experiments.

4 Notes

1. The Opti-MEM and LF3000 mixture can be upscaled for a master mix.
2. Changing the medium at this point is particularly recommended when MEKs are used. The addition of Opti-MEM leads to increased Ca^{2+} levels that will induce MEK differentiation when the medium is not replaced. The medium changing step can be omitted when working with cell lines that are insensitive to high Ca^{2+} concentrations, such as fibroblasts.
3. Many expression vectors use strong promotors, which often lead to high biosensor expression in transient transfections. Note that this can result in overexpression artefacts and complicates microscopy as well as data interpretation. To increase the number of cells with low protein expression, it may be helpful to transfect cells in a 6-well plate (same protocol can be used) and transferring these cells one day after transfection onto a μ-dish. Subsequent cycles of cell division will reduce the copy number of plasmids per transfected cell, leading to lower expression levels.
4. Note that the FAD medium is used in this protocol because MEKs cannot tolerate DMEM medium; Gryphon cells can be maintained in FAD medium for virus production although their appearance does change. The use of FAD medium may be omitted if cells other than MEK are to be transduced.
5. Since the virus titer drops with every freeze and thaw cycle, aliquot virus-containing supernatant in appropriate volumes, for example 7–8 mL for transducing a 10 cm dish of cells or 2–3 mL for a 6 cm dish of cells.
6. Depending on the experimental goals, it may be helpful to use lower cell densities. The isolation of clonal colonies, for instance, may be facilitated if lower cell numbers are used.
7. Tension sensors experiments should aim at physiological expression levels. If the protein expression level of the biosensors should turn out to be too low, which is often the case for large constructs (>250 kDa), cells can be transduced multiple times. Lower expression levels of the tension sensor construct can be obtained by diluting the virus-containing supernatant with medium before addition to the target cells.

8. An alternative procedure is to transduce the second construct 24 h after the first one and select double expressing cells by treating target cells with two antibiotics at the same time.
9. The fastest acquisition mode is used to check quickly whether both sensor constructs are expressed at appropriate levels, without exposing the sample to unnecessary laser light and thereby induce photobleaching. Note that this setting allows for significant bleed-through of mTFP1 emission into the YPet detection channel.
10. Changing the acquisition speed will influence the overall amount of collected photons. Longer pixel dwell times will lead to an increased number of photons at the potential cost of increased photobleaching. Shorter dwell times minimize photobleaching but will result in lower photon counts. We recommend to evaluate the degree of photobleaching at the onset of the experiment and adjust the pixel dwell time to maximize the signal-to-noise ratio.
11. Adjust the zoom factor according to the subcellular structure of interest. As a general rule, try to focus on the signal of interest and exclude background signals.
12. Even though FLIM measurements are largely insensitive to fluorescence intensities, we recommend to use the same laser excitation values throughout the experiment. This ensures that similarly bright cells and therefore samples with comparable expression levels are investigated.
13. The number of cycles will depend on the signal intensity and subcellular structure of interest. In our experience, 15 cycles will collect sufficient amounts of photons when working with cell adhesion proteins that are expressed at physiological expression levels. Acquire sufficient photon numbers and ensure that the fluorescence lifetime data can be fit with high quality.
14. Depending on the experimental needs, the pulse rate can be adjusted. However, be sure to give both fluorophores sufficient time to decay their fluorescent signal before a new laser pulse is generated. As a general rule, the decay time window should be at least four times longer than the fluorescence lifetime of the donor fluorophore.
15. Intensities can be adjusted depending on the experimental needs. However, laser intensities should result in comparable photon numbers for both tension sensor constructs. To ensure reproducibility, do not change the laser intensities in between measurements. Moreover, ensure that the photon count rate does not exceed 10% of the laser repetition rate to avoid pileup effects and incorrect fluorescence lifetime calculations.

16. We strongly discourage including overexpressing cells into tension sensor measurements. Even though it may be tempting to record cells that provide high photon counts, note that overexpressing cells do not represent the physiological condition.
17. It is strongly recommended to analyze three constructs in parallel, as described earlier [28]: the tension sensor construct, the donor-only control, and the no-force FRET control. Acquire at least 10–15 images per construct and perform the experiment on at least 3 independent days to obtain at least 30–45 data points per construct and condition. Prepare fresh samples for each experimental day.
18. The number of OFLIM files will increase with the number of connected FLIM detectors. Take care to choose the correct files.
19. In some images the "Min" value is preset to a value greater than "0"; this happens when at least one photon is acquired in the majority of pixels. In this case, keep the preset "Min" value and use this value plus one for "Max," otherwise the image may appear devoid of any signal.
20. The median lifetime is used because it is less sensitive to outliers than the mean lifetime. Nevertheless, median and mean values should be similar in reproducible experiments.
21. Note that the collected data may not follow a Gaussian distribution. The Kolmogorov-Smirnov test does not require a normal distribution of data points and is often a suitable statistical test for evaluating FRET tension sensor data.
22. In some cases, the FRET efficiency of the no-force FRET control can be lower than the FRET efficiency of the tension sensor sample. This may occur, for instance, when N- or C-terminal tension sensor fusions or protein truncations are used as no-force FRET controls and the overall tension on the protein of interest is low. For certain tension sensor modules, such an effect has been also attributed to compression of the sensor module. Be sure to evaluate this particular effect with suitable control experiments, which may include different tension sensor modules and the use of more than just one no-force FRET control.

References

1. Hoffman BD, Grashoff C, Schwartz MA (2011) Dynamic molecular processes mediate cellular mechanotransduction. Nature 475(7356):316–323. https://doi.org/10.1038/nature10316
2. Petridou NI, Spiro Z, Heisenberg C-P (2017) Multiscale force sensing in development. Nat Cell Biol 19(6):581–588. https://doi.org/10.1038/ncb3524
3. Kanoldt V, Fischer L, Grashoff C (2019) Unforgettable force - crosstalk and memory of mechanosensitive structures. Biol Chem 400(6):687–698. https://doi.org/10.1515/hsz-2018-0328

4. Roca-Cusachs P, Conte V, Trepat X (2017) Quantifying forces in cell biology. Nat Cell Biol 19(7):742–751. https://doi.org/10.1038/ncb3564
5. Fischer LS, Rangarajan S, Sadhanasatish T et al (2021) Molecular force measurement with tension sensors. Annu Rev Biophys 50:595–616. https://doi.org/10.1146/annurev-biophys-101920-064756
6. Grashoff C, Hoffman BD, Brenner MD et al (2010) Measuring mechanical tension across vinculin reveals regulation of focal adhesion dynamics. Nature 466(7303):263–U143. https://doi.org/10.1038/nature09198
7. Austen K, Ringer P, Mehlich A et al (2015) Extracellular rigidity sensing by Talin isoform-specific mechanical linkages. Nat Cell Biol 17(12):1597–1606. https://doi.org/10.1038/ncb3268
8. Freikamp A, Cost A-L, Grashoff C (2016) The Piconewton force awakens: quantifying mechanics in cells. Trends Cell Biol 26(11): 838–847. https://doi.org/10.1016/j.tcb.2016.07.005
9. Kanoldt V, Kluger C, Barz C et al (2020) Metavinculin modulates force transduction in cell adhesion sites. Nat Commun 11(1):6403. https://doi.org/10.1038/s41467-020-20125-z
10. Kumar A, Ouyang M, van den Dries K et al (2016) Talin tension sensor reveals novel features of focal adhesion force transmission and mechanosensitivity. J Cell Biol 213(3): 371–383. https://doi.org/10.1083/jcb.201510012
11. Ringer P, Weißl A, Cost A-L et al (2017) Multiplexing molecular tension sensors reveals piconewton force gradient across Talin-1. Nat Methods 14(11):1090–1096. https://doi.org/10.1038/nmeth.4431
12. Hernández-Varas P, Berge U, Lock JG et al (2015) A plastic relationship between vinculin-mediated tension and adhesion complex area defines adhesion size and lifetime. Nat Commun 6:7524. https://doi.org/10.1038/ncomms8524
13. Rothenberg KE, Scott DW, Christoforou N et al (2018) Vinculin force-sensitive dynamics at focal adhesions enable effective directed cell migration. Biophys J 114(7):1680–1694. https://doi.org/10.1016/j.bpj.2018.02.019
14. Chang C-W, Kumar S (2013) Vinculin tension distributions of individual stress fibers within cell-matrix adhesions. J Cell Sci 126(Pt 14): 3021–3030. https://doi.org/10.1242/jcs.119032
15. Price AJ, Cost A-L, Ungewiß H et al (2018) Mechanical loading of desmosomes depends on the magnitude and orientation of external stress. Nat Commun 9(1):5284. https://doi.org/10.1038/s41467-018-07523-0
16. Borghi N, Sorokina M, Shcherbakova OG et al (2012) E-cadherin is under constitutive actomyosin-generated tension that is increased at cell-cell contacts upon externally applied stretch. Proc Natl Acad Sci U S A 109(31): 12568–12573. https://doi.org/10.1073/pnas.1204390109
17. Conway DE, Breckenridge MT, Hinde E et al (2013) Fluid shear stress on endothelial cells modulates mechanical tension across VE-cadherin and PECAM-1. Curr Biol 23(11):1024–1030. https://doi.org/10.1016/j.cub.2013.04.049
18. Nordenfelt P, Elliott HL, Springer TA (2016) Coordinated integrin activation by actin-dependent force during T-cell migration. Nat Commun 7:13119. https://doi.org/10.1038/ncomms13119
19. Krieg M, Dunn AR, Goodman MB (2014) Mechanical control of the sense of touch by β-spectrin. Nat Cell Biol 16(3):224–233. https://doi.org/10.1038/ncb2915
20. Déjardin T, Carollo PS, Sipieter F et al (2020) Nesprins are mechanotransducers that discriminate epithelial-mesenchymal transition programs. J Cell Biol 219(10). https://doi.org/10.1083/jcb.201908036
21. Carley E, Stewart RM, Zieman A et al (2021) The LINC complex transmits integrin-dependent tension to the nuclear lamina and represses epidermal differentiation. elife 10. https://doi.org/10.7554/eLife.58541
22. Lemke SB, Weidemann T, Cost A-L et al (2019) A small proportion of Talin molecules transmit forces at developing muscle attachments in vivo. PLoS Biol 17(3):e3000057. https://doi.org/10.1371/journal.pbio.3000057
23. Hirano S, Yamamoto T, Michiue T (2018) FRET-based tension measurement across actin-associated mechanotransductive structures using Lima1. Int J Dev Biol 62(9–10): 631–636. https://doi.org/10.1387/ijdb.180110tm
24. Lagendijk AK, Gomez GA, Baek S et al (2017) Live imaging molecular changes in junctional tension upon VE-cadherin in zebrafish. Nat Commun 8(1):1402. https://doi.org/10.1038/s41467-017-01325-6
25. Tao H, Zhu M, Lau K et al (2019) Oscillatory cortical forces promote three dimensional cell intercalations that shape the murine

mandibular arch. Nat Commun 10(1):1703. https://doi.org/10.1038/s41467-019-09540-z

26. Ham TR, Collins KL, Hoffman BD (2019) Molecular tension sensors: moving beyond force. Current Opinion in Biomedical Engineering 12:83–94. https://doi.org/10.1016/j.cobme.2019.10.003

27. Freikamp A, Mehlich A, Klingner C et al (2017) Investigating piconewton forces in cells by FRET-based molecular force microscopy. J Struct Biol 197(1):37–42. https://doi.org/10.1016/j.jsb.2016.03.011

28. Cost A-L, Ringer P, Chrostek-Grashoff A et al (2015) How to measure molecular forces in cells: a guide to evaluating genetically-encoded FRET-based tension sensors. Cell Mol Bioeng 8(1):96–105. https://doi.org/10.1007/s12195-014-0368-1

29. Demeautis C, Sipieter F, Roul J et al (2017) Multiplexing PKA and ERK1 & 2 kinases FRET biosensors in living cells using single excitation wavelength dual colour FLIM. Sci Rep 7. https://doi.org/10.1038/srep41026

30. Hendrix J, Lamb DC (2013) In: Digman MA, Stakic M, Gratton E (eds) Pulsed interleaved excitation: principles and applications, vol 518. Raster Image Correlation Spectroscopy and Number and Brightness Analysis, pp 205–243

31. Müller BK, Zaychikov E, Bräuchle C et al (2005) Pulsed interleaved excitation. Biophys J 89(5):3508–3522. https://doi.org/10.1529/biophysj.105.064766

32. Reissaus CA, Day KH, Mirmira RG et al (2020) PIE-FLIM measurements of two different FRET-based biosensor activities in the same living cells. Biophys J 118(8):1820–1829. https://doi.org/10.1016/j.bpj.2020.03.003

Chapter 16

Visualizing Neurons Under Tension In Vivo with Optogenetic Molecular Force Sensors

Neus Sanfeliu-Cerdán, Li-Chun Lin, Alexander R. Dunn, Miriam B. Goodman, and Michael Krieg

Abstract

The visualization of mechanical stress distribution in specific molecular networks within a living and physiologically active cell or animal remains a formidable challenge in mechanobiology. The advent of fluorescence-resonance energy transfer (FRET)-based molecular tension sensors overcame a significant hurdle that now enables us to address previously technically limited questions. Here, we describe a method that uses genetically encoded FRET tension sensors to visualize the mechanics of cytoskeletal networks in neurons of living animals with sensitized emission FRET and confocal scanning light microscopy. This method uses noninvasive immobilization of living animals to image neuronal β-spectrin cytoskeleton at the diffraction limit, and leverages multiple imaging controls to verify and underline the quality of the measurements. In combination with a semiautomated machine-vision algorithm to identify and trace individual neurites, our analysis performs simultaneous calculation of FRET efficiencies and visualizes statistical uncertainty on a pixel by pixel basis. Our approach is not limited to genetically encoded spectrin tension sensors, but can also be used for any kind of ratiometric imaging in neuronal cells both in vivo and in vitro.

Key words FRET, TSMod, Molecular tension sensors, *C. elegans*, Neuroscience, Mechanobiology, Optogenetics, Mechanosensing, Cytoskeleton

1 Introduction

The ability to sense mechanical forces is a ubiquitous property of all cells including neurons [1]. The mechanical stress is sensed by specialized ionotropic or metabotropic force sensors located in the cell membrane or the intracellular compartments [2, 3]. Often though, these sensors reside at a different location from where it is generated, and thus, mechanical stress needs to be transferred through the cell membrane, cytoskeleton or extracellular matrix, e.g., from the skin to the nerve cell [4, 5]. Due to mechanical anisotropy, nonlinear mechanics, and viscoelasticity,

Ronen Zaidel-Bar (ed.), *Mechanobiology: Methods and Protocols*,
Methods in Molecular Biology, vol. 2600, https://doi.org/10.1007/978-1-0716-2851-5_16,

mechanical stresses distribute within a cell in different directions, which determines the spatial extent to which the cell reacts to them. Whereas methods to apply and measure forces on the cell membrane are available and have been applied to various problems in cell biology [6], the techniques available to study such force transmission pathway inside cells are limited and just emerging [7–9]. Over the last decade, a number of genetically encoded molecular tension sensors were developed that allow researchers to visualize cellular and mechanical properties by live cell microscopy [10–29]. The method is based on a FRET pair of fluorescent proteins that are connected by a molecular spring composed of a flexible protein domain with defined length and mechanics. This flexible domain varies between different sensors, and the most successful approach is based on the flagelliform protein spring from spider silk [30]. A force across the spring separates the fluorophores, which is detectable by a decrease in FRET. Importantly, the spring is calibrated [10, 30], thus allowing researchers to directly convert the average FRET efficiency from a pixel-based measurement into a force acting on a single molecule in vivo. The presented method is versatile, as this FRET tension sensor module, called TSMod, can be incorporated in any molecule of interest, which is expected to be under tension. Imaging can be either performed on a standard confocal microscopy or an epifluorescence microscope available in nearly every lab or imaging facility. Critically, the presented protocol is not limited to TSMod but is equally applicable to any other genetically encoded FRET-based strain sensor employing cpstFRET to infer stress across host molecules [31].

1.1 FRET-Force Measurements

FRET is the non-radiative transfer of energy by near-field dipole–dipole coupling, which occurs after excitation of a fluorescent donor molecule adjacent to a spectrally and spatially overlapping acceptor molecule. The efficiency of the energy transfer is proportional to the inverse of the 6th power of the separation r of the FRET pair and is scaled by the Förster distance R_0 according to:

$$\frac{R_0^6}{R_0^6 + r^6}$$

The Förster distance is a property of a particular FRET pair and can be calculated from the spectral overlap integral J of the emission of the donor and the excitation of the acceptor, the relative orientation of the dipoles of both chromophores κ, the refractive index n, and the quantum donor yield Q_D:

$$R_0 = 0.211\ \left(\kappa^2 n^{-4} Q_D J\right)^{\frac{1}{6}}$$

At a separation of $r = R_0$, the energy transfer is 50%. Since κ^2 is difficult to measure [22] but only slightly subjected to change [32], a constant value of 2/3 is assumed which incorporates the equal likelihood of all possible angles between the emission transition dipole of the donor and the absorption transmission dipole of the acceptor.

The Förster distance of the mTFP1 and mVenus pair is 5.2 nm [33] and hence is most sensitive at a fluorophore separation of ≈5.2 nm. In TSMod, the fluorophores are linked by a 40 aa flagelliform protein from spider silk, which in solution forms an entropic spring. Because each fluorophore has a radius of about 2 nm plus the entropic spring which accumulates to nearly 5 nm, only efficiencies of <50% can be measured theoretically. How does the tension sensor module measure forces? If the TSMod cassette is incorporated into a host molecule which is under tension, the force across the module will pull the two fluorophores apart with a distance proportional to the spring constant of the elastic spider silk domain. The increase in distance can then be measured as a decrease in the fluorescence intensity of the acceptor, indicative as a decrease in FRET. Importantly, the insertion of the TSMod cassette into the native host protein might significantly perturb its structure and thus its function. The placement should thus be carefully considered and several iterations might be necessary to obtain a minimally invasive tension sensor module. If new tension sensors are required, several protocols describe the creation and design of such probes [34, 35]. Ideally, the transgenic TSMod can substitute for the function of the untagged host protein, e.g., rescue a mutant, to facilitate the interpretation of the results. If the tagged protein leads to a dominant effect in the presence of the untagged host, different locations of the TSMod or alternative hosts may need to be considered.

Here, we present a method to map the distribution of mechanical stress in the neuronal spectrin cytoskeleton. We take advantage of a probe that is incorporated in between β-spectrin repeats 8 and 9, as described in detail in Ref. [12]. This method allows for quantitative FRET-force microscopy in neurons of living animals in a dynamic setting, e.g., during laser axotomy [12] and proprioception [36], but also in developing tissues of an early embryo [37]. To enable quantitative FRET-force microscopy based on sensitized emission, several different reference constructs need to be investigated (Fig. 1a, b). These include, but are not limited to, the actual sensor and a sensor domain that cannot be subjected to mechanical stress. This is most commonly achieved by fusing the FRET-TSMod to one of the termini of the host protein (Fig. 1c). We further suggest the construction of pre-calibrated FRET-rulers, in which the elastic spring is replaced by fixed-length linkers such as five amino acids or by large separation domains.

1.2 Organization of the Protocol

We first begin the protocol with a brief description of the nematode *Caenorhabditis elegans* preparation for noninvasive imaging and a protocol to calibrate the collection efficiency of the photodetectors that is necessary to correct the FRET signal. The calibration also

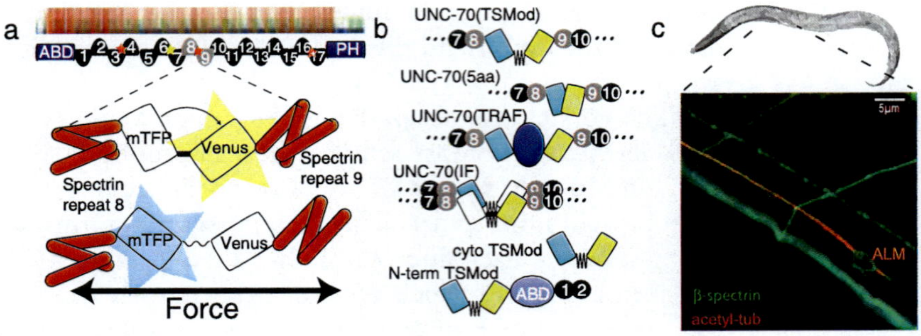

Fig. 1 A β-spectrin tension sensor
(a) Placement of TSMod between β-spectrin repeats 8 and 9 to generate a spectrin force sensor. Red stars indicate regions of high conformational flexibility [38]. **(b)** Suggested experimental and control constructs to validate and verify FRET values obtained from the protocol. From top to bottom: β-spectrin::TSMod force sensor; high FRET (5aa) and low FRET (TRAF) standards; intermolecular FRET control; soluble TSMod and terminal fusion as no-force control. **(c)** Expression of UNC-70::TSMod in adult *C. elegans*. A single ALM touch receptor neuron is highlighted by its coexpression of acetylated α-tubulin

enables us to estimate individual pixel error with the aim to visualize these uncertainties in the final FRET map. We also implement an automatic neurite tracing algorithm, which tracks the boundary of the axon based on its Gaussian intensity distribution. We finally provide a detailed description of the analysis and the assessment of inter- vs. intramolecular energy transfer. This protocol is complemented by several excellent reviews dedicated to facilitate interpretation of the measurements [8, 39, 40].

2 Materials

2.1 Reagents and Consumables

(i) Household bleach.

(ii) KOH.

(iii) NGM Agar plates seeded with *Escherichia coli* OP50.

(iv) Agarose.

(v) Coverslips # 1.5.

(vi) Glass slides.

(vii) Immersion oil for microscopy (Leica; $N = 1.513$).

(viii) Optical lens cleaning paper (Thorlabs).

(ix) Windex.

(x) Lab tape.

(xi) 0.1 μm latex beads, 2.5% w/v (PolySciences 00876-15).

(xii) Na_2HPO_4.

(xiii) KH_2PO_4.

(xiv) NH_4Cl.

(xv) NaCl.

(xvi) NaOH.

(xvii) Fluorescein isothiocyanate (FITC).

(xviii) Ibidi 8-well microslide.

(xix) Distilled H_2O.

2.2 Equipment

(i) Table top centrifuge.

(ii) Worm pick.

(iii) Stereoscopic dissecting microscope.

(iv) Laboratory glassware.

(v) Leica SP5 confocal microscope with argon laser line for excitation of mTFP (458 nm) and Venus (514 nm) with single photon sensitive Hybrid (HyD) photodetectors (or other microscope equipped to visualize mTFP and Venus).

(vi) Vibration isolation table.

(vii) Glass slide holder.

(viii) 63× NA 1.4 or 100× NA 1.4 oil immersion objective.

(ix) 10× NA 0.5 air objective.

(x) Excitation filter for GFP: 480 ± 20.

(xi) Emission filter for GFP 520 ± 20.

(xii) High performance image processing computer.

(xiii) Epifluorescence light source.

2.3 Software

ImageJ; Python 3.10 (see https://gitlab.icfo.net/fret-analysis/fret-analysis).

2.4 Animal Strains

Described in Table 1.

2.5 Equipment Setup

2.5.1 Confocal Laser Scanning Microscope

A Leica SP5 II confocal scanning microscope is assembled around an upright Leica DMi6000 body equipped with an external epifluorescence illumination and GFP excitation and emission filters if dedicated mTFP and mVenus filters are not available. The microscope should be equipped with a holder for a glass slide, a 10× low magnification lens for searching the specimen and a high resolution 63×/1.4 or 100×/1.4 oil immersion objective for data acquisition. If the sample consists of mammalian neurons or other temperature-sensitive specimen, the microscope needs to be equipped with a temperature incubator. The confocal scanning system demands low noise GaAsP hybrid photodetectors with single photon sensitivity, a variable speed galvo scanner, and an argon laser with 458 and 514 nm laser lines. Spectral separation of the emission wavelength is best achieved with an acusto-optical beam splitter or a prism spectrometer detector. Alternatively, emission filters of

Table 1
Inventory of strains for mapping mechanical tension in neuronal spectrin cytoskeleton of living animals

Strain	Purpose	Genotype	Reference
GN517	TSMod	pgEx116 [*unc-70* p::*unc-70* (1-1166)::TSMod::*unc-70* (1167-2267); *myo-3* p::mCherry]	[12]
GN519	High FRET	pgEx131 [*unc-70* p::*unc-70* (1-1166)::mTFP-5aa-Venus:: *unc-70* (1167-2267); *unc-122* p::RFP]	[12]
GN600	No force	pgIs22 [*unc-70* p::*unc-70*::TSMod]; oxIs95[*pdi-2* p::*unc-70* (fl); *myo-2* p::GFP, lin-15] IV	[37]
MSB233/ MSB366	Mutant	mirEx77 [*unc-70* p::*unc-70* (1-1166)::TSMod::*unc-70* (1167-2267) E2008K; *myo-2* p:mCherry]	[36]
MSB339	Low FRET	mirIs23 [*unc-70* p::*unc-70* (1-1166)::mTFP-TRAF-Venus:: *unc-70* (1167-2267]; *unc-70* (s1502); oxIs95[*pdi-2* p:: *unc-70* (fl); *myo-2* p::GFP, lin-15] IV	[36]
GN498	Acceptor crosstalk	pgEx138[*unc-70* p::*unc-70* (1-1166)::TSMod (mTFPG73V)::*unc-70* (1167-2267)::*unc-54* utr; *myo-2* p:: mCherry; *unc-119* p::*unc-119*]	[12]
GN495	Donor bleedthrough	pgEx135 [*unc-70* p::*unc-70* (1-1166)::TSMod(G344V):: *unc-70* (1167-2267)::*unc-54* utr, *myo-2* p::mCherry; *unc-119* p::*unc-119*]	[12]
ARM101	Donor bleedthrough	wamSi101 [*eft-3* p::mTFP::*unc-54* 3′UTR + Cbr-*unc-119* (+)] V.	[41]
GN493	Intermolecular FRET	pgEx133[*unc-70* p::*unc-70* (1-1166)::TSMod (mTFPG73V]::*unc-70* (1167-2267)::*unc-54* utr; *unc-70* p::*unc-70* (1-1166)::TSMod(G344V)::*unc-70* (1167-2267)::*unc-54* utr; *myo-2* p::mCherry	[12]
GN566	Soluble TSMod	pgEx146[*unc-70* p::TSMod::*unc-54* utr]	[12]

490 ± 30 nm for donor emission and 550 ± 30 nm for acceptor emission can be used in combination with a double band pass dichroic mirror (460/520 nm).

2.5.2 Standard Setup

1. Leica DMi6000 with external epifluorescence mercury arc or LED light source and corresponding filter sets.
2. Motorized Piezo z-stage for fast z-stacks.
3. HeNe 633 nm/10 mW, HeNe 594 nm/2.5 mW, HeNe 543/1.0 mW, Ar 458 nm/5 mW, 476 nm/5 mW, 488 nm/20 mW, 514 nm/20 mW or supercontinuum white light laser (covering wavelength from 450 to 590 nm).
4. Leica tandem scanner system.
5. Spectral detection module for adjustable spectral emission windows.

6. HC PL Fluotar 10×, air; HCX PL APO 63×/1.4 OIL.
7. Three standard fluorescent photomultiplier tubes (PMT) and two hybrid avalanche photodetectors (HyD).
8. Mounted on active vibration isolation.

3 Methods

3.1 Calibrating the Collection Efficiency of the Photodetector

Because the detection efficiency of common photodetectors varies with the wavelength of the photon, the linearity and collection efficiency of the hybrid GaAsP photodetectors have to be calibrated at least once for a given setup. Collection efficiency ψ_D and ψ_A can be determined experimentally by collecting photons from a homogeneous fluorescein solution with increasing gain at constant laser power (Fig. 2).

3.1.1 Determination of Collection Efficiency

1. Prepare a 10% FITC solution in PBS.
2. Make successive dilutions to obtain 1, 0.1, 0.01, 0.001% and pipette 30 μL of each dilution into a well of an ibidi "angiogenesis" μ-slide, which contains a coverslip bottom suitable for high-resolution imaging. Cover with an ibidi-cover and seal to prevent leakage of the dye-solution.
3. Proceed to confocal microscope, add a drop of immersion oil on the coverslip and bring 63×/1.4 oil immersion lens into place.

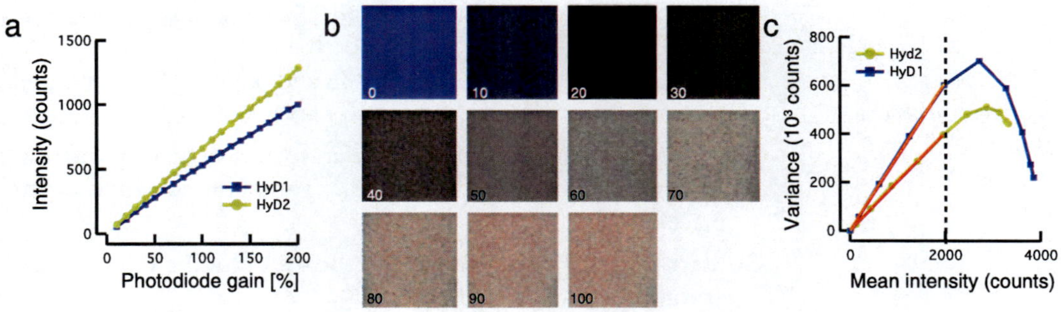

Fig. 2 Calibration of photodetector using a homogeneous FITC solution
(**a**) Intensity counts plotted against increasing gain of each photodiode. Final settings were chosen in regions where the gain curves were perfectly linear, e.g., 150% for HyD1 and 100% for HyD2. (**b**) Representative images for the calibration in the HiLo lookup table highlighting oversaturated pixels in red. (**c**) Variance vs. mean intensity for a given ROI at increasing laser power. Laser power was adjusted in final imaging experiments to minimize saturated pixels. Standard deviation for each pixel as a function of intensity was determined from the line fit to the data. Roll-off above 2000 counts is due to pixel saturation. For a purely shot noise limited detection system a line is expected, hence fit range was limited between 0 and the dotted line

4. Start confocal microscope with the 488 nm line of a standard argon laser activated.
5. In the configuration tab settings for the lasers, the argon laser should normally be set at 25% output power. Prolonged use of this laser at higher output powers will shorten its lifespan.
6. Open the pinhole and adjust the prism spectrometer to between 500 and 600 nm emission wavelength.
7. Adjust imaging parameters such that they are similar to those planned for biological samples, e.g., 1024 × 1024 pixels, 700 Hz linescan rate. Start with a low laser power, e.g., 10% transmission and adjust the gain to 10%.
8. Collect an image and increase the gain successively until 200%.
9. Repeat the same process with both hybrid GaAsP photodetectors.

3.1.2 Analysis

10. Load each image into ImageJ or any other image analysis software.
11. Draw a rectangular ROI in the center of the image and save it.
12. Measure the mean pixel intensity and the standard deviation of the pixels in the ROI and repeat it for all images acquired.
13. Plot the mean fluorescence against the gain for each photodetector (Fig. 2a).
14. Select the gain for subsequent data acquisition for which the curve is linear. The steeper the curve, the more sensitive the diode. Select gains for which the mean fluorescence is similar. In our imaging system, the ideal gain was determined to be 100% for the acceptor diode and 150% for the donor diode.

3.1.3 Optional: Determination of Pixel Noise for Each Photodetector

1. Prepare a 1% FITC solution in PBS and pipette 30 μL into ibidi u-slide. Cover with a coverslip to prevent leaking.
2. Proceed to confocal microscope and add a drop of immersion oil on the coverslip and bring 63×/1.4 oil immersion lens into place.
3. Start confocal microscope with the 488 nm line of a standard argon laser activated.
4. Open the pinhole and adjust the spectral collection window between 500 and 600 nm emission wavelength.
5. Adjust imaging parameters similar to the actual experiment, e.g., 1024 × 1024 pixels, 700 Hz linescan rate. Start with a low laser power, e.g., 0% transmission and adjust the optimal gain settings determine above for the donor and acceptor photodetectors, respectively.
6. If collecting in photon counting mode, gain adjustment will be inactivated.

7. Take an image and increase the laser power successively until 100% or pixels will start to saturate (*see* **Note 1**, Fig. 2b).
8. Repeat the same process with both HyD photodetectors.

3.1.4 Analysis

9. Load each image into ImageJ or any other image analysis software.
10. Draw a rectangular ROI in the center of image and save it.
11. Measure the mean pixel intensity and the standard deviation of the pixels in the ROI and repeat it for all images acquired (using the same ROI).
12. Plot the variance of the pixel intensity of each ROI vs. the mean pixel intensity for each photodetector (Fig. 2c).
13. Fit a line to the data and determine the *y*-intercept and the slope. The square-root of the slope is a measure for the standard deviation of each pixel, while the *y*-intercept is a measure for the dark current (*see* **Note 2**). If a significant number of pixels are saturated, the slope deviates from one for higher counts.

3.2 Age Synchronizing Animals Before Imaging

Keep animals using standard protocols and procedures as described in Ref. [42]. Because mechanical tension is subject to cellular regulation and might depend on animal size, developmental stage, and age, only animals from an age-synchronized population should be imaged.

1. Prepare bleaching solution by combining 0.2 mL household bleach (stored in an opaque or brown glass container), 20 μL 5 M KOH, and 0.8 mL H_2O to obtain 1 mL bleaching solution. Prepare always fresh bleach stock before the experiment and change it every couple of months since it is not stable.
2. Prepare M9 buffer: add 6 g Na_2HPO_4; 3 g KH_2PO_4; 1 g NH_4Cl; 0.5 g NaCl, adjust the volume to 1 L with H_2O and autoclave.
3. Wash worms off from a 6 cm plate using 1 mL M9 buffer (or filtered ddH_2O) and transfer onto a 1.5 mL microcentrifuge tube.
4. Spin down worms for 20 s at 1200 rpm.
5. Optional: Discard supernatant and repeat the wash three times until no bacteria or other contaminants are visible in the supernatant.
6. Resuspend the worms in 1 mL bleaching solution and incubate for 5 min or until most of the adult worms break apart and release their eggs into the solution (*see* **Note 3**).
7. Pellet carcasses and eggs by spinning for 1 min at 150–200 rpm.

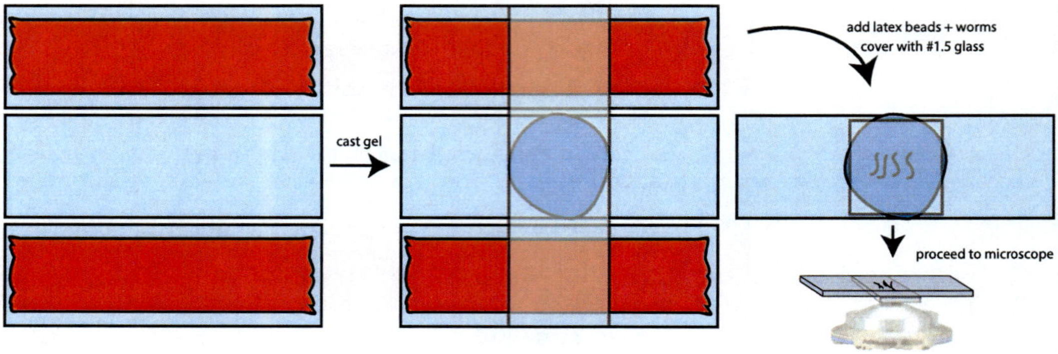

Fig. 3 Preparation of the agar pad
One glass slide is sandwiched between two slides that are modified with lab tape that acts as a spacer. A drop of agar solution is pipetted on top of the middle slide and compressed with another slide to create a flat bed of agar. A droplet of latex beads is pipetted on top of the pad, in which the worms are deposited with a worm pick. The droplet with the worms is covered with a glass coverslip #1.5 thickness

8. Remove supernatant and wash 3 times with M9 buffer.
9. Add 20 μL of M9 and pipette the eggs onto a new agar plate containing a lawn of *E. coli* OP50.
10. Check after 60 h growth at 20 °C and use young adult worms or any age for imaging.

3.3 Mounting Animals for Imaging

Young adult animals or any other age-synchronized stage can be mounted in various ways [43–50] and in different body positions [36, 50, 51]. For the sake of simplicity, we follow here the simple agar pad compressive immobilization with latex beads [12, 43].

11. Prepare a 6% agarose solution in M9 and dissolve by heating in a microwave.
12. Meanwhile, prepare two glass slides with a "spacer-tape" adhered on top of them (*see* Fig. 3). Bring a third glass slide between them. Pipette a drop (50–100 μL) of dissolved agar solution on the middle glass slide and immediately compress the drop with a glass slide. The adhesive tape of the two bordering glass slides will act as a spacer and hence ensure a constant thickness of the agar pad. Store in a humidified chamber to prevent drying and use immediately (*see* **Note 4**).
13. Take a glass slide with an agar pad and pipette 2 μL of latex bead solution onto the middle of the agar pad.
14. Pick brightly fluorescing young adult worms under a stereo microscope into the drop.
15. Cover gently with a glass coverslip and proceed immediately to imaging (*see* **Note 5**).

3.4 Imaging

Critical: The following procedure is optimized for a Leica SP5 point scanning confocal setup described above. If an alternative setup is used, it will be necessary to modify relevant parameters (such as emission filter band, excitation laser intensity, and scan line rate) to achieve the best resolution and FRET signal.

1. Proceed to the microscope immediately after mounting the worms. Prevent drying of the agar pad by keeping it in a humidified chamber. Caution! Drying will increase the salt content and therefore perturb the worms osmotically and lead to dehydration.
2. Look for the worms with a low magnification objective (10× air) and bring into the center of the field of view.
3. Place a drop of immersion oil on top of the coverslip and change to the 63×/1.4 oil immersion lens.
4. Using an attached epifluorescence lamp, observe worms through widefield fluorescence by illuminating with any standard GFP excitation and emission filter, identify the region of interest, and bring the animal into focus (*see* **Note 6**).
5. Switch to confocal mode and select imaging parameter according to **step 11**. Three different images need to be acquired to gain access to the FRET signal using sensitized emission. Four additional images are needed to eliminate spectral contamination (donor bleedthrough and acceptor crosstalk) from the raw FRET intensity image.
6. Define a sequential scan to take the donor and acceptor images sequentially and minimize spectral contamination. The sequential scans will have only one laser active at a time while collecting photons in both the mTFP and Venus channels.
7. Choose 1024 × 1024 pixels to obtain adequate resolution and adjust the zoom so that the pixel size of the image does not cause over- or undersampling. For optimal sampling, *see* **Note 7**.
8. Select the 458 and 514 nm lines of continuous wave argon ion laser to excite the mTFP and Venus fluorophores, respectively. Start with a low laser power and increase if the signal is too poor, e.g., 50% for the donor (≈6 μW) and 11% for the acceptor excitation (≈4 μW). For reproducible measurements, the incident power of the excitation light should be measured with a Thorlabs microscope slide power meter head (S170C) attached to PM101A power meter console.
9. Select bit depth of 16 (preferred) or 12 (adequate) to obtain 64,000 or 4096 gray levels, respectively. Importantly, do not excede these intensities since FRET values cannot be derived from saturated pixels (*see* **Note 1**).

10. Activate two HyD photodetectors and set the emission bandwidth using the prism spectrometer detector to 465–500 nm for the donor channel and 520–570 nm for the acceptor channel.
11. Adjust detector gain of the donor photodiode to 150% and acceptor detector gain to 100%. Gains need to be set according to the calibration to achieve equal collection efficiency (see above). Calibration of the detectors ensures similar collection efficiency under preset gain setting (*see* **Note 8**).
12. Find the animal in "live" mode only in the acceptor channel and choose appropriate zoom level matching optimal sampling (*see* **Note** 7). Reducing the imaging time and power in the acceptor channel avoids unnecessary bleaching of the rather labile mTFP donor fluorophore.

Optional to reduce imaging noise: perform 4× line averaging during acquisition.

13. Switch to line-by-line scan for the final image acquisition. This minimizes motion artefacts and ensure that donor and acceptor pixels match perfectly.
14. Capture four images under the following conditions (*see* **Note 9**):
    (a) Donor excitation → Donor emission = I_D → Donor channel.

    Ex 458 nm.

    Em 465–500 nm

    (b) Donor excitation → Acceptor emission = I_F → FRET channel.

    Ex 458 nm.

    Em 520–570 nm

    (c) Acceptor excitation → Acceptor emission = I_A → Acceptor channel.

    Ex 514 nm.

    Em 520–570 nm

    (X) Acceptor excitation → Donor emission (This channel does not contain any information and can be discarded).

    Ex 514 nm.

    Em 465–500 nm
15. Save them appropriately and proceed to bleedthrough estimation.

3.4.1 Quantifying Spectral Contaminants (Bleedthrough, Crosstalk)

Intensity-based FRET imaging is prone to spectral bleedthrough and crosstalk between individual channels due to the overlapping fluorescence spectra of the donor and acceptor chromophores. Acceptor crosstalk is defined as the direct excitation of the acceptor fluorophore by the donor laser line (Fig. 4a-i). Donor bleedthrough is defined as the donor emission in the FRET channel after direct donor excitation, in the absence of an acceptor (Fig. 4a-ii). Several methods have been introduced that rely on constant or varying correction factors [53–55]. Critically, determination of the bleedthrough factors has to be performed for each experiment. It is not sufficient to do it once for a given setup, especially if imaging parameters such as photodetector gain, laser power, and detection bandwidth are changed during an experiment to improve imaging conditions (*see* **Note 8**).

1. Immobilize a sample that contains just the donor fluorophore, e.g., ARM101 is a worm that expresses only the mTFP [41].

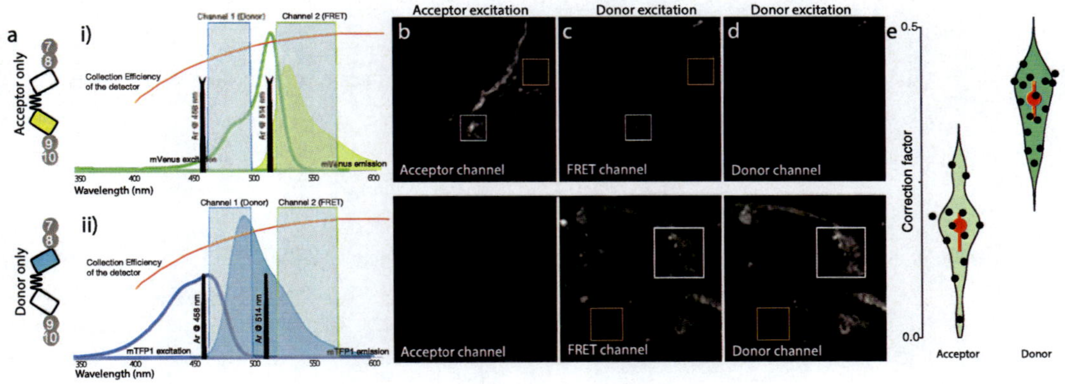

Fig. 4 Determining the spectral contamination in the FRET channel
(**a**) Schematic of the single fluorophore construct to determine spectral crosstalk. (i) Spectral situation of the FRET experiment in the absence of the donor. Residual excitation of the acceptor with the 458 nm line leads to detectable emission in the FRET channel. (ii) Spectral situation of the FRET experiment in the absence of the acceptor. The long-tailed emission of the donor leads to significant signal in the FRET channel. Red line indicates the varying collection efficiency of the photodetector; gray squares represent collection band determined by the spectral detectors; black peaks indicate the two laser lines. Spectra assembled with fpbase.org [52]. (**b**–**d**) Panel of different imaging controls characterizing spectral contamination in the imaging channels. The fluorescent probe consists of either the acceptor (upper row) or the donor (lower row). (**b**) Representative images of the acceptor emission after direct acceptor excitation for the two control constructs. (**c**) Representative images of the acceptor emission after donor excitation for the two different controls. (**d**) Representative images of the donor channel after direct excitation of the donor. The white square indicates the ROI to extract the signal from either spectral contamination in the FRET channel. (**e**) Distribution of bleedthrough factors acquired over the course of 2 years. Red point = median ± confidence interval on the median

2. Take two images after direct excitation of the donor and recording the emission in the donor and the FRET channels to get access to I_D and I_F, respectively.
   (a) Donor excitation → Donor emission = I_D → Donor channel

   Ex 458 nm

   Em 465–500 nm.

   (b) Donor excitation → Acceptor emission = I_F → FRET channel

   Ex 458 nm

   Em 520–570 nm

3. Immobilize a worm that expresses only the acceptor fluorophore, e.g., GN498 is a worm that only expresses mVenus [12].
   (a) Donor excitation → Acceptor emission = I_F → FRET channel

   Ex 458 nm

   Em 520–570 nm

   (b) Acceptor excitation → Acceptor emission = I_A → Acceptor channel

   Ex 514 nm

   Em 520–570 nm

3.5 Analysis

In the following section we provide a detailed description of the FRET analysis protocol and its accompanying analysis scheme (Fig. 5). For the convenience of the experimenter, a Python code with all steps is available under this link, which has been adapted from the original protocol in Ref. [12]. The program contains three modules, one module for the bleedthrough determination (Fig. 5, Module 1) and two modules for the FRET data processing (Fig. 5, Modules 2 and 3). Module 2 elaborates an automated neurite tracing and local background subtraction to determine the FRET efficiency in neurons. This module has advantages for FRET calculation in vivo with spatial varying background fluorescence, e.g., in the nematode. Alternatively, if the ROI is an irregular object, Module 3 provides the user with a manual ROI selection for global background subtraction (Fig. 5). In general, to obtain the FRET values from the recorded data, the following steps need to be performed.

1. *Determine the spectral bleedthrough factors*

   To correct for spectral contamination in the FRET channel from donor bleedthrough and acceptor cross excitation, the

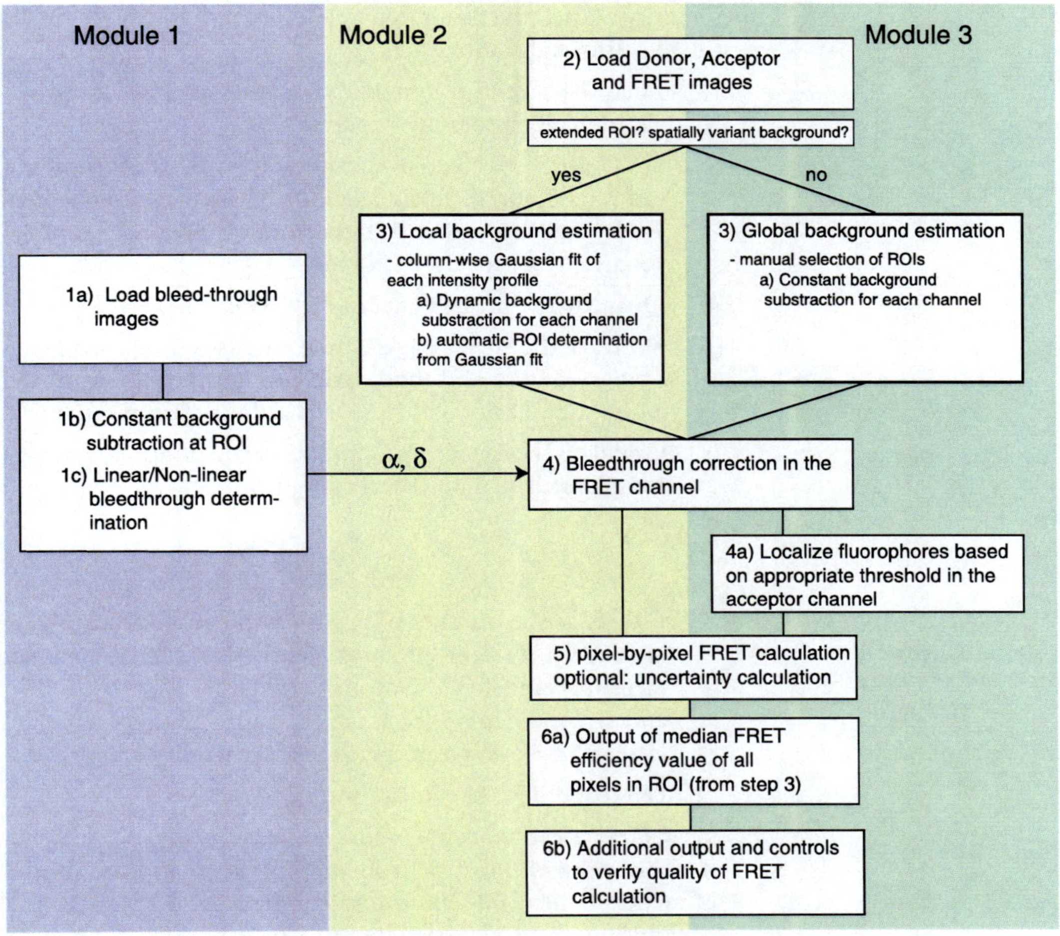

Fig. 5 Overview of the analysis with flowchart

(1) Bleedthrough determination and FRET processing are done on separate sets of images. Bleedthrough factors were loaded into the main program. (2–6) Representative steps in the flow of the FRET calculation

(2) Load images

(3) Global or local background subtraction of a moving window after Gaussian fit to the neurite intensity profile. Simultaneous determination of ROI boundaries from the width of the fit

(4) Bleedthrough correction

(5, 6) FRET calculation and depicting the estimates of the associated, pixelwise error and quality of the seFRET signal

A python code with the three modules and detailed instructions is available under https:/gitlab.icfo.net/fret-analysis/fret-analysis intensity in the FRET channel is divided by the corresponding intensity of the directly illuminated channel, which gives rise to the bleedthrough ratios.

(a) Load the bleedthrough images into ImageJ or any equivalent image processing program.

(b) Select an ROI containing a region with strong but not saturated fluorescent signal in the image pair and select an ROI with background fluorescent (white and orange squares, respectively, in Fig. 4b–d).

(c) Quantify the mean signal in each ROI.

(d) Subtract the mean background intensity (e.g., from the orange square) and the "signal" intensity (e.g., from the white square) in the image pairs as indicated in Fig. 4.

(e) Calculate the bleedthrough ratio of the donor and acceptor channels (δ and α) according to:

$$\delta = \frac{\sum \frac{I_F(x, y)}{I_{D(x, y)}}}{N} \quad \text{and } \alpha = \frac{\sum \frac{I_F(x, y)}{I_{A(x, y)}}}{N}$$

with N as the number of pixels (with the coordinate x, y) in the ROI, I_F is the intensity value in the "FRET" channel (acceptor emission after donor excitation, Fig. 4b), and I_A and I_D are the intensity in the acceptor and donor channel (after direct excitation, Fig. 4a, c), respectively.

(f) Keep α and δ for the linear unmixing.

2. *Local background subtraction and neurite tracing*

The width of most neurons in *C. elegans* approaches the diffraction limit (200–300 nm in diameter) and therefore is a convolution of the Gaussian intensity distribution of the point spread function of the imaging system (Fig. 6a-i). The extension of the neuron, however, spans many hundreds of micrometers in length. Therefore, background fluorescence is not constant and regional variation exists along the length of the neuron through the animal's body (Fig. 6a-i, b, c). To accommodate these special needs, we perform a local background subtraction at each position in the image after fitting the raw intensity profile to a single Gaussian and subtracting the baseline fit parameter I_0 at that position.

(a) Load image stack containing all three channels into your favorite image processing software (e.g., Fiji/ImageJ) (Fig. 7a–c).

(b) Select the neuron of interest with the 20–30 pixel wide segmented line tool and straighten the ROI (Edit > Selection > Straighten). Save selected ROI for reference.

(c) Save the straightened image stack.

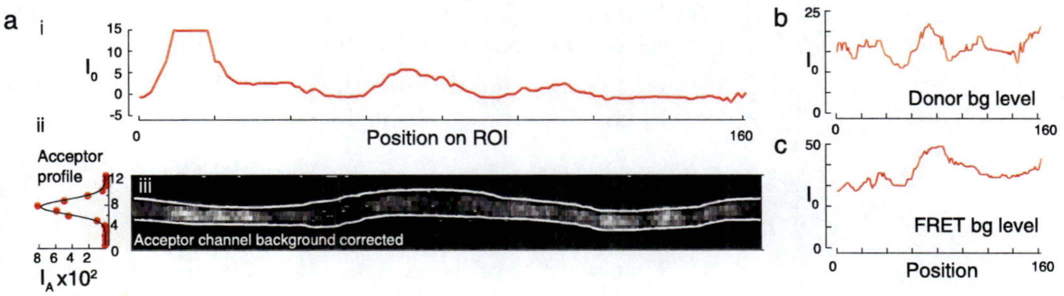

Fig. 6 Background subtraction and neurite tracing
(**a**) Representative background subtraction of a neurite in the acceptor channel after direct excitation of the acceptor. (i) The background level is shown for each row of pixels in the image. (ii) Gaussian distribution of the intensity at a certain pixel position after background subtraction. (iii) Background subtracted acceptor channel intensity image. White lines indicate boundaries of the ROI found by our fitting procedure. (**b**) Background level of the same ROI of the donor channel and (**c**) the raw FRET channel

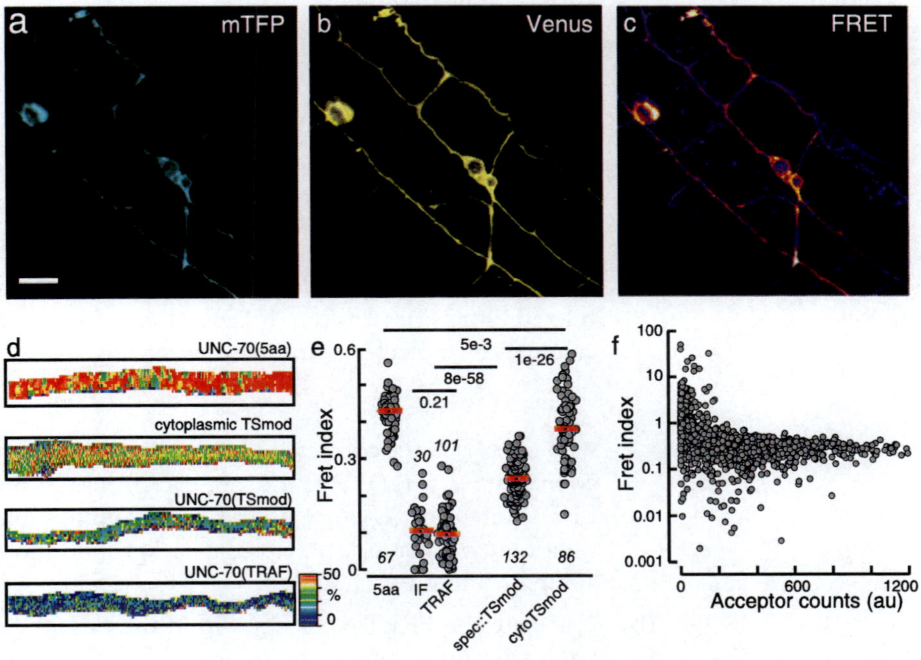

Fig. 7 Representative results from UNC-70 β-spectrin::TSMod and control constructs
(**a**) Donor fluorescence after direct excitation. (**b**) Acceptor fluorescence after direct excitation. (**c**) Acceptor fluorescence after donor excitation (uncorrected FRET signal). (**d**) Processed and corrected FRET efficiency maps of representative axonal ROIs. (**e**) Summary of the observed FRET efficiencies derived from the individual constructs. (**f**) Plot of the FRET efficiency against the acceptor intensity indicates that most FRET is due to intramolecular transfer

(d) Import into data analysis program to determine the ROI selection through iterative-fitting procedures.

(e) Fit a Gaussian to the intensity profile (Fig. 6a-ii), column by column for each channel separately:

$$I\,(x, y) = I_0 + A \cdot \exp\left(\left(-\frac{x - x_0}{\sigma}\right)^2\right)\Bigg|_y$$

(f) Keep width σ of the Gaussian to define the ROI for final calculation of FRET efficiency, E (white line in Fig. 6a-iii). Hence, we restrict the final calculation of the FRET efficiency to the pixels which fall into the range of the intensity distribution of the acceptor channel according to $E(x, y) \in_{x0} (x, y) \pm \sigma(x, y)$. For segmentation and shape analyses, the fit also determines the centerline of the ROI, x_0.

(g) Subtract the I_0, the local background, from each corresponding column:

$$I_D; I_A; I_F(x,\ y) = I(x) - I_0|_y$$

3. *FRET calculation*

(a) Use the empirically determined bleedthrough factors, α and δ, from Subheading 3.4.1 to calculate the bleedthrough corrected FRET signal from background subtracted intensities in the FRET, donor, and acceptor channels on a pixel-by-pixel basis according to.

$$\mathrm{cF}(x, y) = I_F(x, y) - \delta \cdot I_D(x, y) - \alpha \cdot I_A(x, y)$$

with $\mathrm{cF}(x, y)$ as the corrected FRET signal at a certain pixel coordinate x, y, $I_F\,(x, y)$ as the raw, background subtracted FRET intensity, $I_D(x,\ y)$ as the background subtracted donor intensity, and $I_A(x, y)$ as the background subtracted acceptor intensity.

(b) Calculate the FRET efficiency within the neurite boundaries determined above according to [32]:

$$E = \frac{\mathrm{cF}}{I_D + \mathrm{cF}}$$

$$E = 100 \times \left(\frac{(\mathrm{cF} \cdot Q_D \cdot \frac{D}{A})}{\mathrm{qD} + (\mathrm{cF} \cdot Q_D \cdot \frac{D}{A})}\right)$$

with I_D and cF being the donor and bleedthrough corrected FRET channel intensities, respectively, Q_D as the previously determined donor quantum yield [33], qD and cF are the (quenched) donor emission (in the presence of acceptor) and bleedthrough corrected FRET emission, respectively, and the ratio D/A is the collection efficiency/gain ratio of the donor and acceptor photodetector as experimentally determined in Subheading 3.1.

(c) For each neuron, calculate an average FRET value and repeat this analysis for several samples taken from multiple animals. The number of samples needed to detect an effect of a treatment or genetic perturbation will depend on the variance in the biological control samples and the expected effect size.

(d) Repeat this analysis for different imaging control constructs with FRET pairs connected by varying linker length or a "no-force control" (Fig. 1b). A constitutively high FRET construct was engineered by replacing the entropic spring in TSMod by a 5 amino acid linker or a large domain (TRAF domain) that separate the two fretting fluorescent proteins [12, 33, 56] (*see* **Note 10**).

(e) All FRET efficiency values obtained from the force-sensitive TSMod cassette should fall into the dynamic range delineated by the low and high FRET standards (Fig. 7e). In our case, we measured an average FRET value of 22% for the experimental β-spectrin::TSMod construct, whereas the "no-force control," a TSMod without linkers expressed in the cytoplasm, showed an average FRET efficiency of 38%. If the average TSMod FRET is lower than the average of the no-force control construct, the difference indicates the amount of force by which the host protein is under tension (for representative results, interpretation, and more imaging tips to prevent artefacts, *see* **Notes 11–19**).

4. *Optional: Error analysis and estimating the uncertainty of the FRET calculation*

One of the drawbacks of FRET is poor signal-to-noise metrics, especially at low photon counts [57]. To visualize the certainty or uncertainty of the FRET calculation (Fig. 8a–d), we can estimate the error for each pixel inside the region of interest (Fig. 8c). The error for FRET calculation for each pixel can be propagated from the empirically determined pixel error in each channel according to

$$\sigma_{\mathrm{cF}}(x, y) = \frac{1}{\sqrt{N}} \times \sqrt{\left((I_F(x, y) \cdot \sigma_F)^2 + (I_D(x, y) \cdot \sigma_D)^2 + (I_A(x, y) \cdot \sigma_A)^2\right)}$$

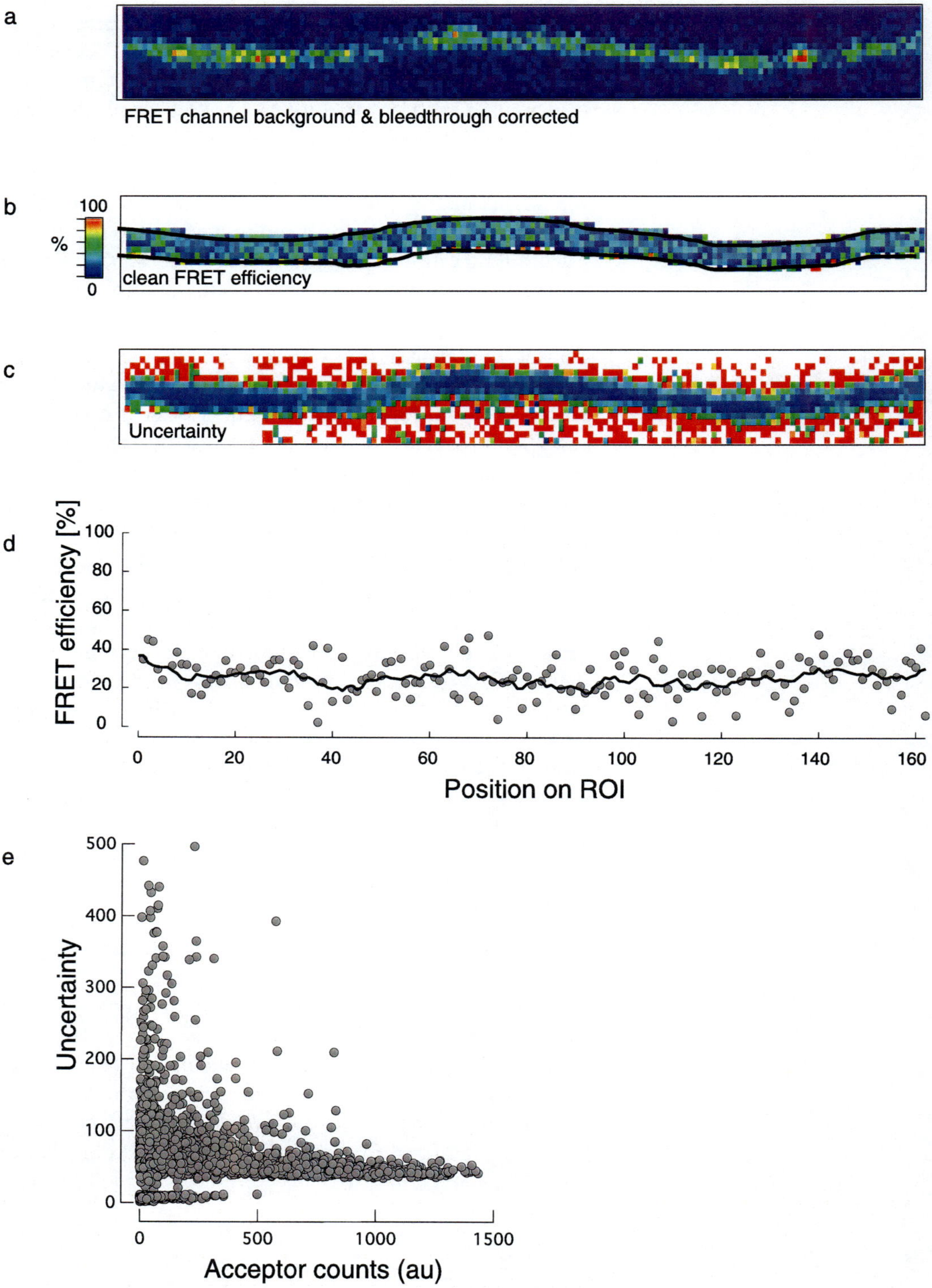

Fig. 8 FRET and error calculation
(**a**) Pseudocolored FRET channel intensity image after bleedthrough correction and background subtraction. Same ROI as in Fig. 6. (**b**) FRET efficiency image after FRET calculation of all pixels within the fitted ROI boundaries. (**c**) Corresponding uncertainty image of all pixels including the pixels outside the fitted ROI. Red

$$\sigma_E(x, y) = \left(\frac{E}{\mathrm{cF}(x,\ y)}\right)^2 \cdot \frac{1}{\sqrt{N}} \times \sqrt{(\mathrm{cF}(x,\ y) \cdot \sigma_D)^2 + (I_D(x,\ y) \cdot \sigma_{\mathrm{cF}})^2}$$

where N is the number of measurements for each pixel; σ_F, σ_A, σ_D are standard deviation of each pixel in the FRET, acceptor, and donor channel (Fig. 8a–c), respectively (as experimentally determined in Subheading 3.1); σ_{cF} is standard deviation of spectrally corrected FRET channel; and σ_E is the standard deviation of the value for FRET in each single pixel, assuming complete independence of noise in the donor and acceptor detectors. Both detectors follow Poisson statistics and variance scales linearly with number of photons counted as tested experimentally (Fig. 2c). We found that pixels with low acceptor signal (e.g., outside the ROI) have a high uncertainty in the FRET channel (Fig. 8e) (*see* **Note 20**).

4 Notes

1. Important for all image acquisition steps. Do not saturate pixels since this causes a deviation from linearity in the response. Reduce laser power during the image acquisition such that pixels are never saturated. Check with HiLo palette if necessary.
2. Avalanche photodiodes and hybrid GaAsP (HyD) detectors can be operated in photon counting mode with all gains on the detection disabled. If possible, repeat the calibration in photon counting mode. Such an analysis should reveal a slope of one, expected for shot noise limited detectors like these that follow Poisson statistics.
3. The egg shell is bleach resistant and therefore embryos can be easily isolated using this procedure. Over-incubation, however, may result in damage to some eggs, reducing the yield of age-synchronized animals.

Fig. 8 (continued) pixels highlight the ones with a high error and are excluded from the statistics. (**d**) Average FRET efficiency as a function of neurite position. (**e**) Uncertainty of the FRET calculation plotted against the acceptor intensity indicates that the largest error is expected for pixels with a low molecular representation. Further analysis shows that the large errors occur outside the respective ROIs

4. High-percentage agarose is needed to completely immobilize the worm. If lower percentage agarose is used, e.g., 2%, the worm will be able to move and eventually crawl out of the field of view. Alternatively, when using 2% agarose, worms can be anesthetized by adding 1 μL of 1 mM levamisole onto the pad prior transferring the animals.
5. The compressive immobilization method in a high-percentage agarose might alter mechanical behavior of the animal. If neither high-percentage agarose nor anesthetics can be used, microfluidic devices have specifically been engineered to trap living animals for imaging [49, 58, 59].
6. Microscopy facilities often do not have the optimal filter combinations for mTFP and Venus in the widefield epi-illumination pathway. In our experience, however, specialized mTFP or Venus filters are not required to see these fluorophores. Take care to limit the illumination time and power to prevent accidental bleaching of the fluorophores before actually acquiring data.
7. Because confocal imaging is a process of converting a continuous space signal into a numeric sequence, it is subjected to artefacts based on the sampling rate. The Nyquist theorem suggests that sampling of an analog signal should be carried out at the Nyquist rate, which is twice the fundamental frequency of the expected signal. In the confocal microscope, the highest fundamental frequency to be recorded is imposed by the optical system and is set by the resolution of the objective and particular wavelength to be imaged [60]. According to Nyquist theorem, the pixel size therefore has an optimum at $p < r/2.3$, with r as the smallest resolvable structure (e.g., the diffraction limit). A much smaller pixel size leads to unnecessary bleaching without additional gain in information.
8. Always keep the laser power and photodetector settings constant during an experiment. If parameters have to be varied, be sure to acquire bleedthrough images at the exact same imaging conditions (Fig. 4e) to eliminate cross excitation of the fluorophores due to the overlapping excitation spectra.
9. Make sure that the two channels are spatially well aligned. This is particularly important for camera-based detection and image splitting optics. Such pixel shift needs to be meticulously corrected for with submicron beads and post-acquisition image processing [34].
10. The bleedthrough ratios are generally assumed to be constant for all images taken with the same microscope setting and laser power. Therefore, the fraction of the donor and acceptor fluorescence in the raw FRET signal was subtracted. Even though the assumption of a constant bleedthrough ratio is generally

valid, photon arrival rate can also be modeled as a Poisson process in which the bleedthrough ratio might not be constant but depend on the actual pixel intensity of the directly illuminated channel. These phenomena play an important role when FRET is based on photon counting and hence signal-to-noise ratio approaches its limits. In such a case, the bleedthrough ratio can be modeled as a single exponential.

11. The comparison of the TSMod construct with the proposed FRET standards provide a reliable means to deduce quantitative values from intensity-based FRET imaging. Since the high FRET (5aa) and low FRET (TRAF) control standards have a fixed linker length and therefore have FRET efficiencies that do not vary with the mechanical state of the host protein. Likewise, the FRET values obtained from these constructs should not change upon different experimental conditions, e.g., drug treatment or mechanical manipulation. Previously, the FRET efficiency of both constructs was quantified by FLI microscopy and determined to be 50% and 10%, respectively [12, 33]. Even though the values derived from these standards can vary depending on the experimental system and specific FRET calculation, any experimental values falling outside of this dynamic range delineated by the fixed length FRET sensors should be treated with caution.

12. Dynamic measurement may be used, whenever possible, to manipulate the molecular tension while reading out FRET signals in real time. In that case, the FRET reporter can be assessed to which degree it is able to report a mechanical stress during a physiological process. For example, if no change of FRET occurs despite an imposed cellular strain, internal force transmission pathways might be disrupted. To a rough approximation, 5% difference in average FRET values correspond to $\approx$1.5 pN of tensile load. A detailed description of the calibration and design of different TSMod variants can be found in Refs. [10, 24].

13. In some circumstances, FRET values of the TSMod can be larger than the relaxed or no-force controls [20, 36]. Not all host molecules need to be under tension and several reports indicate the TSMod can also report whether or not the host proteins are under compression. Glycocalyx proteins experience significant compression during cellular adhesion [20], and β-spectrin is a flexible molecule that can sustain tensile and compressive stresses, depending on the physiological context and the animal's locomotory activity [36].

14. Pitfall: Intermolecular FRET.

    Several factors can make the interpretation of the FRET values challenging. For instance, intermolecular FRET may occur if the TSMod-expressing transgene is over-expressed, which would preclude the force measurements. A straightforward way to probe for the occurrence of intermolecular FRET is to generate doubly transgenic cells or animals in which two different molecules are expressed with either the acceptor or donor fluorophores silenced. Such constructs can be generated by mutating the conserved glycine residue to valine in each fluorophore (Table 1 and [12]). Any significant FRET signal in this transgenic must come from intermolecular FRET (Fig. 7d, e), as no FRET pair exists within the same molecule. Alternatively, if such constructs are not available, the calculated FRET efficiency can be related to the emission intensity of the acceptor channel after direct excitation of the acceptor (Fig. 7f). Under first approximation, this serves as a proxy for protein concentration [61].

15. Pitfall: Autofluorescence.

    Autofluorescence is a common problem when imaging in living tissues. *C. elegans* is known to express a high density of autofluorescent and birefringent vesicles in the intestine [62]. Several mutations of the *glo* genes interfere with the formation of the gut granules and thus lead to a strong reduction of the autofluorescence [62]. If using a genetic intervention is not possible because these genes also interfere with processes that are important for the question under investigation, fluorescence lifetime imaging microscopy can be used to decipher signal that originates from specific and nonspecific fluorophores. Autofluorescent dyes have a much shorter excited state lifetime (<1 ns) than genetically encoded fluorescent proteins (>>1 ns) such that pixels with this contribution can be easily distinguished [12, 63].

16. The current force sensors are built with standard fluorescent proteins well-matched to the reducing environment of the cytoplasm. These sensors might not work as well in extracellular milieu, under strongly oxidizing conditions. To improve force sensors enabling the visualization of stresses in the extracellular matrix, fluorescent proteins optimized for specific cellular environments could be used [64, 65].

17. Even though molecular tension sensors report on the average force per molecule, it represents a bulk measurement, as many molecules within a single pixel measurement contribute to the signal. Thus, individual molecules might experience more or less tension or even be compressed within the imaged volume. As pointed out before [8], the obtained FRET values are also averaged over the acquisition time and, especially, for scanning confocal represents a tiny snapshot within the image. This is even more prevalent in FLI microscopy, which may require tens of seconds to collect enough photons for subsequent fitting routines [33].

18. Sensitized emission FRET can produce known and characterized artefacts, e.g., associated with spectral bleedthrough or differential bleaching of the donor/acceptor fluorophores during the acquisition time. Other methods are available that minimize the impact of these issues, e.g., FLIM-FRET [12, 63], acceptor photobleaching [15], and spectral imaging [66].
19. If FRET values are notoriously high or negative, check bleedthrough constants and/or reevaluate the background subtraction.
20. For low signal, a high uncertainty is expected, as seen in Fig. 8e. Mostly, these pixels fall outside the ROI and can be excluded.

Acknowledgments

We thank the ICFO SLN facility for the generous use of their microscopes. M.K. acknowledges financial support from the Spanish Ministry of Economy and Competitiveness through the Plan Nacional (PGC2018-097882-A-I00), "Severo Ochoa" program for Centres of Excellence in R&D (CEX2019-000910-S; RYC-2016-21062), from Fundació Privada Cellex, Fundació Mir-Puig, and from Generalitat de Catalunya through the CERCA and Research program (2017 SGR 1012), in addition to funding through ERC (MechanoSystems) and HFSP (CDA00023/2018), la Caixa Foundation (ID 100010434, LCF/BQ/DI18/11660035), and MINECO (FPIPRE2019-088840 funded by MCIN/AEI/10.13039/501100011033 and ESF "Investing in your future" to N.S.). M.B.G. is supported by NIH Grant R35105092 and A.R.D. by NIH grant 1R35GM130332-01.

References

1. Kung C (2005) A possible unifying principle for mechanosensation. Nature 436(7051): 647–654
2. Katta S, Krieg M, Goodman MB (2015) Feeling force: physical and physiological principles enabling sensory mechanotransduction. Annu Rev Cell Dev Biol 31(1):347–371. https://doi.org/10.1146/annurev-cellbio-100913-013426
3. Venturini V, Pezzano F, Castro FC et al (2020) The nucleus measures shape changes for cellular proprioception to control dynamic cell behavior. Science 370(6514). https://doi.org/10.1126/science.aba2644
4. Hoffman BD, Grashoff C, Schwartz MA (2011) Dynamic molecular processes mediate cellular mechanotransduction. Nature 475(7356):316–323
5. Krieg M, Dunn AR, Goodman MB (2015) Mechanical systems biology of *C. elegans* touch sensation. BioEssays 37(3):335–344. https://doi.org/10.1002/bies.201400154
6. Krieg M, Fläschner G, Alsteens D et al (2018) Atomic force microscopy-based

mechanobiology. Nat Rev Phys 1:41–57. https://doi.org/10.1038/s42254-018-0001-7
7. Castro F, Venturini V, Ortiz-V'asquez S et al (2021) Direct force measurements of subcellular mechanics in confinement using optical tweezers. J Vis Exp 2021(174):1–35. https://doi.org/10.3791/62865
8. Gayrard C, Borghi N (2016) FRET-based molecular tension microscopy. Methods 94(2016):33–42. https://doi.org/10.1016/j.ymeth.2015.07.010
9. Hurst S, Vos BE, Brandt M et al (2021) Intracellular softening and increased viscoelastic fluidity during division. Nat Phys 17(11): 1270–1276. https://doi.org/10.1038/s41567-021-01368-z
10. Grashoff C, Hoffman BD, Brenner MD et al (2010) Measuring mechanical tension across vinculin reveals regulation of focal adhesion dynamics. Nature 466(7303):263–266
11. Canever H, Carollo PS, Fleurisson R et al (2021) Molecular tension microscopy of E-Cadherin during epithelial- mesenchymal transition. Methods Mol Biol 2179:289–299
12. Krieg M, Dunn AR, Goodman MB (2014) Mechanical control of the sense of touch by β-spectrin. Nat Cell Biol 16(3):224–233. https://doi.org/10.1038/ncb2915
13. Cai D, Chen SC, Prasad M et al (2014) Mechanical feedback through E-cadherin promotes direction sensing during collective cell migration. Cell 157(5):1146–1159. https://doi.org/10.1016/j.cell.2014.03.045
14. Vuong-Brender TTK, Boutillon A, Rodriguez D et al (2018) HMP-1/ α-catenin pro- motes junctional mechanical integrity during morphogenesis. PLoS ONE 13(2):1–21. https://doi.org/10.1371/journal.pone.0193279
15. Ye AA, Cane S, Maresca TJ (2016) Chromosome biorientation produces hundreds of piconewtons at a metazoan kinetochore. Nat Commun 7:1–9. https://doi.org/10.1038/ncomms13221
16. Lemke SB, Weidemann T, Cost AL et al (2018) A small proportion of Talin molecules transmit forces to achieve muscle attachment in vivo. PLoS Biol 310(939):446336. https://doi.org/10.1101/446336
17. Conway DDE, Breckenridge MMT, Hinde E et al (2013) Fluid shear stress on endothelial cells modulates mechanical tension across VE-cadherin and PECAM-1. Curr Biol 23(11):1024–1030. https://doi.org/10.1016/j.cub.2013.04.049.Fluid
18. Aird EJ, Tompkins KJ, Ramirez MP et al (2020) Enhanced molecular tension sensor based on Bioluminescence Resonance Energy Transfer (BRET). ACS Sensors 5(1):34–39. https://doi.org/10.1021/acssensors.9b00796
19. Ringer P, Weißl A, Cost AL et al (2017) Multiplexing molecular tension sensors reveals piconewton force gradient across talin-1. Nat Methods 14(11):1090–1096. https://doi.org/10.1038/nmeth.4431
20. Paszek MJ, DuFort CC, Rossier O et al (2014) The cancer glycocalyx mechanically primes integrin-mediated growth and survival. Nature 511(7509):319–325
21. Guo J, Wang Y, Sachs F et al (2014) Actin stress in cell reprogramming. Proc Natl Acad Sci 111(49):E5252–E5261. https://doi.org/10.1073/pnas.1411683111
22. Meng F, Suchyna TM, Sachs F (2008) A fluorescence energy transfer-based mechanical stress sensor for specific proteins in situ. FEBS J 275(12):3072–3087
23. Ye N, Verma D, Meng F et al (2014) Direct observation of α-actinin tension and recruitment at focal adhesions during contact growth. Exp Cell Res 327(1):57–67. https://doi.org/10.1016/j.yexcr.2014.07.026
24. LaCroix AS, Lynch AD, Berginski ME et al (2018) Tunable molecular tension sensors reveal extension-based control of vinculin loading. elife 7:1–36. https://doi.org/10.7554/elife.33927
25. Arsenovic PT, Ramachandran I, Bathula K et al (2016) Nesprin-2G, a component of the nuclear LINC complex, is subject to myosin-dependent tension. Biophys J 110(1):34–43. https://doi.org/10.1016/j.bpj.2015.11.014
26. Déjardin T, Carollo PS, Sipieter F et al (2020) Nesprins are mechanotransducers that discriminate epithelial-mesenchymal transition programs. J Cell Biol 219(10). https://doi.org/10.1083/JCB.201908036
27. Borghi N, Sorokina M, Shcherbakova OG et al (2012) E-cadherin is under constitutive actomyosin-generated tension that is increased at cell-cell contacts upon externally applied stretch. Proc Natl Acad Sci U S A 109(31): 12568–12,573
28. Price AJ, Cost AL, Ungewiß H et al (2018) Mechanical loading of desmosomes depends on the magnitude and orientation of external stress. Nat Commun 9(1). https://doi.org/10.1038/s41467-018-07523-0
29. Sanfeliu-Cerdán N, Mateos B, Garcia-Cabau C et al (2022) A rigidity phase transition of Stomatin condensates governs a switch from transport to mechanotransduction. BioRxiv.

https://doi.org/10.1101/2022.07.08.499356
30. Brenner MD, Zhou R, Conway DE et al (2016) Spider silk peptide is a compact, linear nano-spring ideal for intracellular tension sensing. Nano Lett 16(3):2096–2102. https://doi.org/10.1021/acs.nanolett.6b00305
31. Meng F, Sachs F (2012) Orientation-based FRET sensor for real-time imaging of cellular forces. J Cell Sci 125(3):743–750
32. Elangovan M, Wallrabe H, Chen Y et al (2003) Characterization of one- and two-photon excitation fluorescence resonance energy transfer microscopy. Methods 29(1):58–73
33. Day RN, Booker CF, Periasamy A (2008) Characterization of an improved donor fluorescent protein for Forster resonance energy transfer microscopy. J Biomed Opt 13(3): 31203
34. LaCroix AS, Rothenberg KE, Berginski ME et al (2015) Construction, imaging, and analysis of FRET-based tension sensors in living cells, vol 125. Elsevier Ltd. https://doi.org/10.1016/bs.mcb.2014.10.033
35. Cost AL, Ringer P, Chrostek-Grashoff A et al (2015) How to measure molecular forces in cells: a guide to evaluating genetically-encoded FRET-based tension sensors. Cell Mol Bioeng 8(1):96–105. https://doi.org/10.1007/s12195-014-0368-1
36. Das R, Lin LC, Catala-Castro F et al (2021) An asymmetric mechanical code ciphers curvature-dependent proprioceptor activity. Sci Adv 7: eabg4617. https://doi.org/10.1126/sciadv.abg4617
37. Kelley M, Yochem J, Krieg M et al (2015) FBN-1, a fibrillin-related protein, is required for resistance of the epidermis to mechanical deformation during c. elegans embryogenesis. elife 2015(4):1–71. https://doi.org/10.7554/eLife.06565
38. Johnson CP, Tang HY, Carag C et al (2007) Forced unfolding of proteins within cells. Science (New York, NY) 317(5838):663–666
39. Gates EM, LaCroix AS, Rothenberg KE et al (2019) Improving quality, reproducibility, and usability of FRET-based tension sensors. Cytometry Part A 95(2):201–213. https://doi.org/10.1002/cyto.a.23688
40. Fischer LS, Rangarajan S, Sadhanasatish T et al (2021) Molecular force measurement with tension sensors. Annu Rev Biophys 50:595–616. https://doi.org/10.1146/annurev-biophys-101920-064756
41. Sands B, Burnaevskiy N, Yun SR et al (2018) A toolkit for DNA assembly, genome engineering and multicolor imaging for C. elegans. Transl Med Aging 2(2018):1–10. https://doi.org/10.1016/j.tma.2018.01.001
42. Porta-de-la Riva M, Fontrodona L, Villanueva A et al (2012) Basic Caenorhabditis elegans methods: synchronization and observation. J Vis Exp 64:e4019. https://doi.org/10.3791/4019
43. Kim E, Sun L, Gabel CV et al (2013) Long-term imaging of Caenorhabditis elegans using nanoparticle-mediated immobilization. PLoS ONE 8(1):e53419
44. Bao Z, Murray JI (2011) Mounting Caenorhabditis elegans embryos for live imaging of embryogenesis. Cold Spring Harb Protoc 6(9):1089–1094. https://doi.org/10.1101/pdb.prot065599
45. Rivera Gomez KA, Schvarzstein M (2018) Immobilization nematodes for live imaging using an agarose pad produced with a Vinyl Record. microPublication Biology. https://doi.org/10.17912/QG0J-VT85
46. Jeong B, Kim SW, Bae YH (2012) Thermosensitive sol-gel reversible hydrogels. Adv Drug Deliv Rev 64(SUPPL):154–162. https://doi.org/10.1016/j.addr.2012.09.012
47. Gilleland CL, Rohde CB, Zeng F et al (2010) Microfluidic immobilization of physiologically active Caenorhabditis elegans. Nat Protoc 5(12):1888–1902
48. Burnett K, Edsinger E, Albrecht DR (2018) Rapid and gentle hydrogel encapsulation of living organisms enables long-term microscopy over multiple hours. Commun Biol 1(1). https://doi.org/10.1038/s42003-018-0079-6
49. Kopito RB, Levine E (2014) Durable spatiotemporal surveillance of Caenorhabditis elegans response to environmental cues. Lab Chip 14(4):764–770. https://doi.org/10.1039/c3lc51061a
50. Wen Q, Po MD, Hulme E et al (2012) Proprioceptive coupling within motor neurons drives C. elegans forward locomotion. Neuron 76(4): 750–761
51. Lockery SR, Lawton KJ, Doll JC et al (2008) Artificial dirt: microfluidic substrates for nematode neurobiology and behavior. J Neurophysiol 99(6):3136–3143. https://doi.org/10.1152/jn.91327.2007
52. Lambert TJ (2019) FPbase: a community-editable fluorescent protein database. Nat Methods 16(4):277–278. https://doi.org/10.1038/s41592-019-0352-8
53. Chen H 3rd, Koushik SV et al (2006) Measurement of FRET efficiency and ratio of donor to acceptor concentration in living cells. Biophys J 91(5):L39–L41

54. Sun Y, Periasamy A (2010) Additional correction for energy transfer efficiency calculation in filter-based Förster resonance energy transfer microscopy for more accurate results. J Biomed Opt 15(2):020513. https://doi.org/10.1117/1.3407655
55. Feige JN, Sage D, Wahli W et al (2005) PixFRET, an ImageJ plug-in for FRET calculation that can accommodate variations in spectral bleed-throughs. Microsc Res Tech 68(1):51–58
56. Koushik SV, Chen H, Thaler C et al (2006) Cerulean, Venus, and VenusY67C FRET reference standards. Biophys J 91(12):L99–L101
57. Esposito A (2020) How many photons are needed for FRET imaging? Biomed Opt Express 11(2):1186. https://doi.org/10.1364/boe.379305
58. Hulme SE, Shevkoplyas SS, Apfeld J et al (2007) A microfabricated array of clamps for immobilizing and imaging C. elegans. Lab Chip 7(11):1515
59. Berger S, Spiri S, DeMello A et al (2021) Microfluidic-based imaging of complete Caenorhabditis elegans larval development. Development (Cambridge) 148(18). https://doi.org/10.1242/DEV.199674
60. Heintzmann R (2006) Band limit and appropriate sampling in microscopy. Cell Biol 3:29–36. https://doi.org/10.1016/B978-012164730-8/50131-3
61. Kenworthy AK, Edidin M (1998) Distribution of a glycosylphosphatidylinositol-anchored protein at the apical surface of MDCK cells examined at a resolution of <100 A using imaging fluorescence resonance energy transfer. J Cell Biol 142(1):69–84
62. Morris C, Foster OK, Handa S et al (2018) Function and regulation of the Caenorhabditis elegans Rab32 family member GLO-1 in lysosome-related organelle biogenesis. PLoS Genet 14(11):1–36. https://doi.org/10.1371/journal.pgen.1007772
63. Datta R, Heaster TM, Sharick JT et al (2020) Fluorescence lifetime imaging microscopy: fundamentals and advances in instrumentation, analysis, and applications. J Biomed Opt 25(07):1. https://doi.org/10.1117/1.jbo.25.7.071203
64. Meiresonne NY, Consoli E, Mertens LM et al (2019) Superfolder mTurquoise2 ox optimized for the bacterial periplasm allows high efficiency in vivo FRET of cell division antibiotic targets. Mol Microbiol 111(4):1025–1038. https://doi.org/10.1111/mmi.14206
65. Costantini LM, Baloban M, Markwardt ML et al (2015) A palette of fluorescent proteins optimized for diverse cellular environments. Nat Commun 6(May). https://doi.org/10.1038/ncomms8670
66. Arsenovic PT, Mayer CR, Conway DE (2017) SensorFRET: a standardless approach to measuring pixel-based spectral bleed-through and FRET efficiency using spectral imaging. Sci Rep 7(1):1–15. https://doi.org/10.1038/s41598-017-15,411-8

Chapter 17

Single-Cell Quantification of the Mechanical Stability of Cell–Cell Adherens Junction Using Glass Micropipettes

Wenmao Huang, Chaoyu Fu, and Jie Yan

Abstract

Micropipette-based methods have been widely used for the manipulation of cells and characterization of the mechanical properties at the cell or tissue level. Here, we introduce the glass micropipette-based mechanical assays for the stability of cell–cell adhesion. A probing microbead coated with specific adhesion ligands, captured by a glass micropipette, is manipulated to form the adhesion complexes with the corresponding receptors on a single cell. Once the cell is moving away from the micropipette, forces are generated from 20 pN to 100 nN to the adhesion complexes, which are quantified in real-time based on the bending of the glass micropipette. We specifically emphasize the principle and method to probe the rupturing forces of the adhesion complexes at controlled force loading rates, the ligand coating on the probe microbeads, the force calibration of the glass micropipette, and the applications of the method to probe the E-cadherin-based cell–cell adhesions. The principles can be broadly applied to other cell adhesions such as cell–matrix adhesions, neuronal synapses, and bacterial–cell adhesions.

Key words Glass micropipette, Cell mechanics, Cell adhesion, Cell, Cell junction, Mechanical stability, Single-cell manipulation, Mechanobiology, Mechanotransduction

1 Introduction

Cell adhesion, the process by which cells interact with and attach to neighboring cells or extracellular matrix through various cell adhesion molecules (CAMs) [1], is the molecular basis of tissue architecture and morphogenesis [2, 3]. Such adhesions are necessary for the transmission of tension, which is generated from the internal actomyosin contraction or the external perturbation, in the cytoskeleton network and across cells in tissue [4]. This leads to the generation of mechanical communications and the regulation of cellular functions based on these CAMs, which enables cells to sense (mechanosensing) and respond (mechanotransduction) to local mechanical cues [5–8]. It has been well known that mechanosensing and mechanotransduction play critical roles in the

Ronen Zaidel-Bar (ed.), *Mechanobiology: Methods and Protocols*,
Methods in Molecular Biology, vol. 2600, https://doi.org/10.1007/978-1-0716-2851-5_17,

regulation of cell differentiation, cell cycle, cell migration, tissue growth, and repair [9–15].

The mechanical stability of the adhesion complexes is a critical determinant of the mechanosensing and mechanotransduction at cell–cell and cell–matrix contacts, dysregulation of which is implicated in many diseases [16–20]. For example, during cancer metastasis, cell–cell junctional stability is downregulated to allow cancer cells to detach from the tumor [16, 17]. The stability of the adhesion complexes is regulated in an inside-out manner depending on the intracellular mechanosensing proteins [21–23]. It is therefore of importance to quantify the mechanical stability of the cell–cell and cell–matrix adhesion complexes and investigate their regulations by intracellular mechanosensing proteins.

The current assays of the adhesion complexes have been mainly focused on their size, dynamics, and the tension levels generated on the complexes under different experimental conditions. Common technologies applied in these studies include atomic force microscopy [24, 25], traction force microscopy [26, 27], and micropillar arrays [28, 29]. However, these assays do not inform us of the mechanical stability of the adhesion complexes, namely the ability of the adhesion complexes to withstand tension under physiologically relevant conditions.

More recently, our group has developed an assay for the mechanical stability of cell–cell adherens junction (AJ) complexes and shown that the mechanical stability of AJ complexes is sensitively regulated by intracellular mechanosensing proteins α-catenin and vinculin [30]. Our assay is based on using a hollow flexible glass micropipette to catch an E-cadherin protein-coated microbead, which forms an E-cadherin/E-cadherin adhesion complex with cells adhered to a surface. The mechanical stability of the adhesion complex is quantified as the tension needed to detach the microbead from the cell when tension is increased at different loading rates. The rupturing tensions could be measured based on the level of the deflection of the tip of the glass micropipette due to micropipette bending at the time of the rupturing of the adhesion complex, using pre-calibrated bending stiffness of the micropipette [30]. Similar glass micropipette mechanical assays were previously applied to investigate chromosomal elasticity, nuclear deformation, and muscle fibril contraction [31–35].

In this chapter, we describe in detail the glass micropipette-based single-cell mechanical manipulation method for the mechanical stability of cell–cell adherens junction complexes. We demonstrate this method using the adhesion complex formed between E-cadherin-coated probing microbead and E-cadherin receptors on a single cell, a work we recently published [30], for example.

2 Materials

1. Thin-wall glass capillaries (1.0 mm outer diameter/0.75 mm inner diameter) and Microfil soft tip (28 g/97 mm long) are used for the glass micropipette preparation (Fig. 1).
2. Prepare 10 μg mL^{-1} E-cadherin human (his-tag) protein solution in 1× PBS buffer.
3. Prepare 10 μg mL^{-1} fibronectin protein solution in 1× PBS buffer.
4. Prepare a blocking buffer (1% (m/v) BSA, 1× PBS, pH = 7.4): weigh 1 g BSA and add 100 mL 1× PBS, pH = 7.4, ultrasound treat for 5 min until the BSA is totally dissolved. Filter the blocking buffer and deionized (DI) water with 0.22 μm filter.
5. Prepare diluted Dynabeads™ M-270 amino beads for the microbeads modification.
6. Prepare cell culture medium with Dulbecco's Modified Eagle Medium (DMEM), supplement with 10% fetal bovine serum (FBS), 100 units mL^{-1} penicillin, and 100 μg mL^{-1} streptomycin.
7. Prepare Polydimethylsiloxane (PDMS) using SYLGARD 184 reagents.

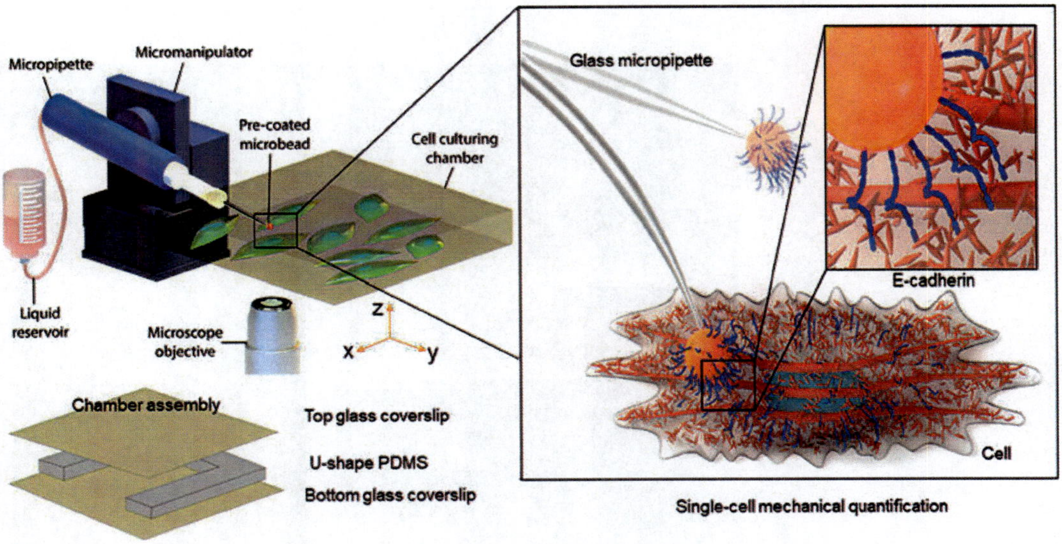

Fig. 1 Schematic diagram of the glass micropipette manipulation system for the stability measurement of single-cell adherens junction. (Reproduced from ref. 30 with permission from Elsevier [30])

3 Methods

3.1 Glass Micropipette Preparation

The micropipettes are made from thin-wall glass capillaries by a micropipette puller (P97, Fig. 2a). The glass micropipettes are subsequently cut by a heat platinum wire to get a round opening tip with a diameter of around 2 μm, which is suitable for capturing 3 μm-diameter microbeads [36, 37]. After that, the glass micropipettes were filled with water to remove all the air bubbles inside. Finally, the micropipettes were immersed in a blocking buffer and ready for use.

3.1.1 Micropipette Pulling

1. Gently put a thin-wall glass capillary into the platinum (Pt) heater of the micropipette puller (P97) and lock the capillary on both sides (Fig. 2b).
2. The program on the P97 flaming brown for 0.75 mm inner diameter (ID) and 1 mm outer diameter (OD) capillaries is normally set as: Pull 250, Velocity 250, Time 25, and Pressure 500. The parameters should be adjusted based on the condition of the Pt heater and the desired geometry of the tip (*see* **Note 1**).
3. Close the safety box of the P97 puller and press "Pull." After a breaking sound, the capillary is pulled apart, producing two glass micropipettes. Gently remove the glass micropipettes from the locker. Be careful with handling the glass micropipette during the whole experiment (*see* **Note 2**).

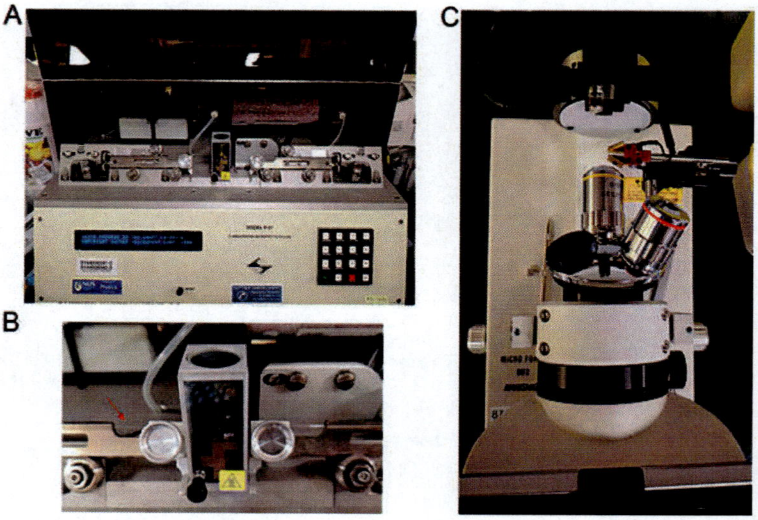

Fig. 2 Glass micropipette preparation. (**a**) micropipette puller, (**b**) a glass capillary (arrow) mounted in the puller, (**c**) micropipette cutter

3.1.2 Micropipette Tip Cutting

1. A commercial micropipette cutter (MF2, Fig. 2c) is used to obtain a smooth round opening tip with the proper diameter of around 2 μm.
2. First, clean the Pt wire using a soft brush. Adjust the voltage applied to the Pt wire, find a working voltage (V_w) under which the Pt wire is heated up to a temperature that is just high enough to melt the glass micropipette (*see* **Note 3**).
3. Adapt the micropipette to a micropipette holder. Adjust the *x*–*y*–*z* stage to move the micropipette and find the tip area in the microscope using a low-magnification objective lens (a 10× lens was used in our assay). Adjust the microscope stage to move the Pt wire to right above the micropipette at the desired cutting place (*see* **Note 4**).
4. Heat the Pt wire at V_w, move the Pt wire toward the micropipette until the glass micropipette tip is cut.
5. Double check the size and shape of the opening cut, and redo **step 4** if necessary. Then remove the micropipette from the holder and keep it in a safe box for future use.

3.1.3 Tip Filling and Blocking

1. Put the micropipette into a Petri-dish and fix it with adhesive tape if needed. Add a 100 uL DI water droplet to the tip of the glass micropipette (*see* **Note 5**). After incubation of 10 min, check the water–air interface inside the pipette under the microscope with a 10× objective lens.
2. Adapt a Microfil soft tip (28 g, 97 mm long) to a syringe filled with PBS. Carefully insert the soft tip into the glass micropipette as close to the tip as possible. Press the syringe to fill the hollow micropipette with water, and then pull out the Microfil soft tip.
3. Adhere the glass micropipette vertically to a desk or vertical wall to let the air bubble trapped inside the tip move up. Gentle finger tapping the glass micropipette can speed up this step.
4. After removing the bubbles, immerse the micropipette tip into the blocking buffer and incubate it overnight.
5. Add some distilled water using Microfil soft tip to compensate for the lost liquid volume due to evaporation overnight before use.

3.2 Force Calibration

The bending stiffness of each micropipette, characterized as the amount of restoring force per micrometer deflection of the micropipette tip from the equilibrium position, needs to be calibrated before experiments. This is done by using a pre-calibrated standard micropipette with a known bending stiffness (e.g., $k_s = 20.0 \pm 2.0$ pN μm^{-1}). The calibration of the standard micropipette is done by using transverse magnetic tweezers that generate a well-controlled force to a superparamagnetic microbead captured at the tip of the standard micropipette with negative pressure, in a direction perpendicular to the micropipette axis. The applied force causes the defection of the micropipette tip, from which the bending stiffness of the micropipette can be obtained with a relative error of ~20%. The force applied to the captured microbead can be controlled by adjusting the distance between the bead and magnet. The force–distance curve needs to be calibrated first [37].

3.2.1 Force Calibration of the Transverse Magnetic Tweezers

1. Prepare 90% glycerol solution (1 mL) mixed with 1 μL of stock M270 superparamagnetic beads. Add the solution into a thin-wall glass tube (1 mm square glass).
2. Put the thin-wall glass tube onto the inverted microscope-based transverse magnetic tweezers (Fig. 3a).
3. Set the reference position of the magnet. Use a micro-manipulator to move a pair of permanent magnets along the x-direction. Record the two x-positions, x_1 and x_2, of the micro-manipulator, one where the magnet front just enters the view area and the other where the magnet fully blocks the view area, using a 10× objective lens. The middle position $x_0 = \frac{x_1+x_2}{2}$ is set to be zero, representing the origin of the magnet in the x-direction. With this reference, the reading of the x-position magnet is the distance of the magnet from the microbead when the microbead is at the center of the view area.

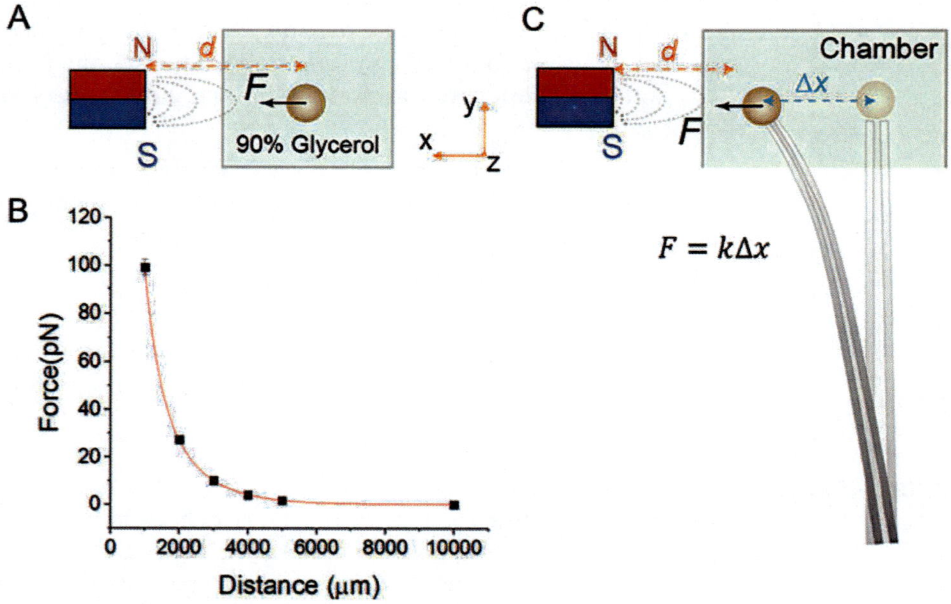

Fig. 3 The calibration of standard glass micropipettes by transverse magnetic tweezer. (**a**) The schematics of the force calibration of the transverse magnetic tweezers. (**b**) A typical force–distance (**d**) curve of a pair of magnets used in the transverse magnetic tweezers. (**c**) Schematic illustration of the standard micropipette calibration by the calibrated transverse magnetic tweezers

4. Visualize the microbeads inside the glycerol solution using a long working distance objective (40×). Track a few microbeads and record their *x–y* positions with a nanometer resolution [38].
5. Stepwise change the *x*-position of the magnet, from 10 μm to 5 μm, 3 μm, 2 μm, 1.5 μm, 1 μm, 0.8 μm, 0.5 μm, and 0.3 μm. At each step, wait for 30 s to record the movements of the beads and calculate the average velocities (v) of the beads at each magnet position. Make sure that beads move in the *x*-direction by adjusting the *y*- and *z*-positions of the magnet.
6. Using Stocks law, the force applied to each microbead is calculated by the equation $F = 6\pi\eta r \cdot v$, where r is the radius of the beads, η is the solution viscosity (e.g., $\eta = 0.21$ Pa s for 90% glycerol [39]).
7. Plot the average force at different magnet positions. Fit the force–distance data with a double-exponential function [38, 40], yielding a calibrating curve $F(d)$ (Fig. 3b).

3.2.2 Calibrate a Standard Glass Micropipette

1. Prepare a standard glass micropipette following Subheading 3.1.
2. Prepare a homemade thin-wall chamber with one side open. Add 1 μL stock of M270 superparamagnetic beads in 1 mL DI water into the chamber, which sink to the bottom of the chamber.
3. Fix the standard glass micropipette into a micropipette holder connected with a water reservoir as a pressure control system (Fig. 1). Test the pressure control system by changing the height of the connector.
4. Insert the glass micropipette into the chamber from the open side and pick a superparamagnetic bead from the bottom surface with negative pressure.
5. Adjust the height of the micropipette so that the tip is slightly above the surface. Adjust the focus to visualize both the micropipette and the microbeads. Move the micropipette holding a microbead to the center of view, and track the bead location and movement.
6. Stepwise change the magnet position from 10 μm to 5 μm, 3 μm, 2 μm, 1.5 μm, 1 μm, 0.8 μm, 0.5 μm, and 0.3 μm. At each step wait for 10 s to record the deflection of the micropipette tip from the resting position (Δx) (Fig. 3c).
7. Repeat **steps 4–6** for multiple microbeads.
8. At each *x*-position d of the magnet, the force applied to the bead on the tip of the micropipette is obtained from the pre-calibrated $F(d)$ curve.

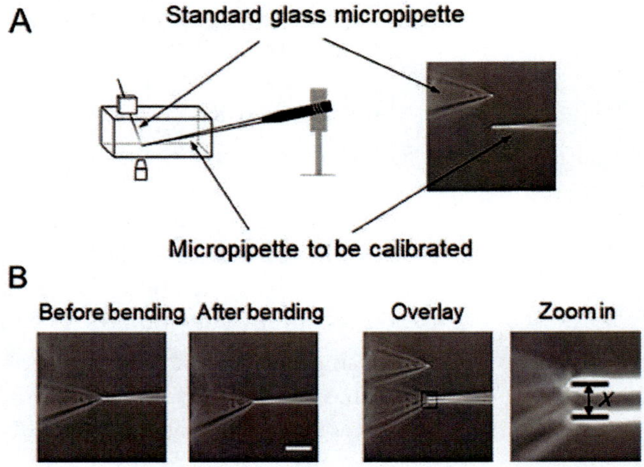

Fig. 4 The calibration of working micropipettes. (**a**) The schematics of the calibration chamber. The images of the standard glass micropipette and the working micropipette to be calibrated are shown on the right. (**b**) Move the standard micropipette to push against the working micropipette and record the resulting defection of the working micropipette tip. Scale bar, 50 μm. (Reproduced from ref. 30 with permission from Elsevier [30])

9. Plot the forces and the corresponding micropipette deflection data on a force– Δx graph. Linear fitting of the data leads to the calibration of the bending stiffness (k_s) of the standard glass micropipette (*see* **Note 6**).

3.2.3 Calibrate the Working Glass Micropipette

1. Each working glass micropipette is calibrated before or after the single-cell micropipette experiments.
2. Position the glass micropipette to be calibrated near the standard micropipette (Fig. 4a).
3. Use a micro-manipulator to move the standard glass micropipette toward the working glass micropipette to be calibrated until both tips of the glass micropipettes contacted each other. Set this position as the reference s_0.
4. Stepwise move the standard micropipette in the direction perpendicular to the axis of the working glass micropipette in the focal plane, at displacements of $\Delta s = 50$ μm, 100 μm, 150 μm, and 200 μm (**Note 7**). Track the position of working micropipette tips at each movement (Fig. 4b).
5. Measure the deflection of the working glass micropipette to be calibrated by the recorded bending (Δx) at each step using ImageJ program.
6. Calculate the average bending stiffness of the working micropipette: $k = k_s \times \frac{\Delta s - \Delta x}{\Delta x}$.

3.3 Prepare Probing Microbeads

1. Add 50 μL amine microbeads (Dynabeads M270) into 1.0 mL 1× PBS buffer. Use a magnet to pull down the beads and remove the supernatant solution. Wash the beads with 1× PBS buffer 5 times.
2. Take 70% glutaraldehyde solution out from −20 °C fridge, and wait in the chemical fume hood for 15 min to raise the temperature to room temperature. Prepare 1% (v/v) glutaraldehyde solution in DI water to a total volume of 1.0 mL.
3. Transfer the 1% glutaraldehyde solution to the test tube containing the washed amine microbeads. Use parafilm to seal the tube and mount the tube to an agitation rotator at room temperature for 2 h.
4. Wash the beads with PBS buffer at least 5 times. Each time use magnet to pull down the beads and remove the supernatant solution.
5. Prepare 10 μg mL^{-1} E-cadherin solution in PBS buffer (*see* **Note 8**). Transfer the E-cadherin solution to the test tube containing the glutaraldehyde-treated beads. Incubate the mixture overnight at 4 °C on an agitation rotator to avoid bead aggregation.
6. Wash the beads using magnet 5 times. Add blocking buffer to the E-cadherin-coated beads and incubate at 4 °C overnight on an agitation rotator to avoid bead aggregation until use.

3.4 Chamber Preparation

A homemade thin-wall chamber with one side open is made of PMDS and cover glasses (Fig. 1). Cells are planted into the chamber and adhere to the fibronectin-coated glass bottom surface for use.

3.4.1 Chamber Assembly

1. Prepare a PDMS membrane with a thickness of ~10 mm. Pour the PDMS mixture with the ratio of 10:1 (w/w) of the base and curing agent into a Petri-dish. Place the Petri-dish in a sealed vacuum chamber to degas the mixture, followed by curing the mixture at 80 °C for 2 h. Cut the PDMS membrane to obtain a U-shape PDMS frame for use.
2. Wash the glass coverslips (22 × 22 mm, 22 × 40 mm) with detergent, DI water (3 times), and 70% ethanol. Dry the coverslips by steam of nitrogen flow.
3. Use a 22 × 40 mm coverslip as the bottom glass, and adhere the U-shape PDMS frame on the top of it. Put another 22 × 22 mm coverslip onto the PDMS as the top glass (Fig. 1). Gently press the chamber on top and bottom to stick them together.
4. Prepare a solution of 10 μg mL^{-1} fibronectin in 1× PBS buffer (*see* **Note 9**). Add 0.5 mL fibronectin solution into the

chamber and incubate at room temperature in the culture fume hood for 30 min. Open the UV light inside the hood to avoid contamination.

5. Wash the chamber with 1× PBS 3 times. The chamber is now ready for planting cells.

3.4.2 Cell Planting

1. The cell lines to be tested were cultured in T75 cell culture flask at 37 °C, 5% CO_2 in cell culture medium.
2. Grow the cells for about 3 days until a confluence of ~90%. Digest the cell with Trypsin for 5 min. Resuspend the cells with medium and centrifuge for 3 min. After removal of the supernatant, resuspend the cells with fresh medium to yield ~10^5 per mL concentration (*see* **Note 10**).
3. Add 0.5 mL medium solution of cells to the chamber. Wait for 30 min and check the adhered cell density under a microscope. Culture the cells for 24–48 h (*see* **Note 10**).
4. Wash 1 μL stock E-cadherin-coated microbeads with culture medium 3 times. Add 0.5 mL fresh medium containing the microbeads into the chamber to change the medium. The cells are now ready for use.

3.5 Single-Cell Adhesion Force Measurement

3.5.1 System Setup

1. An inverted fluorescence microscope with a temperature and CO_2-maintaining box is used in the adhesion force measurement. A 10× lens is used for the initial localization of the micropipette, and a 40× lens is used for the force measurement.
2. A glass micropipette holder mounted on a micromanipulator (MP285) is installed beside the microscopy stage (Fig. 1). The holder is connected to a liquid reservoir, whose height can be adjusted to introduce proper aspiration pressure to hold or release a microbead on the micropipette tip.
3. Before mounting the micropipette, increase the liquid height of the reservoir to apply positive pressure to remove the air bubbles inside the connecting pipe and the holder.
4. Carefully mount the glass micropipette to the micropipette holder, seal all the parts, and check if there is any leaking. Position the micropipette by controlling the manipulator to close the center view of the microscope and adjust the micropipette to an angle of 5–15° above the stage (Fig. 1).
5. Visualize the glass micropipette under 10× lens. Move the microscope stage away from the micropipette to mount the chamber containing the cells to be tested. The opening side of the chamber should be toward the micropipette tip. Move the microscope stage back. Adjust the height of the micropipette to ensure that the micropipette tip is inserted into the chamber (*see* **Note 11**).

3.5.2 Adhesion Force Measurement

1. Visualize the glass micropipette in the chamber under 10× lens. Move down the micropipette tip to the bottom surface. Capture a microbead by applying negative pressure to the glass micropipette.
2. Change the microscope lens to 40×. Move the microbead captured on the micropipette to a single cell until touching the cell surface (*see* **Note 12**).
3. Move up the cell (z-axis) to exert a contact force to the microbead for around 500 pN based on the bending stiffness of the pre-calibrated glass micropipette and wait for 3 min. The waiting time can be altered based on the formation rate of the adhesion protein complex but should be kept consistent throughout the measurement.
4. Subsequently move the cell away from the micropipette along the y-direction (Fig. 1) by controlling the microscope stage at a constant speed v. The speed can be converted to loading rate by $r = kv$ with known bending stiffness k. This will exert a lateral force to the bead–cell adhesion complex.
5. Keep recording the bead position during the movement of the cell, until the bead–cell membrane adhesion is ruptured (Fig. 5a).
6. Change the microbead and repeat the measurements for multiple cells.
7. Analyze the data using ImageJ to determine the micropipette tip deflection distance (Δx) at the rupture point. Convert Δx to the rupture force based on the pre-calibrated bending stiffness k of the working micropipette ($F = k\Delta x$).

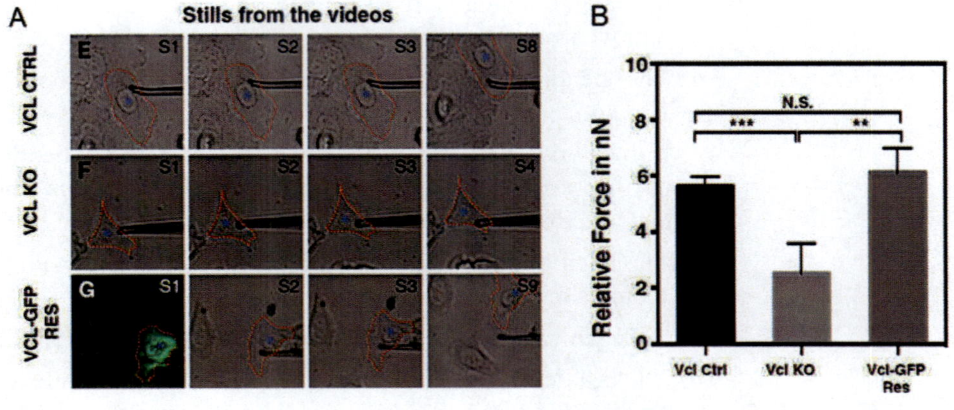

Fig. 5 The adhesion force measurements. (**a**) Stills of the videos from the bead-detachment experiments performed on vinculin (VCL) Ctrl, VCL-KO, VCL-GFP-rescued cells. (**b**) Histograms representing the relative force measurements required for bead displacement for Ctrl, VCL-KO, VCL-GFP-rescued cells ($n = 20$ cells/cell type; mean ± SEM). (Reproduced from ref. 30 with permission from Elsevier [30])

8. Build a histogram of the rupturing forces for further scientific analysis (Fig. 5b, results of the rupture force of vinculin Ctrl and KO cells as an example).

4 Notes

1. The micropipette puller program and parameters in the method are set for capillaries with 0.75 mm ID and 1 mm OD to obtain glass micropipette with diameters ranging between 1 to 3 μm and spring constant in the range from pN μm^{-1} to nN μm^{-1}. Glass micropipette with other sizes and stiffness could be accomplished by adjusting the parameters (for more details, see the manual).
2. Researchers must follow the Standard Operating Procedures (SOP) or Risk Assessment (RA) throughout handling the glass micropipette. Wear all personal protection equipment and use cut-resistance gloves. Never point the glass micropipette tip to anyone. Drop the glass micropipette into a sharp box.
3. The working voltage for micropipette cutting: The too high temperature could damage the tip shape and lead to a bent or blocked tip, while the too low temperature is not enough to adhere and cut the micropipette. In our experience, a working voltage should be chosen right before the Pt wire starts to change red.
4. Choosing the accurate site for cutting: A pre-calibrated eyepiece with a length scale can help to find the cutting site on the glass micropipette.
5. The best way of glass micropipette filling should start with filling the tip side of the micropipette with water or ethanol by capillary rising. Otherwise the bubbles will be difficult to remove from the inner pipette.
6. The bending of the micropipette tip follows Hooks law: $F = k\Delta x$.
7. If the working glass micropipette is too stiff that the bending is not observable during the calibration, researchers should increase the bending step from 50 μm to larger steps.
8. The E-cadherin protein can be replaced by any other adhesion proteins or ligands using a similar coating method.
9. Collagen can also be used for the chamber surface coating.
10. The cell density and incubation time should be adjusted based on the cell types and purpose of the experiment.
11. When the tip of the micropipette is inserted into the chamber, the movement of the stage should be slow. Sometimes because of the high tension on the water–air interface, the soft glass

micropipette tip would bend and adhere to the surface of the chamber. If the tip is not visualized after being inserted into the chamber, we suggest moving away from the chamber and trying to insert it again.

12. How to find the contact point where the microbead just touches the cell membrane: First, move the micropipette capturing a microbead to the top of the cell. Then drop down the microbead slowly by controlling the micropipette and adjust the focal plane to follow the microbead. The contact point is the point where the microbead is not further going down when moving the micropipette.

Acknowledgments

This work was supported by the Singapore Ministry of Education under the Research Scholarship Block, Singapore Ministry of Education Academic Research Fund Tier 2 (MOE-T2EP50220-0015), and the Ministry of Education under the Research Centres of Excellence program.

References

1. Edelman GM (1983) Cell adhesion molecules. Science 219(4584):450–457
2. Gumbiner BM (1996) Cell adhesion: the molecular basis of tissue architecture and morphogenesis. Cell 84(3):345–357
3. Hynes RO (1999) Cell adhesion: old and new questions. Trends Genet 15(12):M33–M37
4. Parsons JT, Horwitz AR, Schwartz MA (2010) Cell adhesion: integrating cytoskeletal dynamics and cellular tension. Nat Rev Mol Cell Biol 11(9):633–643
5. Braga VM (2002) Cell–cell adhesion and signalling. Curr Opin Cell Biol 14(5):546–556
6. Ladoux B et al (2015) The mechanotransduction machinery at work at adherens junctions. Integr Biol 7(10):1109–1119
7. Iskratsch T, Wolfenson H, Sheetz MP (2014) Appreciating force and shape—the rise of mechanotransduction in cell biology. Nat Rev Mol Cell Biol 15(12):825–833
8. Ladoux B, Mège R-M (2017) Mechanobiology of collective cell behaviours. Nat Rev Mol Cell Biol 18(12):743–757
9. Sumigray KD, Lechler T (2015) In: Yap AS (ed) Chapter twelve – cell adhesion in epidermal development and barrier formation, in current topics in developmental biology. Academic Press, pp 383–414
10. Frixen UH et al (1991) E-cadherin-mediated cell-cell adhesion prevents invasiveness of human carcinoma cells. J Cell Biol 113(1): 173–185
11. Guadagno TM et al (1993) A link between cyclin A expression and adhesion-dependent cell cycle progression. Science 262(5139): 1572–1575
12. Wight TN, Kinsella MG, Qwarnström EE (1992) The role of proteoglycans in cell adhesion, migration and proliferation. Curr Opin Cell Biol 4(5):793–801
13. Aplin AE, Howe AK, Juliano R (1999) Cell adhesion molecules, signal transduction and cell growth. Curr Opin Cell Biol 11(6): 737–744
14. Bacakova L et al (2011) Modulation of cell adhesion, proliferation and differentiation on materials designed for body implants. Biotechnol Adv 29(6):739–767
15. Kiss JZ et al (2001) The role of neural cell adhesion molecules in plasticity and repair. Brain Res Rev 36(2–3):175–184
16. Hirohashi S, Kanai Y (2003) Cell adhesion system and human cancer morphogenesis. Cancer Sci 94(7):575–581

17. Cavallaro U, Christofori G (2004) Cell adhesion and signalling by cadherins and Ig-CAMs in cancer. Nat Rev Cancer 4(2):118–132
18. Jang Y et al (1994) Cell adhesion molecules in coronary artery disease. J Am Coll Cardiol 24(7):1591–1601
19. Amagai M, Klaus-Kovtun V, Stanley JR (1991) Autoantibodies against a novel epithelial cadherin in pemphigus vulgaris, a disease of cell adhesion. Cell 67(5):869–877
20. Szekanecz Z, Koch AE (2000) Endothelial cells and immune cell migration. Arthritis Res Ther 2(5):1–6
21. Kim C, Ye F, Ginsberg MH (2011) Regulation of integrin activation. Annu Rev Cell Dev Biol 27:321–345
22. Ellis SJ et al (2014) The Talin head domain reinforces integrin-mediated adhesion by promoting adhesion complex stability and clustering. PLoS Genet 10(11):e1004756
23. Gumbiner BM (2005) Regulation of cadherin-mediated adhesion in morphogenesis. Nat Rev Mol Cell Biol 6(8):622–634
24. Sagvolden G et al (1999) Cell adhesion force microscopy. Proc Natl Acad Sci 96(2):471–476
25. Friedrichs J, Helenius J, Muller DJ (2010) Quantifying cellular adhesion to extracellular matrix components by single-cell force spectroscopy. Nat Protoc 5(7):1353–1361
26. Style RW et al (2014) Traction force microscopy in physics and biology. Soft Matter 10(23):4047–4055
27. Mertz Aaron F et al (2013) Cadherin-based intercellular adhesions organize epithelial cell–matrix traction forces. Proc Natl Acad Sci 110(3):842–847
28. Saez A et al (2007) Rigidity-driven growth and migration of epithelial cells on microstructured anisotropic substrates. Proc Natl Acad Sci 104(20):8281–8286
29. Vedula Sri Ram K et al (2012) Emerging modes of collective cell migration induced by geometrical constraints. Proc Natl Acad Sci 109(32):12974–12979
30. Biswas R et al (2021) Mechanical instability of adherens junctions overrides intrinsic quiescence of hair follicle stem cells. Developmental Cell 56(6):761–780.e7
31. Francis GW et al (1987) Direct measurement of cell detachment force on single cells using a new electromechanical method. J Cell Sci 87(4):519–523
32. Backholm M, Bäumchen O (2019) Micropipette force sensors for in vivo force measurements on single cells and multicellular microorganisms. Nat Protoc 14(2):594–615
33. Poirier MG, Monhait T, Marko JF (2002) Reversible hypercondensation and decondensation of mitotic chromosomes studied using combined chemical–micromechanical techniques. J Cell Biochem 85(2):422–434
34. Maniotis Andrew J, Christopher SC, Donald EI (1997) Demonstration of mechanical connections between integrins, cytoskeletal filaments, and nucleoplasm that stabilize nuclear structure. Proc Natl Acad Sci 94(3):849–854
35. Knoblauch M et al (2003) ATP-independent contractile proteins from plants. Nat Mater 2(9):600–603
36. Poirier MG, Marko JF (2002) Mitotic chromosomes are chromatin networks without a mechanically contiguous protein scaffold. Proc Natl Acad Sci 99(24):15393–15397
37. Yan J, Skoko D, Marko JF (2004) Near-field-magnetic-tweezer manipulation of single DNA molecules. Phys Rev E 70(1):011905
38. Yu M et al (2017) mDia1 senses both force and torque during F-actin filament polymerization. Nat Commun 8(1):1–9
39. Cheng N-S (2008) Formula for the viscosity of a glycerol– water mixture. Ind Eng Chem Res 47(9):3285–3288
40. Yu M et al (2018) Effects of mechanical stimuli on profilin- and formin-mediated actin polymerization. Nano Lett 18(8):5239–5247

Part VI

Probing Cell Signaling and Gene Expression in Response to Force

Chapter 18

Using Micropatterned Supported Lipid Bilayers to Probe the Mechanosensitivity of Signaling Receptors

Zhongwen Chen

Abstract

The ligand–receptor complexes in the cell–cell and cell–ECM interfaces mediate the mechanical coupling of cells to their microenvironment and transduce downstream signals to modulate cell functions. In this chapter I describe a microfabrication strategy to prepare a substrate of spatially segregated supported lipid bilayers and ECM components on which cells can form juxtacrine receptor signaling complexes and integrin adhesions simultaneously. This platform is specifically applicable for microscopically monitoring the signal transduction of each receptor, as well as the modulatory effects of receptor signaling on integrin adhesions and cell behaviors.

Key words Supported lipid bilayer, Microfabrication, Cell receptor, ECM, Receptor tyrosine kinase, Juxtacrine signaling, Integrin adhesion

1 Introduction

Communications between cells are mediated through cell surface receptors, of which physical parameters such as geometry and force play a crucial role in regulating the ensuing signaling transductions [1–3]. The membrane receptors that are activated in the cell–cell and cell–ECM interfaces enable sensing of the spatial and mechanical properties of cellular microenvironment, and affect the overall cell behaviors [4]. For example, the ephrinA1 ligand and EphA2 receptor juxtacrine signaling system mediates the mechanical coupling of two neighboring cells and transduces signals downstream to regulate integrin adhesions on the ECM [5–7]. Thus, the modulatory effect of EphA2 signaling over integrins should be considered in the context of spatial organization to correctly understand its role in regulating cell migration. Here I introduce a micropatterned supported lipid bilayer (SLB) platform to study the spatial–mechanical regulation of EphA2–integrin signaling crosstalk [8, 9]. The hybrid substrate is composed of fluid ephrinA1 ligands

Ronen Zaidel-Bar (ed.), *Mechanobiology: Methods and Protocols*,
Methods in Molecular Biology, vol. 2600, https://doi.org/10.1007/978-1-0716-2851-5_18,

presented on SLB patches embedded within regions of immobilized Arg-Gly-Asp (RGD) peptides. The regions of immobilized RGD function both as integrin ligands and lipid bilayer diffusion barriers. This configuration permits fluid rearrangement, clustering, and activation of EphA2 receptors in the membrane regions. Simultaneously, mechanically mediated focal adhesions formation occurs on rigid regions of the substrate, enabling cell spreading. The sites of ephrinA1–EphA2 interactions and integrin adhesions are thus separated to mimic in vivo organization. This system allows live monitoring of cognate receptor activation at the cell membrane interface and signaling crosstalk between the two signaling systems with high spatial–temporal resolution. While in the current protocol I specifically describe the procedures to probe EphA2 and integrin signaling, other modified versions of this method have been applied for the studies of mechanically modulated integrin endocytosis [10], filopodia formation [11], and molecular mechanisms of virus–host interactions [12].

This technique requires preparation of (1) small unilamellar vesicles (SUVs), (2) micropatterned substrate of polymer and SLB on a glass surface, (3) differential protein functionalization, (4) and live-cell imaging. Briefly, PLL-(g)-PEG-biotin polymers are first coated on a glass coverslip, followed by selective deep UV etching with a photomask to generate patterned substrate of polymers and bare glass surface. Lipid SUVs are then deposited onto the glass coverslip and lipid bilayers are self-assembled on the bare surface only. The ephrinA1 ligands are then functionalized to the lipid bilayers, while RGD are functionalized to the polymers (Fig. 1). The substrate is then ready for a live-cell imaging to monitor the activation of EphA2 and integrin receptors and cell behaviors.

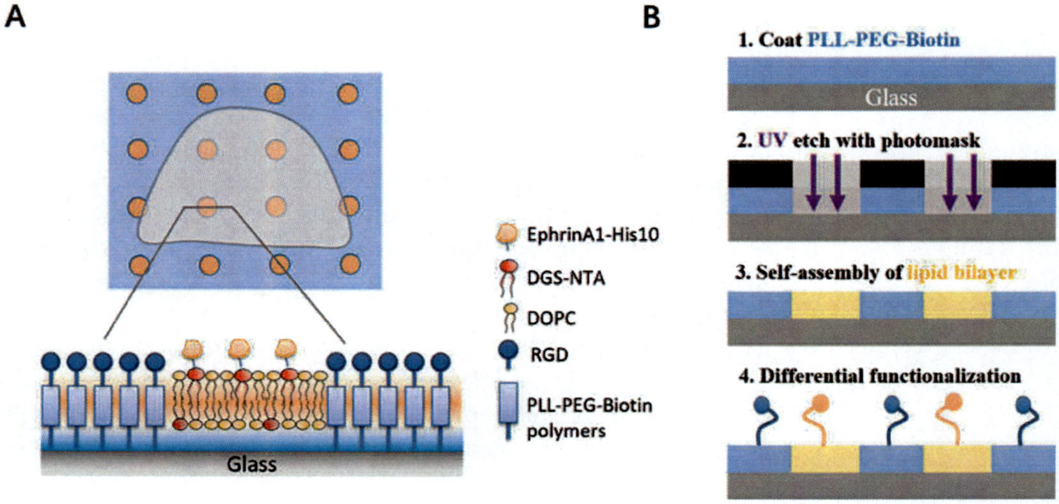

Fig. 1 Schematic illustrations of (**a**) the micropatterned SLBs to probe EphA2 and integrin signaling and (**b**) the microfabrication procedures

2 Materials

2.1 SUV Preparation

1. 18:1 (Δ9-Cis) PC (DOPC) lipid, stocked in chloroform at 5 mg/mL.
2. 18:1 DGS-NTA(Ni) lipid, stocked in chloroform at 1 mg/mL.
3. Chloroform.
4. Round bottom flask (10–25 mL).
5. Rotary evaporator, equipped with a water bath, a dry ice condenser, and a vacuum pump.
6. Tip sonicator.

2.2 Substrate Micropatterning

1. Glass coverslips (round, #1.5H Thickness, Ø25 mm diameter).
2. PBS buffer: 137 mM NaCl, 2.7 mM KCl, 10 mM Na_2HPO_4, 1.8 mM KH_2PO_4.
3. PLL(20)-g[3.5]-PEG(2)/PEG(3.4)-biotin(50%), stocked in PBS buffer at 1 mg/mL.
4. UVO cleaner.
5. Chromium-quartz photomask.
6. NeutrAvidin-Dylight405, stocked in PBS buffer at 1 mg/mL.
7. RGD-biotin, stocked in PBS buffer at 1 mg/mL.
8. EphrinA1 protein, with a 10-histidine tag in the C-terminal.
9. Attofluor™ cell imaging chamber.
10. TBS buffer: 25 mM Tris, 150 mM NaCl, 3 mM KCl.

2.3 Live-Cell Imaging

1. Hanks-based enzyme-free cell dissociation buffer.
2. Imaging buffer: 25 mM Tris, 140 mM NaCl, 3 mM KCl, 2 mM $CaCl_2$, 1 mM $MgCl_2$, 5.5 mM D-glucose.

3 Methods

Carry out all procedures at room temperature unless otherwise specified.

3.1 Substrate Micropatterning

1. Mix lipid molecules (96% DOPC +4% DGS-NTA(Ni), with a total mass of 2 mg) in excess chloroform (~10 mL) in a round bottom flask. Mount the round bottom flask in a rotary evaporator to allow the evaporation of chloroform under vacuum pumping and 50 °C water bath (*see* **Note 1**), leading to the formation of a thin lipid film in the bottom of the flask.
2. Dismount the round bottom flask from the rotary evaporator, immediately flow in N_2 jet for 5 min (*see* **Note 2**).

3. Resuspend the lipid film in 2 mL of deionized water by pipetting, allowing the formation of large, often multilamellar vesicles (1 mg/mL) (*see* **Note 3**).
4. Transfer the vesicle suspensions to a sonication tube. Sonicate it in ice by a probe tip sonicator to generate SUVs. Typically use a program of "10 sec on, 5 sec off" for 8–10 cycles for sonication (*see* **Note 4**).
5. Transfer the SUV suspensions to a fresh tube and centrifuge at 20,000 × *g* and 4 °C for 2 h to remove any debris, and then transfer the supernatant to a fresh tube and store at 4 °C until further use (*see* **Note 5**).

3.2 Prepare Micropatterned Substrate of Polymers and SLBs

1. Clean glass coverslips by sonication in a 1:1 mixture of isopropanol and water for 30 min, followed by overnight (or longer) incubation in 50% H_2SO_4 solution. Before usage, take the coverslips out of H_2SO_4, rinse with water, and expose the glass coverslips to UV light in an enclosed UVO cleaner for 10 min, rinse with water and dry by N_2 jet (*see* **Note 6**).
2. Incubate the glass coverslips with PLL(20)-g[3.5]-PEG(2)/PEG(3.4)-biotin(50%) at a concentration of 0.1 mg/mL for 2 h or overnight (*see* **Note 7**).
3. Rinse the substrates with water to remove excess polymers and dry by N_2 jet.
4. Drop ~1.5 μL of water on the micropatterned area (1 cm × 1 cm square) of a chromium-quartz photomask and lay the polymer-coated glass substrate on top of it to make a close contact (*see* **Note 8**). Expose the substrate to deep UV light for 9–12 min in a UVO cleaner. The UV light exposed polymer will be degraded (*see* **Note 9**).
5. Take the glass coverslip out of the photomask by flowing with water. Rinse the glass substrates with excess water to remove any degraded polymers (*see* **Note 10**).
6. Mix SUVs with TBS buffer in 1:1 ratio. Incubate the glass substrates with SUV solution for 5 min to allow the self-assembly of lipid bilayers in regions of the substrates that were UV etched in the previous step (*see* **Note 11**).
7. Assemble the glass coverslip into a cell culture imaging chamber under an aqueous environment (*see* **Note 12**).
8. Rinse the substrate with excess TBS buffer.

3.3 Protein Functionalization

1. Block the substrate with bovine serum albumin (BSA) solution (0.05–0.1 mg/mL in TBS buffer) for 2 h at room temperature or overnight at 4 °C.

2. Wash the substrate with TBS buffer and incubate with NeutrAvidin-Dylight 405 (1.5 μg/mL) for 15 min to bind surface biotin groups.
3. Wash excess NeutrAvidin-Dylight 405 with TBS buffer and incubate with ephrinA1-His 10 (purified ephrinA1 protein with a 10-histidine tag in the C-terminus, labeled with Alexa Fluor™ 680 dye) and RGD-biotin for an additional 90 min (*see* **Note 13**). The ephrinA1-His 10 will bind to DGS-NTA(Ni) on SLBs and RGD-biotin will bind to NeutrAvidin on the polymers.
4. Remove unbound proteins by washing with imaging buffer.
5. Check the fluorescence of ephrinA1 and NeutrAvidin-Dylight 405 under a microscope. The substrate is then ready for use in a live-cell experiment (*see* **Note 14**).

3.4 Live-Cell Imaging

1. Wash cells with PBS buffer for 2 times. Detach cells by incubating with enzyme-free, Hanks-based cell disassociation buffer for 5–15 min at 37 °C (*see* **Note 15**).
2. Gently pipette the detached cells and transfer to a falcon tube. Centrifuge the cell suspensions at 1000 rpm for 4 min, remove the supernatant, and resuspend cells in the imaging buffer.
3. Seed a low density of cells into pre-warmed (37 °C) cell chambers and allow to interact with the substrate.
4. Take time-laps images as soon as cells land on the substrate (*see* **Note 16**). The cell integrins can bind RGD peptides to form focal adhesions and allow cell spreading, during which cells interact with multiple ephrinA1-functionalized SLBs to form ephrinA1–EphA2 clusters (Fig. 2).

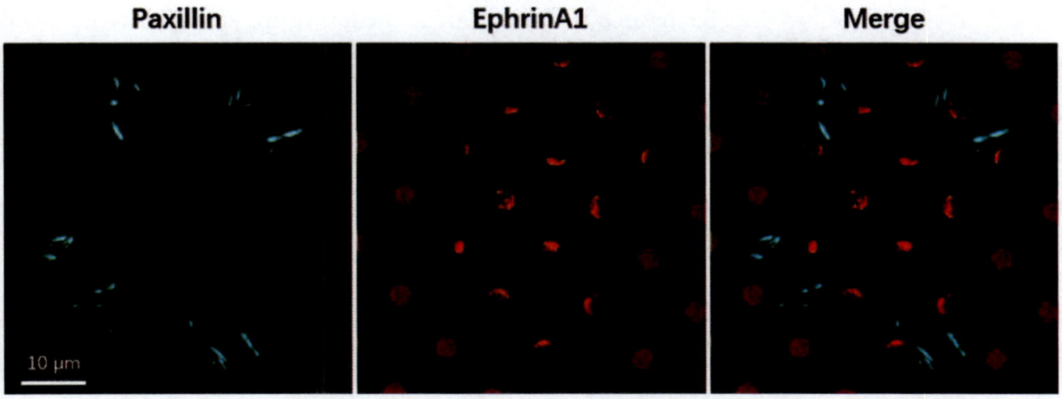

Fig. 2 The fluorescence images of a paxillin-EGFP transfected cell spread on the ephrinA1/RGD substrate for 90 min

4 Notes

1. To avoid chloroform boiling, lower the round bottom flask slowly to just above the water level. After the evaporation of almost all chloroform, immerse the flask into water for ~1 min for complete removal of chloroform.
2. The lipid films can be oxidized when exposed to air.
3. Optionally, immerse the flask into water bath at up to 60 °C to help dissolve the lipid films into water.
4. The program should be optimized based on the power of the tip sonicator. The "off-time" is necessary to avoid overheating of the SUV solution during sonication.
5. The SUVs can be stored for up to 2 weeks to form a good lipid bilayer.
6. The cleaned coverslips should be used for the next step immediately.
7. To reduce the usage of reagents, drop ~30 μL of the incubation solution to a parafilm and lay the glass coverslips on top of it.
8. Water help absorb the glass coverslip onto the photomask. Use a minimal amount of water to keep the gap as tight as possible.
9. This is the most critical step. If the polymers are not degraded completely, it would result in a non-fluid lipid bilayer in the later step. Users should optimize the exposure time based on the distance of the photomask and UV lamp, and the lamp power in their equipment settings. Try not to prolong the exposure too much to avoid the complete evaporation of water underneath the coverslip.
10. The glass substrates should be used for lipid vesicles deposition immediately.
11. To reduce the usage of reagents, drop ~30 μL of the SUV solution to a parafilm and lay the glass coverslips on top of it.
12. Always keep SLBs in aqueous solutions. The lipid bilayer structure will be damaged if it is dried.
13. The protein is recommended to contain more than 8 histidine for a stable binding to the DGS-NTA(Ni) lipids. The biotin–NeutrAvidin–biotin binding is very stable.
14. The mobility of ephrinA1 on lipid bilayers was checked by fluorescence recovery after photobleaching (FRAP). The ephrinA1 will dissociate from SLB gradually because of the dynamic binding of His-tag and NTA(Ni). The cell experiments should be done within a few hours after washing off the unbound ephrinA1.

15. Trypsin-treated cells take a few hours to recover and spread on RGD surface, which is not suitable for this experiment. The enzyme-free dissociation buffer works well for lightly adherent cell lines, but it may take a significant prolonged incubation to detach strongly adherent cells.
16. An Eclipse Ti inverted microscope (Nikon) with a TIRF system and Evolve electron multiplying charge-coupled device (EMCCD) camera (Photometrics) was used for most imaging experiments. TIRF microscopy was performed with a 100× TIRF objective with a numerical aperture of 1.49 (Nikon) and an iChrome MLE-L multilaser engine as a laser source (Toptica Photonics). Throughout live-cell imaging, the cell chamber was kept at 37 °C.

Acknowledgments

Z.C. was supported by Shanghai Pujiang Program (20PJ1400800) and Shanghai Municipal Science and Technology Major Project (2019SHZDZX02).

References

1. Wang XF, Ha T (2013) Defining single molecular forces required to activate integrin and notch signaling. Science 340(6135):991–994
2. Basu R, Whitlock BM, Husson J, Le Floc'h A, Jin W, Oyler-Yaniv A, Dotiwala F, Giannone G, Hivroz C, Biais N et al (2016) Cytotoxic T cells use mechanical force to potentiate target cell killing. Cell 165(1):100–110
3. Natkanski E, Lee W-Y, Mistry B, Casal A, Molloy JE, Tolar P (2013) B cells use mechanical energy to discriminate antigen affinities. Science 340(6140):1587–1590
4. Casaletto JB, McClatchey AI (2012) Spatial regulation of receptor tyrosine kinases in development and cancer. Nat Rev Cancer 12(6): 386–399
5. Salaita K, Nair PM, Petit RS, Neve RM, Das D, Gray JW, Groves JT (2010) Restriction of receptor movement alters cellular response: physical force sensing by EphA2. Science 327(5971):1380–1385
6. Parri M, Buricchi F, Giannoni E, Grimaldi G, Mello T, Raugei G, Ramponi G, Chiarugi P (2007) EphrinA1 activates a Src/focal adhesion kinase-mediated motility response leading to rho-dependent actino/myosin contractility. J Biol Chem 282(27):19619–19628
7. Lisabeth EM, Falivelli G, Pasquale EB (2013) Eph receptor signaling and Ephrins. Cold Spring Harb Perspect Biol 5(9):a009159
8. Chen Z, Oh D, Biswas KH, Yu C-H, Zaidel-Bar R, Groves JT (2018) Spatially modulated ephrinA1:EphA2 signaling increases local contractility and global focal adhesion dynamics to promote cell motility. Proc Natl Acad Sci 115(25):E5696–E5705
9. Chen Z, Oh D, Biswas KH, Zaidel-Bar R, Groves JT (2021) Probing the effect of clustering on EphA2 receptor signaling efficiency by subcellular control of ligand-receptor mobility. elife 10:20210820
10. Yu C-h, Rafiq NBM, Cao F, Zhou Y, Krishnasamy A, Biswas KH, Ravasio A, Chen Z, Wang Y-H, Kawauchi K et al (2015) Integrin-beta3 clusters recruit clathrin-mediated endocytic machinery in the absence of traction force. Nat Commun 6:8672
11. Alieva NO, Efremov AK, Hu S, Oh D, Chen Z, Natarajan M, Ong HT, Jégou A, Romet-Lemonne G, Groves JT et al (2019) Myosin IIA and formin dependent mechanosensitivity of filopodia adhesion. Nat Commun 10(1): 3593
12. Wong JJ, Chen Z, Chung JK, Groves JT, Jardetzky TS (2021) EphrinB2 clustering by Nipah virus G is required to activate and trap F intermediates at supported lipid bilayer–cell interfaces. Sci Adv 7(5):eabe1235

Chapter 19

Monitoring Mechano-Regulation of Gene Expression by RNA Sequencing

Yekaterina A. Miroshnikova

Abstract

The advent of high-throughput sequencing techniques has revolutionized biological research. One such method is RNA sequencing, which has become a relatively affordable and routine method for quantifying and comparing gene expression changes over desired experimental conditions. Along with the popularity of the method, a myriad of user-friendly, open-source computational tools have also emerged for differential gene expression analyses. Correspondingly, decades of mechanobiology research have established that mechanical cues, both alone and/or in combination with biochemical signals, can be powerful regulators of transcriptional programs and consequently cell state/fate transitions. Thus, it has become possible to investigate both universal and specific temporally resolved transcriptional responses upon mechanical stimulation genome-wide. This chapter will describe methods to analyze transcriptional changes in response to extrinsic mechanical stretch.

Key words Mechanotransduction, RNA sequencing, Transcription, Nuclear mechanics

1 Introduction

RNA sequencing was developed more than a decade ago and has become an indispensable tool with which to study RNA biology and genome function [1–4]. Due to the emergence of next generation sequencing (NGS) [3, 5], RNA sequencing can now not only be utilized to access differential gene expression (DGE) transcriptome-wide but also be applied to quantify mRNA splicing, noncoding RNAs, intermolecular RNA interactions, enhancer RNAs, translation, spatial transcriptomics, as well transcriptome analyses at single cell resolution. Despite the wide range of application, DGE analyses remain one of the most central applications and is the subject of this chapter. Mechanical perturbations bring new technical challenges and considerations to this otherwise well-established technique and these aspects will be discussed throughout the chapter.

Ronen Zaidel-Bar (ed.), *Mechanobiology: Methods and Protocols*,
Methods in Molecular Biology, vol. 2600, https://doi.org/10.1007/978-1-0716-2851-5_19,

With the emergence of the field of mechanotransduction, the study of how mechanical forces are converted into biological signals, a large number of both commercially available and custom-built bioreactors to mechanically stimulate cells and tissues in two- and three-dimensions have been developed. Custom-built reactors are generally more cost-effective and offer greater flexibility with respect to sample holder design, material properties of adhesive substrates, etc. but are often also more laborious and require specialist skills. Commercial systems make mechanobiology accessible to non-expertism, which is valuable as laboratories from diverse expertise are able to now study the role of mechanics in many processes. Here, a commercially available Flexcell tension system (Flexcell International Corporation) will be used as an example of setting up a mechanical manipulation experiment for downstream analysis for differential gene expression with RNA sequencing.

1.1 Method Overview

Flexcell tension system utilizes positive and negative air pressure to apply either static or cyclic mechanical strain to cells plated on two-dimensional, commercially available silicon elastomer membrane culture plates. Importantly, culture plates can be purchased pre-coated or with covalently bonded extracellular matrix proteins or be conjugated with desired ligand of interest (described below). One additional important consideration is the stiffness of the material, which should be optimized appropriately for a given cell type. For instance, currently Flexcell offers silicone elastomer with moduli stiffness ranging from 1 to 60 kPa.

It is first important to establish the appropriate mechanical manipulation regime for cellular system of choice. For any stretch experiment, it is first important to establish (1) whether the stretch regime would be equibiaxial or uniaxial and (2) whether the regime would be cyclic or static. Further, it is important to establish the necessary amount of strain to elicit desired effects (i.e., magnitude of elongation required, for Flexcell tension system the maximum achievable elongation is about 21%). For cyclic stretch it is also important to consider the effect of frequency of stimulation on desired effect, as well as the total time of mechanical manipulation, which is closely related to chosen frequency regime. Finally, Flexcell tension system offers static, sinusoidal, triangular, square, heart stimulation-like (as well as custom waveforms), and the choice of waveform is application-dependent and should be chosen to best replicate desired in vivo or experimental conditions under investigation.

Prior to setting up a stretch experiment, one must consider appropriate surface coating of the flexible membranes. In general, it is helpful to utilize 4–5 times higher ligand concentrations than those used for coating glass (for instance 50 μg/mL concentration of fibronectin versus 10 μg/mL for glass). Further, due to the design of the Flexcell tension system, where proper stimulation is only experienced in the central portion of the well at the location of the 25 mm loading post, it is critical to use plating guides (either homemade or purchased from Flexcell International Corporation) to restrict cellular adhesion to the area of proper manipulation.

Another critical consideration for the stretch experiment is utilization of proper controls. In the case of Flexcell tension system, it is possible to block pressure transduction with rubber stoppers, allowing control plates to be embedded within identical experimental setup as those plates that are being stretched. Final consideration is experimental design. Flexcell tension system allows simultaneous stimulation of up to four multi-well plates, thus a maximum of four conditions, including control non-stretched plate. As for any experiment, it is critical that loading and unloading of the plates is performed by the same operator, especially as it relates to the removal of plates from the Flexcell tension system upon completion of stimulation and it may be beneficial to perform RNA lysis (described below) while the plates are still embedded within the stretch device, if possible, to avoid extra stimulation. It is also advisable to install the plates within the Flexcell tension system at least 1 h prior to performing stimulation, as the act of placing the plates onto greased stretch posts and sealing the system may already perturb the flexible membranes.

Briefly, upon completion of mechanical manipulation experiment (minimum of triplicates for all conditions) and all the necessary controls, the following workflow is employed:

1. RNA is extracted immediately post experimental perturbations.
   - If possible, matched experimental conditions should be collected together on the same day/in replicate sequence with controls (i.e., control 1, mechanical manipulation 1, control 2, mechanical manipulation 2, control 3, mechanical manipulation 3 versus controls 1–3, then mechanical manipulations 1–3; thus avoiding sequential sample collection one condition after another).
2. RNA is enriched for mRNA (or any RNA species of interest) or processed for ribosomal RNA depletion.
3. cDNA is synthesized,
4. Adaptor-ligated sequencing library is prepared.
5. Samples are sequenced on desired platform.
6. Quality control (QC) computational analysis is performed.
7. DGE computational analysis is performed.

2 Materials

1. Relevant mechanical manipulator (here Flexcell tension system from Flexcell International Corporation).
2. Relevant cell or tissue system(s).
3. Relevant substrates.
4. Phosphate Buffered Saline (PBS), pH 7.4.
5. UltraPure DNase/RNase-Free Distilled Water.
6. RNA extraction kit (in general, Trizol extraction followed with Zymo RNA cleaning with concentrator kit with DNAse leads to higher yields versus column-based purification methods; however, proper cleaning step post Trizon method is required as this method has higher level of protein and organics contamination, which can inhibit library prep. If RNA amount is not limiting, Qiagen RNeasy or similar kit would be best; depending on starting cell numbers it may be necessary to adjust the kit to best match starting cellular material amount).
7. Agilent tape station and High Sensitivity RNA ScreenTape.
8. ERCC RNA Spike-In Mix.
9. Library preparation kit, if preparing by self – this is platform specific (e.g., TruSeq Total RNA-Seq library prep kit from Illumina).

3 Methods

1. RNA is extracted following specific kit/reagent instructions and QCed for purity, quality, and concentration.
   1.1. RNA sequencing is a powerful tool but only as good as the starting material, thus it is critical to perform QC tests with mechanical manipulations to ensure high-quality sequencing data can be produced. In general, a minimal RIN (RNA integrity value) value for RNA is 7–8, ideally closer to 10, which is maximum quality RNA. With respect to quantity, this is slightly platform-specific but in general a minimum of 500 ng per sample is required for all QC and library preparations steps.
   1.2. As a general rule, despite aiming for best quality RNA, all samples should have relatively constant RIN values, i.e., deviation of a maximum of 1–1.5 points at absolute maximum.
   1.3. If rRNA depletion is to be performed (see also below), it is recommended to check total RNA amount and integrity post depletion step. If this is not performed by the facility, it is important to reQC and quantify RNA amount and integrity of sample post depletion.

2. To control for extrinsic sources of variability, synthetic spike-in standards [6] have developed by the External RNA Control Consortium (ERCC) and should be added to experimental RNA.
   2.1. Depending on the sequencing facility, this is a point at which RNA can be submitted to a facility for all downstream analysis.
3. RNA is enriched for mRNA (or any RNA species of interest) or processed for ribosomal RNA (rRNA) depletion.
   3.1. This depends on the regime and time of mechanical perturbation (seconds to minutes vs. hours to days) and the species of interest where mechanical manipulation is expected to elicit an effect.
   3.2. It is important to consider RNA species turnover times and lifetimes when designing experiments to ensure changes can be captured on the given mechanical manipulation timeline.
4. cDNA is synthesized by RT and second strand sequencing using preferred reagents/kits.
5. Adaptor-ligated sequencing library is prepared (generally this can be done by the sequencing facility).
6. Sequencing library is then sequenced to specified read depth (10–50 million reads per sample, paired end for higher quality alignment as well as the ability to detect indel variants) on desired NGS platform (Roche 454, Illumina, Helicos, and PacBio which use DNA Polymerase for their synthesis or SOLiD which utilizes DNA Ligase; Helicos and PacBio being single molecule-based while all others ensemble-based). Many considerations must go into the choice of the sequencing platform as they have different caveats and advantages – for instance, some do not amplify starting material, while others do which leads to differential error rates, etc. (for more refer to [7]).
7. Computational analysis relating to QC: aligning and assembling sequencing reads to species-specific reference genome of platform-appropriate version, quantification of read spanning transcripts, filtering and normalization between samples.
8. Computational analysis relating to DGE: normalized read count data is then used to perform statistical analysis first for baseline expression of fragments per kilobase of transcript per million mapped reads (FPKMs), which are calculated for quantile normalization and averaged for all biological replicates and then DESeq2/EdgeR (negative binomial distribution approaches) or baySeq/EBSeq (Bayesian approaches) are used for DGE analysis.

4 Notes

1. Minimal RNA amount: 500 ng (100–1000 ng range) if can aim for 1000 ng/sample.

   1.1. 260/280 and 260/230 > 1.8

2. Coverage depth: higher sequencing depth provides more statistical power to detect DGE for genes with low expression levels – this may be particularly important for mechanical perturbations that are of low magnitude or applied for short periods of time.

3. Spike-in utilization: a number of previously published work [8–10] indicate that mechanical forces and master transcription regulators can alter gene expression patterns globally. Thus in order to perform meaningful DGE analyses on the background of global expression changes, it is critical to include spike-in control reference RNAs so that changes can be compared systematically between conditions. Spike-in standards should be utilized for any experimental conditions that may lead to differential amounts of RNA/cell.

Acknowledgments

This work was supported by the Intramural Research Program at the National Institutes of Health (NIH)/the National Institute of Diabetes and Digestive and Kidney Diseases (NIDDK).

References

1. Emrich SJ, Barbazuk WB, Li L, Schnable PS (2007) Gene discovery and annotation using LCM-454 transcriptome sequencing. Genome Res 17:69–73
2. Lister R et al (2008) Highly integrated single-base resolution maps of the epigenome in Arabidopsis. Cell 133:523–536
3. Nagalakshmi U et al (2008) The transcriptional landscape of the yeast genome defined by RNA sequencing. Science 320:1344–1349
4. Mortazavi A, Williams BA, McCue K, Schaeffer L, Wold B (2008) Mapping and quantifying mammalian transcriptomes by RNA-Seq. Nat Methods 5:621–628
5. Wilhelm BT et al (2008) Dynamic repertoire of a eukaryotic transcriptome surveyed at single-nucleotide resolution. Nature 453:1239–1243
6. Jiang L et al (2011) Synthetic spike-in standards for RNA-seq experiments. Genome Res 21:1543–1551
7. Chu Y, Corey DR (2012) RNA sequencing: platform selection, experimental design, and data interpretation. Nucleic Acid Ther 22: 271–274
8. Le HQ et al (2016) Mechanical regulation of transcription controls Polycomb-mediated gene silencing during lineage commitment. Nat Cell Biol 18:864–875
9. Lin CY et al (2012) Transcriptional amplification in tumor cells with elevated c-Myc. Cell 151:56–67
10. Nie Z et al (2012) c-Myc is a universal amplifier of expressed genes in lymphocytes and embryonic stem cells. Cell 151:68–79

Chapter 20

Testing the Role of Focal Adhesion Kinase (FAK) in Topography-Mediated Stem Cell Differentiation by Inhibiting FAK Phosphorylation

Yuan Yao and Evelyn K. F. Yim

Abstract

Stem cell differentiation can be modulated by the substrate topographies. Focal adhesion kinase (FAK) has been identified as a key regulator in topography-induced stem cell mechanotransduction. This chapter will describe a protocol to study the effect of FAK phosphorylation inhibition on topography-mediated stem cell differentiation. The FAK phosphorylation was inhibited using a FAK inhibitor and the effects on stem cell differentiation were examined using western blot and immunofluorescence staining.

Key words Focal adhesion kinase, Stem cell differentiation, Focal adhesion kinase phosphorylation inhibition, Immunofluorescence staining, Western blot

1 Introduction

Stem cells are important cell sources for regenerative medicine due to their capability of self-renewal and differentiation. In native environment, stem cells are exposed to biophysical and biochemical stimuli from their extracellular matrix, which regulate the stem cell differentiation. Biomaterials that are fabricated to mimic the native biophysical cues have shown success in directing cell differentiation. Numerous studies have demonstrated that micro- and nano-sized surface topographies could drive specific cell differentiation [1, 2]. Neuronal differentiation could be induced by both grooves and nanopillars [3–5], and grooves have also shown success in promoting osteogenic differentiation [6]. While the modulation effect of surface topography on stem cell differentiation is widely proved, the underlying mechanism still requires further investigation.

Focal adhesion kinase (FAK) is a central regulator of cell behaviors. Activation of FAK is a critical event in cellular responses to topographical features. Knockout of FAK has significantly reduced

Ronen Zaidel-Bar (ed.), *Mechanobiology: Methods and Protocols*,
Methods in Molecular Biology, vol. 2600, https://doi.org/10.1007/978-1-0716-2851-5_20,

cell migration and adhesion [7, 8]. Many researchers suggested that FAK phosphorylation is also essential in topography-induced stem cell differentiation [9–11]. Human mesenchymal stem cells (hMSCs) cultured on nanograting structures had increased FAK phosphorylation and promoted neuronal differentiation [9, 12]. MSC osteogenic differentiation was also regulated by FAK [13]. These studies indicated the direct effect of phosphorylated FAK (pFAK) in topography-guided stem cell sensing.

In this chapter, we present a method to understand the role of FAK in topography-mediated cell mechanosensing by FAK phosphorylation inhibition. Here, we use human mesenchymal stem cells (hMSCs) as an example and examine the neuronal differentiation of hMSCs based on the expression of microtubule-associated protein 2 (MAP2). We describe the pFAK inhibition protocol and procedures to determine the optimal concentration and incubation time of FAK inhibitor. We also provide the detailed procedures and required materials and instruments to examine the effect of pFAK inhibition on stem cell differentiation using western blot and immunofluorescence staining.

2 Materials

2.1 Sample Preparation

1. Patterned silicon wafer (*see* **Note 1**).
2. Trichloro(1H, 1H, 2H, 2H-perfluorooctyl)silane.
3. Surfactant: 0.01% Triton X-100 in deionized (DI) water.
4. SYLGARD 184 Silicone Elastomer kit.
5. Absolute and 70% ethanol.
6. Collagen type I from rat tail.

2.2 FAK Inhibition

1. Serum-free cell culture medium.
2. 1× phosphate buffered saline (PBS), pH 7.4.
3. FAK inhibitor: Y15 or PF-573228. Prepare 20 μM Y15 or PF-573228 stock solution in 1× PBS and store the stock solution in −20 °C. Thaw the solution before adding to cell culture.

2.3 Western Blot

1. Ice.
2. Buffers: 1× Tris-buffered saline (TBS), pH 7.4; 1× PBS.
3. RIPA protein extraction solution: 10 mM Tris–HCl, pH 8.0, 1 mM EDTA, 0.5 mM EGTA, 1% Triton X-100, 0.1% sodium deoxycholate, 0.1% SDS, 140 mM NaCl.
4. Protease and phosphatase inhibitor cocktail.
5. BCA protein assay kit.

6. Laemmli 2× buffer: 65.8 mM Tris–HCl, pH 6.8, 2.1% SDS, 26.3% (w/v) glycerol, 0.01% bromophenol blue.
7. β-Mercaptoethanol.
8. DI water.
9. 4–15% Mini-PROTEAN® TGX™ Precast Protein Gels (*see* **Note 2**).
10. Protein standards.
11. 1× SDS-PAGE running buffer: 25 mM Tris, 192 mM glycine, 0.1% SDS.
12. Transfer buffer: 1× Tris–glycine, 20% methanol.
13. Polyvinylidene fluoride (PVDF) transfer membrane.
14. Filter paper.
15. 1× TBST buffer: 1× TBS, 0.1% Tween 20.
16. Blocking buffer: 5% bovine serum albumin (BSA) and 1% glycine in TBST buffer.
17. Primary antibodies: rabbit anti-FAK primary antibody, rabbit anti-pFAK primary antibody, and mouse anti-GAPDH (*see* **Note 3**).
18. Horseradish (HRP)-conjugated secondary antibodies.
19. ECL plus western blot detection kit.
20. Strip buffer: 0.2 M glycine, 0.1% SDS, 1% Tween 20, pH 2.2.

2.4 Immunofluorescence Staining

1. Fixation solution: 4% paraformaldehyde (PFA), pH 7.4.
2. Buffers: 1× TBS; 1× PBS.
3. Permeabilization reagent: 0.1% Triton X-100 in 1× PBS.
4. Blocking reagent: 1% BSA and 5% goat serum in 1× TBS.
5. Markers: rabbit anti-pFAK antibody, rabbit anti-Paxillin, and mouse anti-MAP2 (*see* **Note 4**).
6. Secondary antibodies: Alexa-Fluor 488 goat anti-rabbit and Alexa-Fluor 546 goat anti-mouse secondary antibodies.
7. 4′,6-Diamidino-2-phenylindole (DAPI) and Alexa Fluor™ 546 Phalloidin.
8. Aqueous anti-fade mounting medium: e.g. Fluoromount.
9. Fluorescence microscope.

2.5 Equipment and Consumables

1. Plasma surface treatment machine.
2. Vacuum desiccator and vacuum pump.
3. Laminar flow cabinet/biosafety cabinet (BSC).
4. 37 °C, 5% CO_2 cell culture incubator.
5. 37 °C water bath.

6. Confocal microscope.
7. Orbital shaker.
8. Scale and weighing dishes.
9. Petri dishes.
10. Dry block heater.
11. ChemiDoc MP imaging system.
12. Trans-Blot Turbo System.
13. Centrifuge and centrifuge tubes.
14. Tissue culture coated 24-well plates.
15. Pipettes (adjustable volume: 0.5–10 μL, 20–200 μL, 100–1000 μL) and pipette tips.
16. Microscopy glass coverslips.

3 Methods

3.1 Preparation of Samples

3.1.1 Surface Treatment of Patterned Molds for Soft Lithography

1. Move a vacuum desiccator in a fume hood and connect the vacuum desiccator with a vacuum pump. Place the silicon wafer in the vacuum desiccator.
2. Place a small plastic weighing dish in the vacuum desiccator beside the silicon wafer. Add ~25 μL of trichloro(1H, 1H, 2H, 2H-perfluorooctyl)silane in the weighing dish.
3. Close the desiccator lid, turn on the vacuum pump for 30 min, and keep the lid on overnight.
4. Take the silicon wafer out from the desiccator. Wash with 0.01% Triton X-100. Clean with DI water and blow dry with nitrogen gas (*see* **Note 5**).

3.1.2 Preparation of Patterned Polydimethylsiloxane (PDMS) Samples

1. Weigh polydimethylsiloxane (PDMS) base polymer from SYLGARD 184 Silicone Elastomer kit and curing agent at a ratio of 10:1. Mix the base polymer with the curing agent thoroughly for 2–3 min. Place the mixture in a vacuum desiccator and degas for 30 min.
2. Place the silanized silicon wafer in a Petri dish inside a vacuum desiccator. Pour the PDMS mixture on the silanized silicon wafer (*see* **Note 6**). Degas for 30 min.
3. Cure the PDMS in a 60 °C oven overnight.
4. Carefully peel the PDMS samples off from the silicon wafers.
5. Repeat the above **steps 1–4** in Subheading 3.1.2 to make additional patterned PDMS replicas. Avoid using the first five PDMS replicas from a patterned mold after surface treatment for cell culture (*see* **Note 7**).

3.1.3 Preparation of Samples for Cell Culture

1. Cut the samples into disc shape and load the samples in a sterile tissue culture-coated 24-well plate (*see* **Note 8**). Add one sample into each well.
2. Treat samples with air plasma for 1 min, add 500 μL of 70% ethanol in each well, and UV sterilize for 30 min. Remove the 70% ethanol after UV treatment.
3. Coat the PDMS samples with collagen type I solution at a coating density of 2.5 $\mu g/cm^2$ and incubate for 1 h at 37 °C.
4. Aspirate excess collagen solution and wash the samples with 1× PBS. Aspirate the PBS and seed cells directly on the samples.

3.2 Treatment of Cells with Focal Adhesion Kinase Inhibitor

1. Separate samples into four groups: patterned/−pFAKi, unpatterned/−pFAKi, patterned/+pFAKi, and unpatterned/+pFAKi.
2. Detach cells and prepare cell suspension with desirable cell density.
3. For +pFAKi groups, add FAK inhibitor in cell suspension and mix well. For 1 mL of cell suspension, add 5 μL FAK inhibitor stock solution. For −pFAKi groups, add 5 μL of 1× PBS in the cell suspension.
4. Add 500 μL of +pFAKi or − pFAKi cell suspension on each sample (*see* **Note 9**).
5. Culture the cells for 7 days in a 37 °C, 5% CO_2 cell culture incubator. For +pFAKi groups, change media every day using cell culture medium containing FAK inhibitor. For −pFAKi groups, change media using normal cell culture medium (*see* **Note 10**).

3.3 Determination of Optimal Focal Adhesion Kinase Inhibitor Concentration and Incubation Time

The optimal concentration of FAK inhibitor and incubation time are dependent on cell type. The parameters in Subheadings 3.1 and 3.2 were designed for hMSCs. We recommend trying different FAK inhibitor concentrations and incubation times based on the following optimization procedures. The selected concentration and incubation of FAK inhibitor should reduce the pFAK level but should not detach cells from the substrate materials.

3.3.1 Concentration of pFAK Inhibitor

1. Separate samples into several parallel groups with an increasing concentration of FAK inhibitor from 0 to 1 μM.
2. Detach cells and prepare cell suspensions with desirable cell density.
3. Add 0–1 μM, for example, 0, 0.02, 0.1, 0.5, and 1 μM of FAK inhibitor in cell suspensions and mix well.
4. Add 500 μL of cell suspensions containing 0–1 μM of FAK inhibitor on each group of samples.

5. Culture cells for 7 days in a 37 °C, 5% CO_2 cell culture incubator. Change media every day with cell culture media containing 0–1 μM FAK inhibitor. Observe cells with a microscope to check cell adhesion. After 7 days, measure the pFAK/FAK level using western blot (*see* **Note 11**).

3.3.2 Incubation Time with pFAK Inhibitor

1. Separate samples into parallel groups with increasing incubation time with FAK inhibitor from 0 to 7 days.
2. Detach cells and prepare cell suspension with desirable cell density.
3. Add 5 μL FAK inhibitor stock solution per 1 mL of cell suspension and mix well. For the group with 0 day of incubation with FAK inhibitor, add 5 μL of 1× PBS per 1 mL cell suspension.
4. Add 500 μL of cell suspensions on each group of samples.
5. Culture cells for 7 days in a 37 °C, 5% CO_2 cell culture incubator and change media every day. Use FAK inhibitor-containing medium when changing culture medium. Replace the FAK inhibitor-containing medium with normal medium after the stipulated days of pFAK inhibition. For the group of 0-day incubation with FAK inhibitor, use normal medium for the whole culture period. Observe cells with a microscope to check cell adhesion. After 7 days, measure the pFAK/FAK level using western blot.

3.4 Western Blot

3.4.1 Sample Preparation

1. Dilute phosphatase and protease inhibitor in RIPA lysis buffer to produce 1× solution. Keep the solution on ice.
2. Remove cell culture media from adherent cells.
3. Wash cells twice with ice-cold 1× PBS.
4. Add ice-cold lysis buffer to the cells. Use 1 mL buffer per 5×10^6 cells. Incubate the cells with lysis buffer for at least 5 min.
5. Scrape the cells with a sterile ice-cold cell scraper. Transfer the cell suspension into a precooled centrifuge tube. Place the centrifuge tube on ice in a Styrofoam box. Incubate the samples in the lysis buffer for 30 min with constant agitation.
6. Centrifuge the samples at 12,000 rpm for 20 min at 4 °C. Collect the supernatant into a clean precooled tube and discard the pellet.
7. Measure the protein concentration using a BCA assay. After determination of the protein concentration, samples can be stored at −80 °C or used for electrophoresis.

3.4.2 Electrophoresis

1. Prepare sample buffer by adding 50 μL of β-mercaptoethanol to 950 μL of 2× Laemli buffer.
2. Mix protein samples with sample buffer 1:1 (*see* **Note 12**).
3. Denature the proteins by heating the samples to 95 °C for 5 min or 70 °C for 10 min.
4. Set up the gels. Remove the comb and tap from the gel. Load two gel cassettes or one gel cassette and one buffer dam into the electrode assembly. Assemble the electrophoresis cell and add running buffer in the inner chamber until overflow (*see* **Note 13**). Add running buffer in the outer chamber. Rinse the sample wells with running buffer.
5. Load protein samples (20 μL) and protein standards (5 μL standard and 15 μL sample buffer).
6. Run electrophoresis at 300 V and 400 mA for 15 min.

3.4.3 Protein Transfer

1. Pre-wet the PVDF membrane in methanol.
2. Equilibrate gel, membrane, and filter papers in the transfer buffers for at least 30 min.
3. Assemble the transfer sandwich in the order: cathode, filter paper, gel, membrane, filter paper, anode. Gently roll out bubbles.
4. Place the transfer sandwich in the instrument. Transfer for 7 min.
5. After transfer, take out the membrane and wash 3 times with TBST.

3.4.4 Immuno-Blot

1. Block the membrane in blocking buffer for 1 h at 4 °C with agitation.
2. Rinse the membrane with TBST 3 times for 5 min.
3. Cut the membrane horizontally between 75 and 50 kDa with a blade and stain FAK and GAPDH separately (Fig. 1).
4. Dilute the pFAK and GAPDH primary antibodies in blocking buffer and incubate the membrane in primary antibody dilution overnight at 4 °C with agitation (*see* **Note 14**).
5. Rinse the membrane with TBST 3 times for 5 min.
6. Dilute the secondary antibodies in TBST and incubate the membrane in secondary antibody dilution for 1 h at room temperature with agitation.
7. Rinse the membrane 3 times with TBST.
8. Prepare the ECL detection solution according to manufacturer's protocol.
9. Incubate the membrane with the detection solution for 5 min. The recommended volume detection solution is 0.1 mL/cm^2.

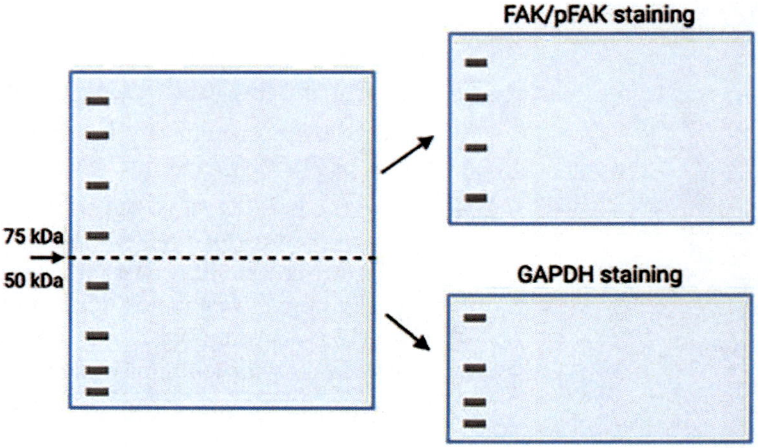

Fig. 1 Schematic diagram showing the cutting of the membrane between 75 and 50 kDa

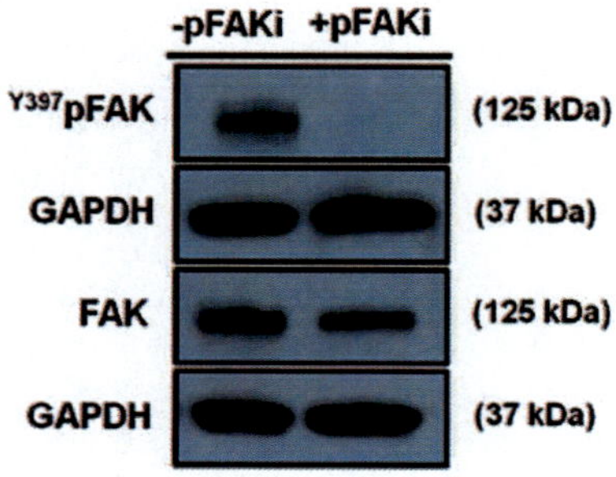

Fig. 2 Western blot of FAK and pFAK protein levels of human mesenchymal stem cells (hMSCs) with pFAK inhibition (+pFAKi) and without pFAK inhibition (−pFAKi). (Adapted with permission from Ref. [10]. Copyright 2013 American Chemical Society)

10. Run chemiluminescence imaging.
11. Strip the membrane. Add enough stripping buffer to cover the membrane and incubate at room temperature for 10 min with gentle agitation. Discard the buffer and repeat incubation for 10 min with fresh stripping buffer. Wash the membrane 3 times with TBST with gentle agitation (*see* **Note 15**).
12. Redo the blocking and staining. Stain the membrane with FAK and GAPDH primary antibodies (*see* **Note 16**).
13. Run chemiluminescence imaging (Fig. 2). Normalize the results to GAPDH and quantify the pFAK and FAK protein levels.

3.5 Immunofluorescence Staining

The volume of reagents used here is for samples with an area of 1 cm^2.

1. Remove culture media and gently wash the samples with 500 μL 1× filtered PBS for 3 times.
2. Fix cells with 4% PFA for 15 min at room temperature. After fixation, remove PFA solution and wash the cells with 500 μL 1× filtered PBS for 3 times.
3. Permeabilize cells with 500 μL 0.1% Triton X-100 for 15 min. After permeabilization, remove Triton X-100 solution and wash the cells with 500 μL sterile 1× filtered TBS for 3 times.
4. Block the cells with 500 μL blocking solution and incubate for 1 h at room temperature or overnight at 4 °C.
5. Primary antibody incubation: Prepare primary antibody solution by diluting the primary antibodies in 1× filtered TBS containing 1% goat serum. Add a drop of primary antibody solution on a piece of clean parafilm inside a Petri dish (*see* **Note 17**). Take out the sample from the culture plate, invert the sample so that the cells are facing down, and put the sample on the drop of the antibody solution. Cover the Petri dish and seal with parafilm to prevent evaporation. Incubate the cells with primary antibody solution overnight at 4 °C. After incubation, wash the samples 3 times with 1× sterile TBS before secondary antibody incubation.
6. Secondary antibody incubation: Prepare secondary antibody solution by diluting secondary antibodies in 1× filtered TBS. Incubate the samples with secondary antibodies as described in **step 5**. Cover the Petri dish with aluminum foil paper and incubate at room temperature for 1 h. After incubation with secondary antibodies, wash the samples for 3 times with 1× sterile TBS.
7. Nuclei and F-actin staining: Prepare DAPI and phalloidin solution by diluting in TBS with a ratio of 1:2000 and 1:500, respectively. Incubate samples with DAPI and phalloidin solution as described in **step 5**. Wash the samples for 3 times with 1× sterile TBS.
8. Mount samples: Add 1 drop of mounting medium on a clean glass cover slip and put the sample on the drop of the mounting medium with the cell side facing down. Add another glass coverslip on top of the sample. Wait for the mounting solution to settle down before imaging.
9. Image the cells with a confocal microscope (Fig. 3).

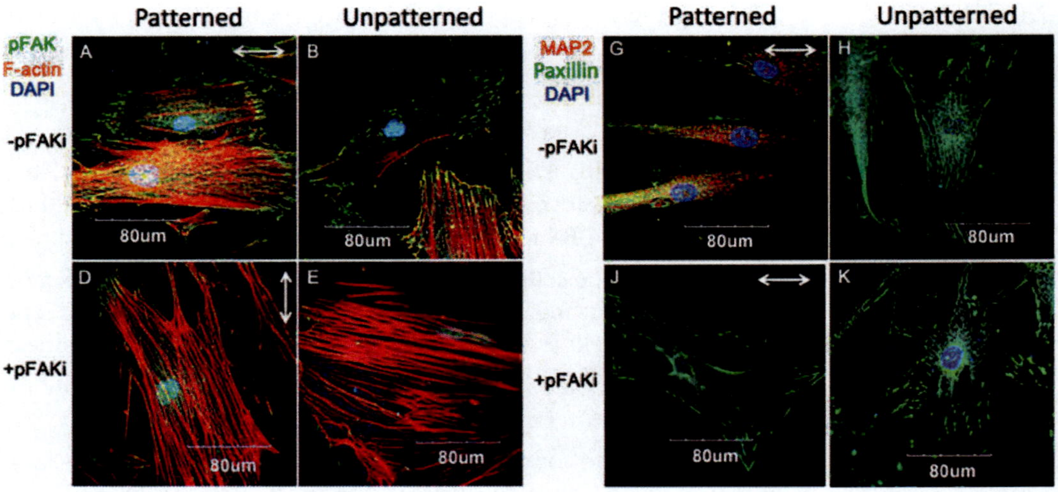

Fig. 3 Immunofluorescence staining images of human mesenchymal stem cells (hMSCs) cultured on unpatterned tissue culture polystyrene (TCPS) and 250 nm grating TCPS with pFAK inhibitor (+pFAKi) or without pFAK inhibitor (−pFAKi). Arrow shows the direction of the 250 nm grating pattern. (Adapted with permission from Ref. [10]. Copyright 2013 American Chemical Society)

4 Notes

1. Patterns on silicon wafers can be fabricated by photolithography or electron-beam lithography; users shall select fabrication methods depending on desired feature size and topography patterns.
2. We used a ready-made gel. Select a gel that has sufficient space between 75 and 50 kDa for cutting.
3. Rabbit anti-FAK primary antibody was purchased from Cell Signaling Technology (catalogue number 3285), rabbit anti-pFAK primary antibody was purchased from Abcam (catalogue number ab39967), and mouse anti-GAPDH was purchased from Santa Cruz (catalogue number sc-47724).
4. Rabbit anti-pFAK antibody was purchased from Abcam (catalogue number ab39967), rabbit anti-Paxillin was purchased from Abcam (catalogue number ab32084), and mouse anti-MAP2 was purchased from Abcam (catalogue number ab11267). MAP2 is used here as an indicator of cell differentiation.
5. The steps in Subheading 3.1.1 only need to be done once for each silicon wafer. After the steps in Subheading 3.1.1, the silicon wafers can be used to fabricate patterned PDMS sample directly or stored in a clean and dry environment at room temperature.

6. We poured degassed PDMS in the Petri dish to a height of 2–3 mm to fabricate PDMS samples for cell culture.
7. Starting from the sixth PDMS replica after the surface treatment of a new mold, the samples can be used directly or stored in a container.
8. If samples float in cell culture medium, sterile Teflon rings can be used to hold the samples down.
9. We used a seeding density of 5000 cells/cm^2 on a sample with an area of 1 cm^2.
10. The 7-day culture period was chosen for examining the expression of MAP2 by hMSCs.
11. Users should choose the lowest concentration that would be essential to inhibit the pFAK. If desired, test the cell differentiation with immunofluorescence staining according to the steps in Subheading 3.4.
12. The protein concentration used in this protocol is 100 μg/mL. It is recommended to try a series of concentrations to test the sensitivity of the antibodies.
13. Wait for 5 min to check leaking after adding running buffer in the inner chamber. Reassemble the gel cassettes if noticeable leaking was observed.
14. Refer to supplier's recommendation for the dilution ratio of antibodies.
15. To make sure the primary antibody was completely removed, incubate the membrane with only secondary antibodies. No signal should be detectable if the primary antibody was removed.
16. Alternatively, run two parallel gels at the same stain, one with FAK and GAPDH and the other with pFAK and GAPDH. Normalize the results to GAPDH.
17. The drop of the antibody solution should be enough to cover the whole sample. A total of 100 μL antibody solution was used for samples with an area of 1 cm^2.

Acknowledgments

This work was supported by National Institutes of Health (NIH R01 HL130274-01A1) and NSERC-CREATE, Training in Global Biomedical Technology Research and Innovation at the University of Waterloo (CREATE Funding 401207296). We acknowledge Benjamin Kim Kiat Teo for generating the western blot and immunofluorescence staining data.

References

1. Jain D, Mattiassi S, Goh EL, Yim EKF (2020) Extracellular matrix and biomimetic engineering microenvironment for neuronal differentiation. Neural Regen Res 15(4):573–585. https://doi.org/10.4103/1673-5374.266907
2. Anselme K, Bigerelle M (2011) Role of materials surface topography on mammalian cell response. Int Mater Rev 56(4):243–266. https://doi.org/10.1179/1743280411Y.0000000001
3. Lu D, Chen C-S, Lai C-S, Soni S, Lam T, Le C, Chen EYT, Nguyen T, Chin W-C (2016) Microgrooved surface modulates neuron differentiation in human embryonic stem cells. In: Turksen K (ed) Human embryonic stem cell protocols. Springer, New York, pp 281–287. https://doi.org/10.1007/7651_2014_81
4. Ankam S, Lim CK, Yim EKF (2015) Actomyosin contractility plays a role in MAP2 expression during nanotopography-directed neuronal differentiation of human embryonic stem cells. Biomaterials 47:20–28. https://doi.org/10.1016/j.biomaterials.2015.01.003
5. Lee JM, Kang WS, Lee KG, Cho H-Y, Conley B, Ahrberg CD, Lim JH, Mo SJ, Mun SG, Kim E-J, Choi J-W, Lee K-B, Lee SJ, Chung BG (2020) Combinatorial biophysical cue sensor array for controlling neural stem cell fate. Biosens Bioelectron 156:112125. https://doi.org/10.1016/j.bios.2020.112125
6. Ding Z, Lu G, Cheng W, Xu G, Zuo B, Lu Q, Kaplan DL (2020) Tough anisotropic silk nanofiber hydrogels with Osteoinductive capacity. ACS Biomater Sci Eng 6(4):2357–2367. https://doi.org/10.1021/acsbiomaterials.0c00143
7. Frey MT, Tsai IY, Russell TP, Hanks SK, Wang Y-l (2006) Cellular responses to substrate topography: role of myosin II and focal adhesion kinase. Biophys J 90(10):3774–3782. https://doi.org/10.1529/biophysj.105.074526
8. Yu H, Gao M, Ma Y, Wang L, Shen Y, Liu X (2018) Inhibition of cell migration by focal adhesion kinase: time-dependent difference in integrin-induced signaling between endothelial and hepatoblastoma cells. Int J Mol Med 41(5):2573–2588. https://doi.org/10.3892/ijmm.2018.3512
9. Sathe SR, Jain D, Koh C-G, Yim EKF (2021) POPX2 phosphatase enhances topographical contact guidance for cell morphology and migration. Biomed Mater 16(2):025020. https://doi.org/10.1088/1748-605x/abd3b5
10. Teo BKK, Wong ST, Lim CK, Kung TYS, Yap CH, Ramagopal Y, Romer LH, Yim EKF (2013) Nanotopography modulates mechanotransduction of stem cells and induces differentiation through focal adhesion kinase. ACS Nano 7(6):4785–4798. https://doi.org/10.1021/nn304966z
11. Cassidy JW, Roberts JN, Smith C-A, Robertson M, White K, Biggs MJ, Oreffo ROC, Dalby MJ (2014) Osteogenic lineage restriction by osteoprogenitors cultured on nanometric grooved surfaces: the role of focal adhesion maturation. Acta Biomater 10(2):651–660. https://doi.org/10.1016/j.actbio.2013.11.008
12. Yang K, Yu SJ, Lee JS, Lee H-R, Chang G-E, Seo J, Lee T, Cheong E, Im SG, Cho S-W (2017) Electroconductive nanoscale topography for enhanced neuronal differentiation and electrophysiological maturation of human neural stem cells. Nanoscale 9(47):18737–18752. https://doi.org/10.1039/C7NR05446G
13. Niu H, Lin D, Tang W, Ma Y, Duan B, Yuan Y, Liu C (2017) Surface topography regulates osteogenic differentiation of MSCs via crosstalk between FAK/MAPK and ILK/β-catenin pathways in a hierarchically porous environment. ACS Biomater Sci Eng 3(12):3161–3175. https://doi.org/10.1021/acsbiomaterials.7b00315

Part VII

Biophysical Modeling of Cell Shape and Protein Dynamics

Check for updates

Chapter 21

Long-Term Fluorescence Recovery After Photobleaching (FRAP)

Takumi Saito, Daiki Matsunaga, and Shinji Deguchi

Abstract

Numerous models have been developed for the analysis of fluorescence recovery after photobleaching (FRAP), by which intracellular diffusion and turnover rate are quantitatively evaluated. FRAP analyses typically focus on such events that occur within several minutes, but to precisely evaluate a slow turnover rate of particularly actin stress fibers, achieving long-term FRAP observations of more than 10 min is necessary. In such long-term observations, the effect of intracellular advection is no longer ignored, which motivated us to develop a novel method to decouple the multiple factors associated with the long FRAP response. This method allows us to distinguish the origin of mechanobiological responses of stress fibers that come from either the level of individual actin filaments or that of actin monomers.

Key words Fluorescence recovery after photobleaching, Chemical reaction, Turnover, Advection, Cytoskeleton, Stress fiber, Live-cell imaging, Confocal laser scanning microscopy

1 Introduction

Fluorescence recovery after photobleaching (FRAP) is a widely used technique to evaluate kinetics of intracellular molecules tagged with fluorescent proteins (Fig. 1). In FRAP experiments, the time course of intensity in a photobleached region is obtained by typically using a conventional confocal laser scanning microscope (CLSM). This FRAP curve is categorized into three phases: pre-bleach, photobleaching, and recovery. Since fluorescence intensity is increased with the number of fluorophore [1–3], the FRAP response corresponds to a replacement of molecules, the so-called molecular turnover. The recovery phase is often fitted with a single exponential curve: $F = F_0(1 - e^{-kt})$, where k and F_0 represent molecular turnover rate and mobile fraction, representing how frequently and how many molecules are exchanged, respectively [4].

Ronen Zaidel-Bar (ed.), *Mechanobiology: Methods and Protocols*,
Methods in Molecular Biology, vol. 2600, https://doi.org/10.1007/978-1-0716-2851-5_21,

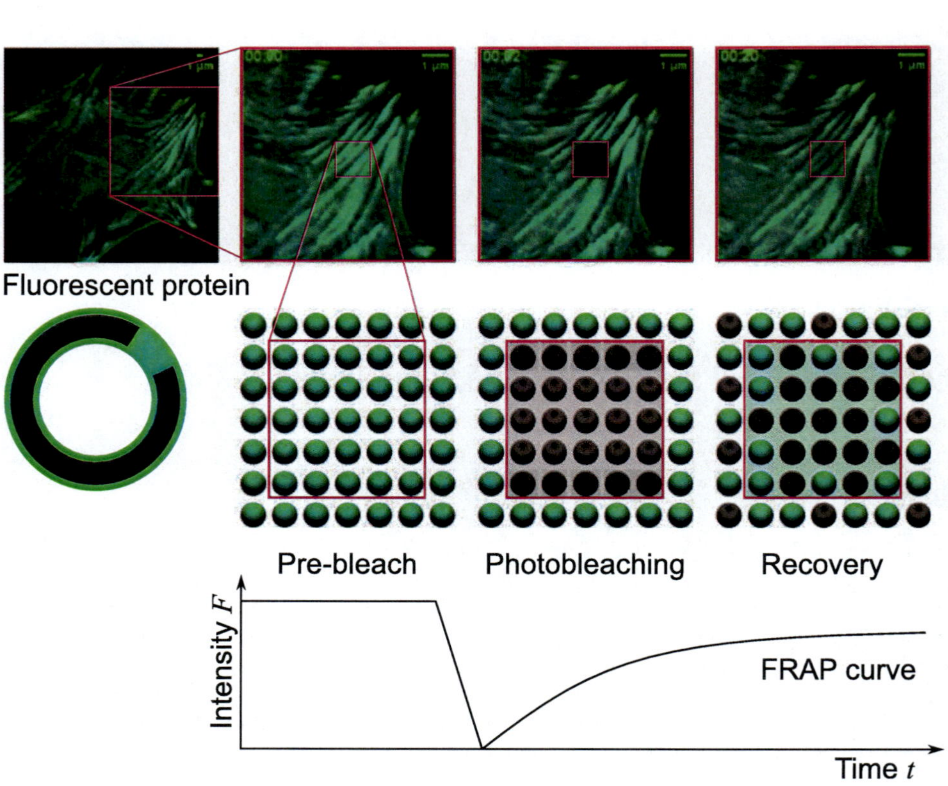

Fig. 1 Schematic of fluorescence recovery after photobleaching (FRAP). Expression plasmids of a target protein tagged with green fluorescent protein (GFP) are constructed and transfected into cells. A region of interest is labeled by photobleaching. Time course of intensity in the labeled region represents a replacement of bleached and fluorescent molecules. The curve is often fitted with a single exponential function to determine the recovery rate

First, we describe the interpretation of the turnover rate as a physicochemical factor in two simple cases. In the case of chemical reaction-dominant turnover, k represents a dissociation rate of labeled molecules from their binding partners [5]. In this case, the intensity in an initially bleached region is increased without a spatial gradient. In a diffusion-dominant case where labeled molecules are transported by the Brownian motion, on the other hand, the intensity distribution exhibits a spatial gradient unlike the reaction-driven case. Fast scanning, thanks to the current CLSM technology, allows us to capture spatiotemporal patterns of the gradient of labeled molecules. Therefore, k, determined from the analysis of the gradient instead of using single exponential functions, estimates the diffusion coefficient [6].

For more complicated diffusion mode known as anomalous diffusion, time-dependent diffusion coefficient $D(t) = D_0\ t^{\alpha-1}$ is substituted for the normal diffusion equation, which is often called a fractional diffusion equation [7]. Here, α represents a scale factor, which characterizes the anomalous diffusion into sub-diffusion

($\alpha < 1$), super-diffusion ($\alpha > 1$), and pure-diffusion ($\alpha = 1$). Intracellular crowded environments reduce the mobility of diffusive molecules in the sub-diffusion case [8–11], whereas diffusive molecules are subjected to active transports such as directed flow and active motion of motor proteins in the super-diffusion case [12, 13].

Moreover, intracellular diffusive molecules are associated and dissociated with their reaction partners. In this reaction–diffusion turnover case, an effective diffusion coefficient – determined by a mean-square displacement of diffusive particles – is slowed by stochastic binding and unbinding events [14]. This effective diffusion coefficient, composed of both pure-diffusion and chemical reaction parameters, is determined using a continuum reaction–diffusion model [5, 15]. To distinguish the mixed behavior of labeled molecules, reaction-free molecules such as green fluorescent proteins [5] and domain-deletion mutants of target molecules [14] are used as a control. The effective diffusion coefficient of these control molecules is regarded to be the pure-diffusion coefficient. Thus, not only pure-diffusion coefficient but also on−/off-rates or equilibrium constants are determined from the analysis of the spatiotemporal intensity distribution of FRAP response. Overall, numerous FRAP models and experimental techniques have been developed to evaluate the intracellular molecular turnover in multiple cases [16].

In this chapter, we introduce a new approach to evaluate the mechanism of relatively "long-term" molecular turnover, which was difficult to be analyzed with previous methods (Fig. 2). Characteristic time estimated by k^{-1} can exceed several tens of minutes in a long-term FRAP curve. Consequently, directed spatial movements such as intracellular bulk-like flow, but not fast diffusion that occurs in a short period of time, dominate the change in the FRAP response. Specific examples of slow turnover include that of actin molecules associated with stress fibers, namely a subcellular

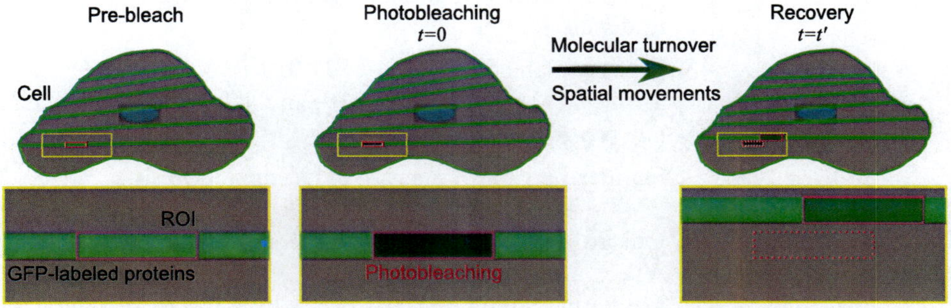

Fig. 2 Schematic of a long-term FRAP measurement in a cell expressing GFP-tagged molecules. The process of FRAP observation is categorized into pre-bleach, photobleaching, and recovering phases. In long-term FRAP, a region labeled by photobleaching is spatially transported because of intracellular advection; meanwhile, a chemical reaction-driven molecular turnover recovers the intensity within the region

tension-bearing structure known to play a role in mechanosensing [14]. Time course of intensity in such long-term FRAP curves often exhibits an abnormal curve distinct from single exponential functions. To assess such curves, double exponential functions may be used for determining two characteristic time constants [17, 18]. However, the relevance of the two characteristic time constants to actual physicochemical factors such as advection rate and on−/off-rate is unclear. Moreover, an additional difficulty includes the involvement of intracellular advection transports of labeled molecules and/or their binding partners, by which the chemical equilibrium distribution is spatially shifted. To overcome these difficulties, we developed a new FRAP model based on a reaction–advection equation that allows us to determine both the long-term advection rate and the dissociation rate of labeled molecules [19]. Combining this approach with typical cytoskeleton inhibitors gives insight into the mechanobiological aspects of the advection-related molecular turnover in living cells.

2 Materials

2.1 Plasmids Construction

1. EYFP-actin vector (#6902–1, Clontech) and mClover2-C1 vector (#54,577, Addgene).
2. 2% (w/v) of template DNA, 2% (v/v) of individual pair restriction enzymes (e.g., 2% (v/v) of XhoI and 2% (v/v) of BamHI), and 10% of CutSmart® Buffer in distilled water.
3. Wizard® SV Gel and PCR Clean-Up System.
4. 50% (v/v) of Ligation high Ver. 2 and 50% of digested and collected DNA solution.
5. Competent Quick DH5a.
6. GenElute™ Plasmid miniprep kit.
7. LB media: 1% (w/v) of Tryptone, 1% (w/v) of NaCl, 0.5% (w/v) of Yeast extract, and 100 μg/mL of Kanamycin (*see* **Note 1**) in distilled water.
8. LB agar plate: 2% (w/v) of BD BBL™ Select APS™ LB broth, 1.5% (w/v) of Bactro Agar™, and 100 μg/mL of Kanamycin (*see* **Note 1**) in distilled water.
9. Standard setup for agarose gel electrophoresis.

2.2 Cell Culture and Transfection

1. Culture media for rat aortic smooth muscle cell lines (A7r5, ATCC): Low-glucose (1.0 g/L) Dulbecco's Modified Eagle Medium containing 10% (v/v) heat-inactivated fetal bovine serum and 1% penicillin–streptomycin.
2. Lipofectamine® LTX and Plus Reagent are used in Opti-MEM® according to the manufacturer's protocol.

2.2.1 Extra Materials

1. (−)-Blebbistatin (final concentration: 10 μM): myosin II ATPase inhibitor.
2. Latrunculin A (final concentration: 10 nM): actin polymerization inhibitor.
3. Dimethyl sulfoxide.

2.3 Fluorescence Recovery After Photobleaching

FV1000 confocal laser scanning microscope (Olympus) equipped with a 60× oil immersion objective lens (NA = 1.42), a laser unit for imaging with 488 nm, laser unit for photobleaching, and camera (FV10-MCPSU laser combiner, IX2-UCB). Image sequences are analyzed by FIJI/ImageJ (NIH) and MATLAB (MathWorks).

3 Methods

3.1 Sample Preparation

3.1.1 Plasmids Construction

First, plasmids of a target protein labeled with green fluorescent protein (GFP) are designed. We used human β-actin for the validation of our method. Here we describe a standard method regarding the insertion of mClover2 (a variant of GFP) into the human β-actin plasmids by using restriction enzymes.

1. Expression plasmids encoding EYFP-actin and mClover2-C1 vectors are separately digested with XhoI and BamHI by using CutSmart® Buffer according to the manufacturer's protocol (*see* Subheading 2.1, **item 2**).
2. Digested plasmids are separated by using a typical agarose gel electrophoresis.
3. DNA fragments of mClover2 and actin are extracted from mClover2-C1 vector and EYFP-actin vector, respectively, by using Wizard® SV Gel and PCR Clean-Up System according to the manufacturer's protocol.
4. These fragments are assembled by using Ligation high Ver. 2 (*see* Subheading 2.1, **item 4**) at 16 °C for 1 h.
5. Transform the plasmid into competent cells (DH5a) by the heat-shock method, and culture them on an LB agar plate with Kanamycin at 37 °C for 16 h.
6. Pick up a positive colony on the agar plate and shake it in LB medium with Kanamycin at 37 °C for 16 h.
7. Elute the plasmid by using GenElute™ Plasmid miniprep kit according to the manufacturer's protocol and appropriately dilute it to 0.5–5 μg/μL.

3.1.2 Cell Culture

1. Rat aortic smooth muscle cell lines (A7r5, ATCC) are cultured in the media (*see* Subheading 2.2, **item 1**) in a humidified 5% CO_2 incubator at 37 °C on a glass-bottom dish for 1–2 days (to prepare 60–80% confluent cells).

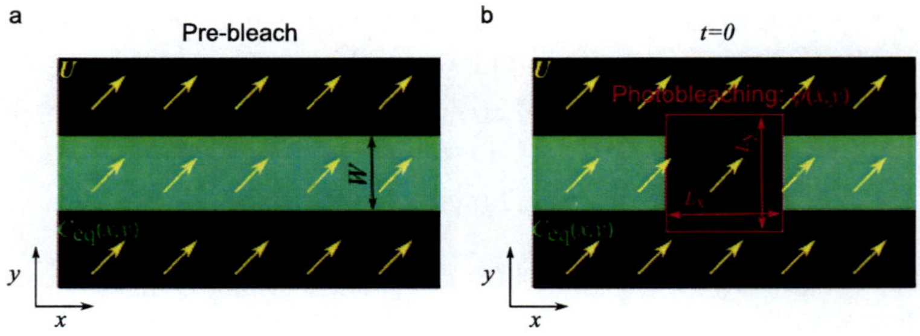

Fig. 3 Schematic of intensity distributions used for the advection–reaction model. (**a**) An example of chemical equilibrium distribution corresponding to an intensity distribution of a single stress fiber. $C_{eq}(x, y) = 1$ for arbitrary x and $-W/2 < y < W/2$, where W represents the width of the stress fiber, while $C_{eq}(x, y) = 0$ at the other region. (**b**) An initial condition ($t = 0$) consists of the chemical equilibrium distribution and photobleaching distribution. $\psi(x, y) = 0$ for $-L_x/2 \leq x \leq L_x/2$ and $-L_y/2 \leq y \leq L_y/2$, while $\psi(x, y) = 1$ at the other region

2. Two micrograms of the diluted plasmids (0.5–5 μg/μL) are transfected into cells by using Lipofectamine® LTX with Plus Reagent and Opti-MEM medium according to the manufacturer's protocol (incubation time, 24 h).
3. The medium is replaced with new one after 24 h of transfection, and the dish is placed into a stage incubator mounted on a confocal microscope.
4. For the use of inhibitors, add premixed inhibitors into the dish with a final concentration of 10 μM (−)-Blebbistatin and 10 nM Latrunculin A.

3.2 FRAP

3.2.1 Reaction–Advection Model for Long-Term FRAP

In long-term FRAP experiments of more than 10 min, a region labeled by photobleaching tends to spatially shift due to intracellular advection, contributing to a part of the recovery of the intensity (Fig. 3). To address cases where such advection is not ignored, the time rate of change of bound molecule concentration, C, is described by

$$\frac{\partial C}{\partial t} = -k_{\text{off}} C + k_{\text{on}} F_{\text{eq}} - \boldsymbol{U} \nabla C \tag{21.1}$$

where F_{eq} represents the concentration of free molecules at unbound state, k_{off} and k_{on} represent the dissociation and association rate, respectively, $\boldsymbol{U}$ represents the velocity vector of the advection, and ∇C represents the gradient. The free molecules are assumed to be at the equilibrium (i.e., $F(t) = F_{eq}$) due to the pool of unbound molecules in cells. The analytical solution of Eq. 21.1 is described by

$$C(x, y, t) = C_{\text{eq}}(x - U_x t, y - U_y t)\left(1 - e^{-k_{\text{off}} t}\right) + \phi(x - U_x t, y - U_y t)\, e^{-k_{\text{off}} t} \tag{21.2}$$

where $C_{eq}(x, y)$ represents the concentration distribution at the prebleach state considered to be at the equilibrium (Fig. 3a), $\phi(x, y)$ represents the initial condition corresponding to the intensity distribution just after photobleaching, and U_x and U_y represent the velocity in x and y direction, respectively.

To determine the specific form of $\phi(x, y)$, first, a spatial profile of photobleaching is required (*see* **Note 2**). In the case of rectangular photobleaching for instance (Fig. 3), a spatiotemporal recovery profile is described by

$$\psi(x, y, t) = 1 - \frac{K_0}{4}\left[\mathrm{erf}\left(\frac{x + \frac{L_x}{2}}{\sqrt{r^2 + 4D_{\mathrm{eff}}t}}\right) - \mathrm{erf}\left(\frac{x - \frac{L_x}{2}}{\sqrt{r^2 + 4D_{\mathrm{eff}}t}}\right)\right]\left[\mathrm{erf}\left(\frac{y + \frac{L_y}{2}}{\sqrt{r^2 + 4D_{\mathrm{eff}}t}}\right) - \mathrm{erf}\left(\frac{y - \frac{L_y}{2}}{\sqrt{r^2 + 4D_{\mathrm{eff}}t}}\right)\right] \tag{21.3}$$

where K_0, r, L_x, L_y, and erf represent the photobleaching efficiency, laser resolution, width and height of the bleached rectangular area, and the error function (i.e., $\mathrm{erf}(z) = \frac{2}{\sqrt{\pi}}\int_0^z e^{-x^2}\mathrm{d}x$), respectively [20], and D_{eff} represents the effective diffusion coefficient defined as $D_{\mathrm{eff}} = D_{\mathrm{pure}}/(1 + K^{-1})$, where D_{pure} and K represent the pure-diffusion coefficient of free actin molecules at unbound state and equilibrium dissociation constant (i.e., $K = k_{\mathrm{off}}/k_{\mathrm{on}}$), respectively [5, 14, 15, 21]. Further details of the effective diffusion in FRAP for reaction–diffusion-dominant molecules are described in our previous research [14]. In long-term FRAP of typically more than 10 min, target molecules are subjected to slow turnover and are predominantly at bound state; hence, $K \gg 1$ and $D_{\mathrm{eff}} \approx 0$. Equation 21.3 is thus rewritten as

$$\psi(x, y) = 1 - \frac{K_0}{4}\left[\mathrm{erf}\left(\frac{x + \frac{L_x}{2}}{\sqrt{r^2}}\right) - \mathrm{erf}\left(\frac{x - \frac{L_x}{2}}{\sqrt{r^2}}\right)\right]\left[\mathrm{erf}\left(\frac{y + \frac{L_y}{2}}{\sqrt{r^2}}\right) - \mathrm{erf}\left(\frac{y - \frac{L_y}{2}}{\sqrt{r^2}}\right)\right] \tag{21.4}$$

indicating that $\psi(x, y) = 0$ for $-L_x/2 < x < L_x/2$ and $-L_y/2 < y < L_y/2$ that correspond to bleached region; meanwhile, $\psi(x, y) = 1$ at the outside.

In the case of rectangular equilibrium distribution for instance (Fig. 3a), we describe $C_{eq}(x, y)$ with error functions based on Eq. 21.4 as

$$C_{eq}(x, y) = \frac{1}{2}\left[\operatorname{erf}\left(\frac{y + \frac{W}{2}}{r}\right) - \operatorname{erf}\left(\frac{y - \frac{W}{2}}{r}\right)\right] \tag{21.5}$$

where W represents the width of the equilibrium distribution; here, $C_{eq}(x, y) = 1$ for arbitrary x and $-W/2 < y < W/2$, where the stress fiber associated with the target actin molecules is present, while $C_{eq}(x, y) = 0$ at the other region (*see* **Note 2**). The geometric shape of $\psi(x, y)$ and $C_{eq}(x, y)$, currently shown by Eqs. 21.4 and 21.5, must actually be appropriately described according to the specific experimental setup for photobleaching and spatial distribution of target molecules at bound state (i.e., pre-bleach distribution), respectively. The initial condition is finally described as the product of the equilibrium distribution and photobleaching distribution (Fig. 3b):

$$\phi(x, y) = C_{eq}(x, y)\psi(x, y) \tag{21.6}$$

3.2.2 Imaging

The fluorescence intensity profiles are obtained by using a confocal laser scanning microscope.

1. Use a time-lapse 2D-scanning with a 488 nm wavelength imaging laser for 2D-scanning.
2. Set imaging parameters of scanning speed, interval, and duration for which we used 2 ns/pixel, 10–15 s, and 15–30 min for FRAP in actin stress fibers, respectively (*see* **Note 3**).
3. Set a rectangular region of interest with $L_x \times L_y$ for photobleaching that is performed from the second (or later) frame for 1–2 frames by using 405/440 nm wavelength lasers (*see* **Note 4**).

3.2.3 FRAP Data Analysis

Images are analyzed by using FIJI/ImageJ software (NIH) and MATLAB (MathWorks) according to the following processes where IJ and ML indicate the use of FIJI/ImageJ and MATLAB, respectively.

(IJ1) Crop XY–T stack images with a rectangular region containing a bleached region (Fig. 4).

(IJ2) If the cropped images contain regions other than those assumed in the model (Fig. 3), delete such regions to reduce fitting errors.

(IJ3) Extract the width W within the first pre-bleach image by using the tool in FIJI/ImageJ named "Measure."

(ML1) Normalize spatiotemporal intensity profiles within the cropped images by

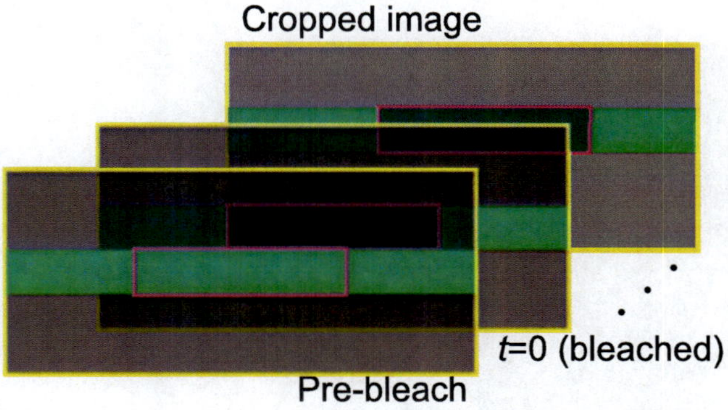

Fig. 4 XY–T stack images are cropped in a rectangle shape including the bleached region for the analysis of spatiotemporal local FRAP response. The time is defined to be zero just after photobleaching

$$\tilde{I}(x, y, t) = \frac{I(x, y, t)}{I_{\max}} \tag{21.7}$$

where $I_{\max}$ represents the maximum intensity in the pre-bleach frame.

(ML2) Fit the cropped spatiotemporal images to Eq. 21.2 consisting of Eqs. 21.4, 21.5, and 21.6 to determine the set of model parameters $\theta = (K_0, k_{\text{off}}, U_x, U_y)$ with the least-square method (*see* **Note 5**) (Fig. 5).

While we have focused on the technical aspect in this article, we found with this method that the actin–myosin interaction and actin polymerization differently affect the advection and chemical dissociation within living cells, thus revealing unknown mechanobiological processes [19]. Specifically, the origin of the fluorescence recovery was investigated using inhibitors (−)-Blebbistatin and Latrunculin A to show that the turnover of actin molecules along individual stress fibers, a tension-bearing structure, is predominantly caused at the level of actin filaments but not that of actin monomers).

4 Notes

1. Add Kanamycin after cooling down the autoclaved LB media. Otherwise antibodies become inactivated.
2. Equations 21.4 and 21.5 depend on readers' respective experimental setup. Therefore, readers are supposed to appropriately define their specific forms regarding the photobleaching distribution and the chemical equilibrium distribution.
3. Imaging parameters of the interval and duration time must be properly adjusted to achieve appropriate photobleaching with

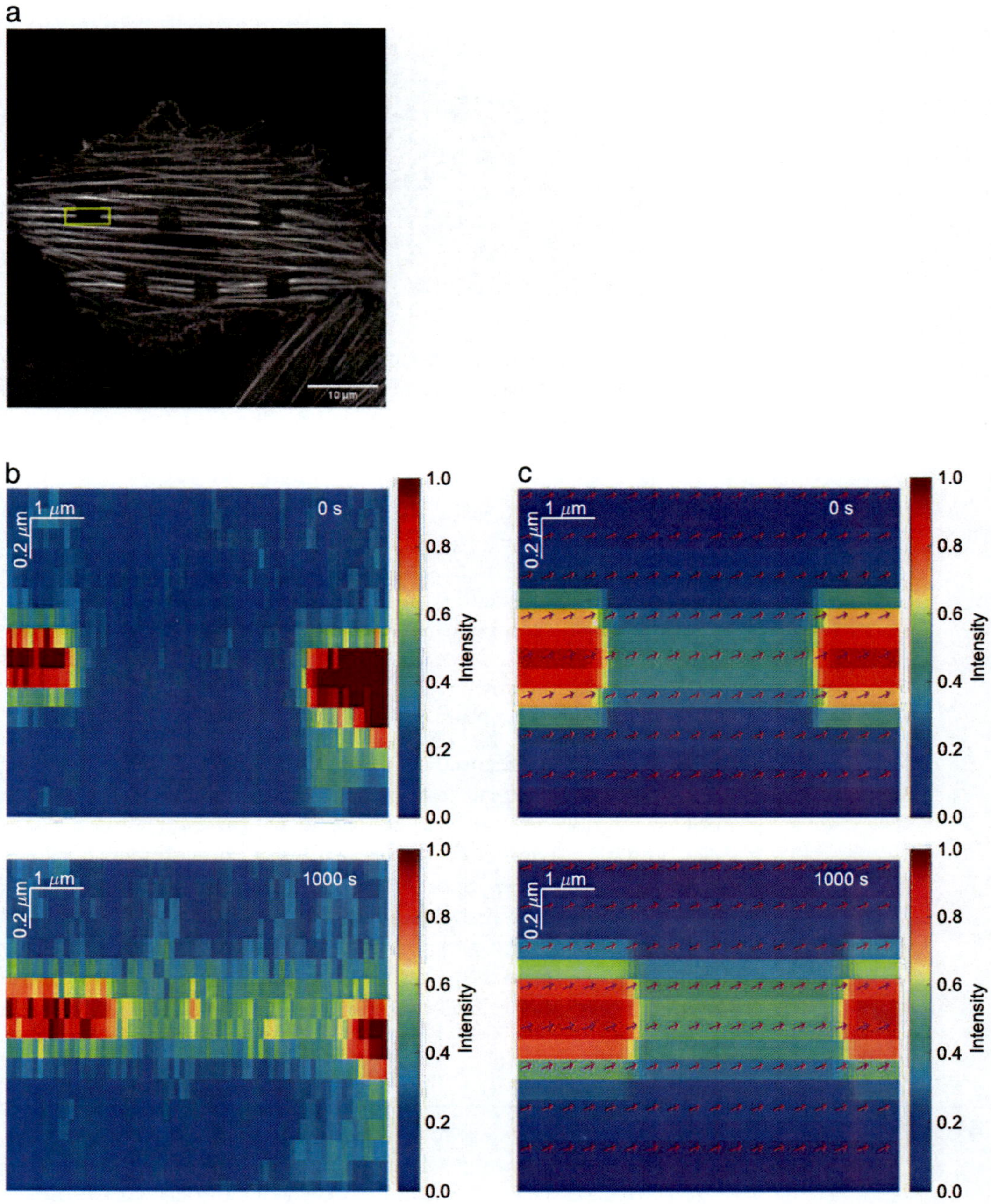

Fig. 5 Long-term FRAP experiment on actin stress fibers determines the advection velocity and the actin dissociation rate from the force-bearing stress fibers. (**a**) An A7r5 cell expressing mClover2- β-actin. (**b**) Experimentally observed time evolution of a rectangular region containing a part of a single stress fiber with a photobleached part. (**c**) Model analysis of the spatiotemporal profile fitted with the least-square method. Arrows represent local advection vectors

the use of a specific imaging laser used. Readers are referred to the article describing how to correct intensity decay during imaging [22].

4. The efficiency of photobleaching depends on the number of fluorophores.

5. Initial values and ranges of the individual model parameters $\theta = (K_0, k_{\text{off}}, U_x, U_y)$ must be appropriately chosen in the least-square method-based fitting to find their reasonable values. For instance, K_0 of the photobleaching parameter must be chosen within a range of [0, 1].

Acknowledgments

T.S. is supported by the Japan Society for the Promotion of Science (JSPS). This work was partly supported by JSPS KAKENHI Grants (18H03518, 19 K22967, and 20 J10828). The authors thank Tsubasa S. Matsui for helpful discussions.

References

1. Fink C, Morgan F, Loew LM (1998) Intracellular fluorescent probe concentrations by confocal microscopy. Biophys J 75:1648–1658. https://doi.org/10.1016/S0006-3495(98)77607-6
2. Piston DW, Patterson GH, Knobel SM (1998) Chapter 3: quantitative imaging of the Green Fluorescent Protein (GFP). Methods Cell Biol 58:31–48. https://doi.org/10.1016/S0091-679X(08)61947-0
3. Lo CA, Kays I, Emran F et al (2015) Quantification of protein levels in single living cells. Cell Rep 13:2634–2644. https://doi.org/10.1016/J.CELREP.2015.11.048
4. Kapitza HG, McGregor G, Jacobson KA (1985) Direct measurement of lateral transport in membranes by using time-resolved spatial photometry. Proc Natl Acad Sci U S A 82:4122–4126. https://doi.org/10.1073/pnas.82.12.4122
5. Sprague BL, Pego RL, Stavreva DA, McNally JG (2004) Analysis of binding reactions by fluorescence recovery after photobleaching. Biophys J 86:3473–3495. https://doi.org/10.1529/biophysj.103.026765
6. Axelrod D, Koppel DE, Schlessinger J et al (1976) Mobility measurement by analysis of fluorescence photobleaching recovery kinetics. Biophys J 16:1055–1069. https://doi.org/10.1016/S0006-3495(76)85755-4
7. Lubelski A, Klafter J (2009) Fluorescence correlation spectroscopy: the case of subdiffusion. Biophys J 96:2055–2063. https://doi.org/10.1016/j.bpj.2008.10.069
8. Daddysman MK, Fecko CJ (2013) Revisiting point FRAP to quantitatively characterize anomalous diffusion in live cells. J Phys Chem B 117:1241–1251. https://doi.org/10.1021/jp310348s
9. Leung WKC, Ching AKK, Chan AWH et al (2011) A novel interplay between oncogenic PFTK1 protein kinase and tumor suppressor TAGLN2 in the control of liver cancer cell motility. Oncogene 30:4464–4475. https://doi.org/10.1038/onc.2011.161
10. Geiger AC, Smith CJ, Takanti N et al (2020) Anomalous diffusion characterization by Fourier transform-FRAP with patterned illumination. Biophys J 119:737–748. https://doi.org/10.1016/J.BPJ.2020.07.013
11. Kang M, Day CA, Kenworthy AK (2019) A novel computational framework for D(t) from fluorescence recovery after photobleaching data reveals various anomalous diffusion types in live cell membranes. Traffic 20:867–880. https://doi.org/10.1111/TRA.12690
12. Hallen MA, Liang ZY, Endow SA (2008) Ncd motor binding and transport in the spindle. J Cell Sci 121:3834–3841. https://doi.org/10.1242/JCS.038497

13. Ciocanel MV, Kreiling JA, Gagnon JA et al (2017) Analysis of active transport by fluorescence recovery after photobleaching. Biophys J 112:1714–1725. https://doi.org/10.1016/j.bpj.2017.02.042
14. Saito T, Matsunaga D, Matsui TS et al (2021) Determining the domain-level reaction-diffusion properties of an actin-binding protein transgelin-2 within cells. Exp Cell Res 404. https://doi.org/10.1016/j.yexcr.2021.112619
15. Ait-Haddou R, Kurachi Y, Nomura T (2010) On calcium-buffer dynamics within the excess buffer regime. J Theor Biol 264:55–65. https://doi.org/10.1016/j.jtbi.2010.01.017
16. Lorén N, Hagman J, Jonasson JK et al (2015) Fluorescence recovery after photobleaching in material and life sciences: putting theory into practice. Q Rev Biophys 48:323–387. https://doi.org/10.1017/S0033583515000013
17. Sakurai-Yageta M, Maruyama T, Suzuki T et al (2015) Dynamic regulation of a cell adhesion protein complex including CADM1 by combinatorial analysis of FRAP with exponential curve-fitting. PLoS One 10. https://doi.org/10.1371/journal.pone.0116637
18. Campbell JJ, Knight MM (2007) An improved confocal FRAP technique for the measurement of long-term actin dynamics in individual stress fibers. Microsc Res Tech 70:1034–1040. https://doi.org/10.1002/jemt.20513
19. Saito T, Matsunaga D, Matsui TS, Deguchi S (2021) Long-term molecular turnover of actin stress fibers revealed by advection-reaction analysis in fluorescence recovery after photobleaching. bioRxiv 2021.06.19.449123. https://doi.org/10.1101/2021.06.19.449123
20. Deschout H, Hagman J, Fransson S et al (2010) Straightforward FRAP for quantitative diffusion measurements with a laser scanning microscope. Opt Express 18:22886. https://doi.org/10.1364/oe.18.022886
21. Crank J (1975) The mathematics of diffusion, 414 p
22. Carnell M, Macmillan A, Whan R (2015) Fluorescence recovery after photobleaching (FRAP): acquisition, analysis, and applications. Methods Mol Biol 1232:255–271. https://doi.org/10.1007/978-1-4939-1752-5_18

Chapter 22

Simulating 3D Cell Shape with the Cellular Potts Model

Rabea Link and Ulrich S. Schwarz

Abstract

Computer simulations have become a widely used method for the field of mechanobiology. An important question is whether one can predict the shape and forces of cells as a function of the extracellular environment. Different types of models have been described before to simulate cell and tissue shapes in structured environments. In this chapter, we give a brief overview of commonly used models and then describe the Cellular Potts Model, a lattice-based modelling framework, in more detail. We provide a hands-on guide on how to build a model that simulates the shape of a single cell on a micropattern in three dimensions in different open source software packages using the Cellular Potts framework. A simulation is set up with an initial configuration of generalized cells that change shape and position due to an energy function that incorporates cellular volume and surface area constraints as well as interaction energies between the generalized cells.

Key words Cell adhesion, Cell shape, Micropatterns, Modelling, Cellular Potts model, Computer simulations

1 Introduction

The emergence of mechanobiology as an own scientific field was triggered by the insight that shape and forces play much more important roles for the behavior and fate of cells than formerly appreciated in the fields of cell biology and cell biophysics [1–3]. Right from its genesis in the late 1990s, advances in mechanobiology were strongly shaped by new experimental technologies, such as adhesive micropatterning [4] or traction force microscopy [5]. In addition, mathematical models and computer simulations also played an important part in understanding how cell shape and forces emerge from the interplay of biochemical signaling, cytoskeleton rearrangement and interactions with the extracellular environment [6]. Here we focus on the question how cell shape can be predicted as a function of the geometry of the extracellular environment. This central question to mechanobiology has been addressed before with many different types of models, but mainly

Ronen Zaidel-Bar (ed.), *Mechanobiology: Methods and Protocols*,
Methods in Molecular Biology, vol. 2600, https://doi.org/10.1007/978-1-0716-2851-5_22,

in two-dimensional ones [7]. Here we not only address three dimensions but also provide a practical guide for the general reader how to implement such procedures in common software packages.

There are several cell-based types of models that can be used to describe the shape dynamics of single cells [8, 9] or cell clusters [10, 11]. In these models, the cell is described on a mesoscopic scale with coarse-grained parameters such as surface tension or interaction energy, which arise from integrating out microscopic degrees of freedom. An energy function that depends on the boundary of the cell is minimized in the standard models to describe cell shape and dynamics. Differences in how the cell boundary is described and the energy is minimized lead to different models. Phase field models use a continuous field ϕ that differentiates between inside ($\phi = 1$) and outside of the cell ($\phi = 0$) [12, 13]. Cells can also be described by polygons in vertex models [14, 15], where vertices are shifted during the simulation to minimize an energy function and, therefore, the cells change their shape, position, and usually also neighborhood relations during the simulation.

In this chapter, we focus on the Cellular Potts Model (CPM) [16, 17]. In contrast to the continuum models mentioned before, the CPM is a lattice model, which means that a simulated cell cannot change its shape continuously in space but occupies a number of pixels (in 2D) or voxels (in 3D) on a lattice, thus leading to discrete shapes. The discrete approach decreases computation time compared to continuum models. A simulation with the CPM is easy to set up and can be adapted to different types of experiments. The framework has been used to simulate many different biological phenomena. The first experiments that were described with a CPM were cell sorting experiments [18, 19]. Other simulations include tumor growth, invasion and progression [20], single cell motility [21, 22], migration of single cells and cell clusters [23, 24] cells on micropatterns [25, 26] and cell competition in epithelia [27]. A disadvantage of the lattice-based formalism is that the calculation of bending energies is computationally expensive, thus other simulation frameworks are better suited when bending energies are important, as, for example, during endocytosis. Here, however, we are concerned only with the shapes of whole cells, which to first order are determined by membrane and cortical tensions of the cell envelope, a situation which is well described by the CPM.

2 Materials

There are many open source software available which can be used to implement CPM-simulations. In Table 1, we give a selected overview of CPM software that is freely available, namely CompuCell3D [28], Morpheus [29], Artistoo [30], Tissue Simulation

Table 1
Open source software packages that implement the Cellular Potts Model and the corresponding websites

Software name	URL
CompuCell3D	https://compucell3d.org/
Morpheus	https://morpheus.gitlab.io/
Artistoo	https://artistoo.net/
Tissue simulation toolkit	https://sourceforge.net/projects/tst/
Chaste	https://www.cs.ox.ac.uk/chaste/
Simmune	https://www.niaid.nih.gov/research/simmune-project

Toolkit [31], CHASTE [32], and Simmune [33]. In recent years, several approaches to parallelize the Cellular Potts Model have been published [34–36], which enable large-scale simulations. We start this chapter with an overview over the CPM-concept in Subheading 3. In Subheading 4 we then explain how to use CompuCell3D and Morpheus to simulate the three-dimensional shape of a cell on a micropattern.

3 Theory

In the CPM, cell shape is defined on a lattice, *see* Fig. 1. Sites of the lattice with the same index σ belong to the same generalized cell, which can be a biological cell, a sub-cellular compartment, or a physical object such as a substrate or a pillar. Sites that belong to the surrounding medium usually have the index $\sigma = 0$. The CPM is energy-based, and the effective energies that determine the dynamics of the simulation can be assigned to quantities such as volume or surface conservation, interaction between different generalized cells or chemotaxis. This description allows the inclusion of any biological mechanism to the model as long as it can be described with an energetic formalism.

The energy function which is used in the CPM is often referred to as Hamiltonian. It assigns an energy to any given configuration of cells on a lattice. Here we discuss the CPM in 3D. In its basic version, the Hamiltonian consists of an interaction term and a term to conserve volume:

$$\begin{aligned}\mathcal{H} = &\sum_{\mathbf{i},\mathbf{j}\ \text{neighbors}} \left(1 - \delta_{\sigma(\mathbf{i}),\sigma(\mathbf{j})}\right) J(\tau(\sigma(\mathbf{i})),\tau(\sigma(\mathbf{j}))) \\ &+ \sum_{\sigma} \lambda_V(\tau(\sigma))(v(\sigma) - V_{\tau(\sigma)})^2\end{aligned} \tag{1}$$

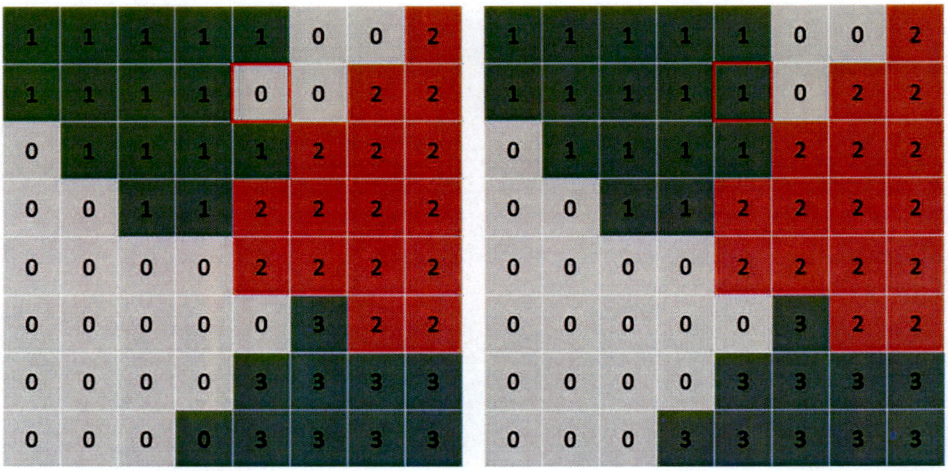

Fig. 1 2D Cellular Potts Model with three cells $\sigma = 1, 2, 3$ and medium $\sigma = 0$. Two cells have the same cell type τ and are thus depicted with the same color. A lattice site is picked at random (red) and its index is changed to that of the neighboring cell. Then the energy difference $\Delta\mathcal{H}$ due to this index change is calculated and the Metropolis algorithm determines if the spin flip is accepted

The first term sums over all neighboring voxels **i**, **j**. The Kronecker Delta $\delta_{\sigma(\mathbf{i}),\sigma(\mathbf{j})}$ is defined as 1, if the cells at **i** and **j** are the same, and 0 if they are different. This ensures that only if **i** and **j** belong to different cells the interaction term is different from 0. The interaction energy J then depends on the cell types of the cells at positions **i** and **j**. The second term of the Hamiltonian sums over all cells in the simulation. It has the form of an elastic energy, which means that the term is 0 if the cell has the target volume $V_{\tau(\sigma)}$ and the energy increases quadratically for differences between the actual volume $v(\sigma)$ and the target volume $V_{\tau(\sigma)}$. The strength with which this constraint is enforced is given by the rigidity $\lambda_V(\tau(\sigma))$, which depends on the cell type. The higher the value of λ_V, the more energy it costs the cell to have a volume that differs from the target value. The volume energy is needed to avoid that the cells shrink and disappear when the total energy is minimized during the simulation.

The Hamiltonian in Eq. 1 can be extended to include other important aspects like surface area constraints and chemotaxis. The surface area of a biological cell is fixed by the amount of cell membrane. This can be described with an elastic area constraint, similar to the volume constraint in Eq. 1:

$$\mathcal{H}_{\text{area}} = \sum_{\sigma} \lambda_A(\tau(\sigma))(a(\sigma) - A_{\tau(\sigma)})^2. \qquad (2)$$

Adding the area constraint Eq. 2 to the Hamiltonian allows us to use negative interaction energies J. Using negative interaction energies J without constraining the area would lead to cells that maximize their surface area to reduce the total energy of the system, possibly even causing the cells to break into smaller parts.

A simple form of chemotaxis has the form of a potential energy

$$\mathcal{H}_{\text{chem}} = \mu(\tau(\sigma))C(\mathbf{i}), \tag{3}$$

where $C(\mathbf{i})$ is the chemical field at the target site $\mathbf{i}$ and $\mu(\tau(\sigma))$ describes the strength of the cellular response to the chemical field.

Depending on the biological system, other forms of the Hamiltonian can be better suited. For example, Albert and Schwarz [26] developed a 2D CPM based on the geometrical Hamiltonian introduced by Vianay et al. [25] to describe cell shape and forces on micropatterns. The Hamiltonian was chosen to be

$$\mathcal{H} = \sigma A + \lambda_s l + \sum_{\text{arc } i} \frac{EA}{2L_{0,i}} (L_i - L_{0,i})^2 - \frac{E_0}{A_{\text{ref}} + A_{\text{ad}}} A_{\text{Ad}}. \tag{4}$$

In contrast to the original Hamiltonian in Eqs. 1 and 2, the cell area A and perimeter l of the cell are not constrained by an elastic energy term but can vary in response to the adhesive environment. The main model parameters are now surface tension σ and line tension λ_s. An additional elastic line tension $\lambda_e = EA(L - L_0)/L_0$ is introduced to describe reinforced actin edge fibers that form between adhesive parts of the micropattern. EA is the one-dimensional elastic modulus of the fiber and L and L_0 are the contour length and the rest length, respectively. The last term describes the energy gain from cell adhesion, which is linear during initial cell spreading and then plateaus. A_{ad} is the adhesive area covered by the cell and E_0 and A_{ref} are parameters to control the strength and saturation of the adhesive energy. This model predicts the dynamics of cell shapes on micropatterned substrates but does not work in an unstructured environment.

A CPM simulation is set up with an initial configuration of the generalized cells, which is shown in Fig. 1 for illustrative purposes in 2D. A modified version of the Metropolis algorithm [16, 38] is used to evolve the cells and find the minimal energy configuration. First a site that is at the boundary of a generalized cell is chosen at random. The index of this site is then changed to the index of a randomly chosen neighboring site. Using the Hamiltonian from Eq. 1, the energy difference $\Delta\mathcal{H} = \mathcal{H}_{\text{after}} - \mathcal{H}_{\text{before}}$ between the old and the new configuration is calculated. If the new position has the same or lower energy, $\Delta\mathcal{H} \leq 0$ the index change is accepted. In case the new configuration has a higher energy than the old one, $\Delta\mathcal{H} > 0$, the index change is accepted with the probability

$$p = e^{-\frac{\Delta\mathcal{H}}{T}}. \tag{5}$$

T is called the simulation temperature, because it is the physical temperature when using the Metropolis algorithm to simulate the Boltzmann ensemble with microscopic degrees of freedom. In the CPM, the simulation temperature T does not correspond to the physical temperature, but a higher T value increases the probability to accept a spin flip and thus leads to more fluctuations in the simulation. At $T=0$, only spin flips that lower the energy in the system are accepted, which prevents the simulation from overcoming energy barriers. Therefore, all simulations use an intermediate simulation temperature as compromise to generate dynamic shape changes while still keeping clear cell identities. One Monte Carlo step consists of n spin flip attempts, where n is the number of lattice sites. Note that only the indices of neighboring cells can be chosen for the spin flip. This modification of the Metropolis algorithm breaks detailed balance, which would be required if one simulated on a microscopic level [39].

4 Methods

In the following, we will discuss how to simulate a cell on a micropattern in 3D with different CPM software. We will simulate a single cell with volume and surface constraints on an H-shaped micropattern, which often is used to immobilize a single cell or a cell doublet without polarization [40], in contrast to the crossbow shape, which leads to polarization [41], or the square, which leads to rotation [40]. In the simulations, the H-pattern is implemented with fixed generalized cell types, which we call `Adhesive` and `NonAdhesive`. To build a simulation, many parameters must be set. We will suggest parameters throughout the instructions. To set up a simulation, parameter scans are a helpful tool to determine the parameter range that best corresponds to the experiment.

To simulate the cell shape on a micropattern, a good lattice size is $100 \times 100 \times 50$. The micropattern is in the x-y plane at $z=0$. We choose the thickness of the adhesive part to be 4 voxels wide and 70 voxels long. Then, the cell is initialized as a half sphere with radius 28, sitting at the center of the micropattern. The target volume of the cell is $V=50,000$ voxels, which is close to the initial cell volume.

4.1 CompuCell3D

The CompuCell3D software is provided by the group of James Glazier at Indiana University, USA. To download the software, follow the instructions at https://compucell3d.org/SrcBin#MostRecent. After installation, double-click the `twedit++.command` file to open the editor. The editor is customized to simplify the definition of new CompuCell3D simulations.

1. You can initialize a new simulation by choosing `CC3D Project` → `New CC3D Project...` in the menu bar.
2. The simulation wizard appears. You can enter the name and directory of your simulation. For the simulation type, choose `Python+XML`.

3. Next, you have to decide the dimensions of your simulation. A good choice for the micropattern simulation is 100×100×50. For the boundary condition, choose `NoFlux`. Change the `Average Membrane Fluctuations`, which corresponds to the temperature T in Eq. 5 to 700. In addition, change the `Pixel Copy Range (NeighborOrder)` to 3 and the number of steps to 1000. Because you want to start with a very specific initial cell layout, you will customize this after the simulation wizard is complete. Select the `Continue` button.
4. Add `Cell`, `Adhesive`, and `NonAdhesive` as generalized cell types. Because you only want the `Cell` to change shape during the simulation, freeze the other two cell types by selecting the corresponding check boxes.
5. In the next part, you are asked to add chemical fields, which you do not need for your simulation.
6. Then, you can specify cell properties and behaviors, *see* Fig. 2. The cellular behaviors and the constraints and forces add terms to the Hamiltonian. For the simulation you want to create, adhesion, volume, and surface energies are needed. These terms are added to the Hamiltonian when you select the check boxes for `Contact`, `VolumeFlex`, and `SurfaceFlex`.
7. After this, the simulation wizard is complete and code corresponding to your selected settings is automatically

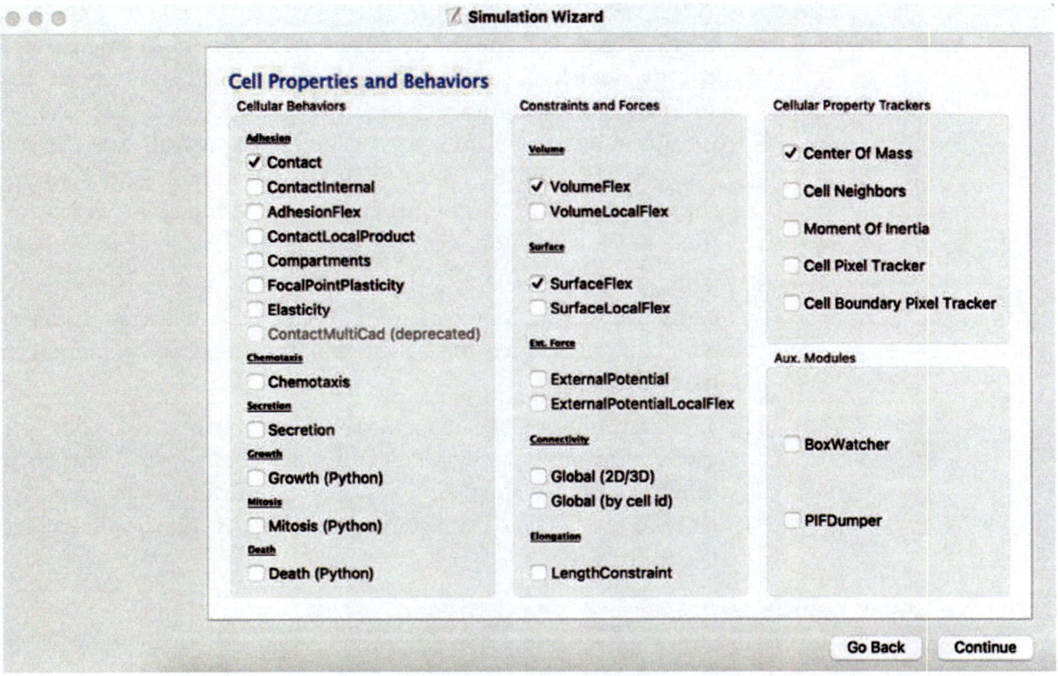

Fig. 2 Graphical user interface of the `Simulation Wizard` in CompuCell3D

generated. The code is distributed in three files, the `Main Python Script` which you do not need to change, the `XML Script`, where properties that stay the same throughout the simulation can be set, and the `Python Steppables` file, where you can use the Python programming language (Python Software Foundation, https://www.python.org/) to describe more complicated starting conditions, changes during the simulation and finishing steps.

8. Open the `XML Script`. In the first section of the `XML Script`, you can find the `Metadata` of the simulation, which you do not need to change. Then, the `Potts` section specifies the parameters that are important for the Cellular Potts Model algorithm, which you already specified in the simulation wizard. Then, several sections follow that specify which plugins are included in the simulation. In the `<Plugin Name="CellType">`, you can find the cell types, as defined in the wizard.
9. The `VolumeFlex` energy is included in the `<Plugin Name="Volume">` section and corresponds to the elastic volume constraint in Eq. 1. Here, you need to change the parameters assigned by the wizard: For the generalized cell type which you called `Cell` the target volume V must be changed to 50,000, and the parameter describing the strength of the elastic volume energy λ_V must be set to 2.0. It is called `LambdaVolume` in the simulation. Since the other two cell types are fixed you do not need volume energies for these cell types and can optionally delete the respective lines in the `XML Script`.
10. You can do the same in the next section for the elastic surface energy, which adds Eq. 2 to the Hamiltonian used in the simulation. You should change the parameters in the section `<Plugin Name="Surface">`. The target surface A of the cell must be changed to 13,000 and the parameter λ_A describing the strength to 1.0. The other lines for `Adhesive` and `NonAdhesive` cell types can be deleted.
11. The `<Plugin Name="CenterOfMass">` allows the user to easily access the center of mass of the cells in the simulation, which is not needed for your simulation but can be helpful in many cases.
12. The last plugin `<Plugin Name="Contact">` specifies the interaction energies between cell types. Here you define how much energy is gained or lost when two different cell types are close to each other. The interaction energies you should use are the following:

```
<Energy Type1="Medium" Type2="Medium">0.0</Energy>
<Energy Type1="Medium" Type2="Cell">20.0</Energy>
<Energy Type1="Medium" Type2="Adhesive">0.0</Energy>
<Energy Type1="Medium" Type2="NonAdhesive">0.0</Energy>
```

```
<Energy Type1="Cell" Type2="Cell">0.0</Energy>
<Energy Type1="Cell" Type2="Adhesive">-150.0</Energy>
<Energy Type1="Cell" Type2="NonAdhesive">-1.0</Energy>
<Energy Type1="Adhesive" Type2="Adhesive">0.0</Energy>
<Energy Type1="Adhesive" Type2="NonAdhesive">0.0</Energy>
<Energy Type1="NonAdhesive" Type2="NonAdhesive">0.0</Energy>
```

These interaction energies ensure that increasing the interface between cell and medium increases the energy and is thus unfavorable. What is more important in our context, increasing the interface between cell and adhesive decreases the overall energy of the system and ensures that the cell attaches to the micropattern. Because the `Cell` has a large volume, the `NeighborOrder` in the `<Plugin Name="Contact">` can be increased to 60.

13. Finally, you need to delete the `Steppable` that details the initial configuration of the simulation. We will write our own initial cell configuration in the `Python Steppables` file. Save the changes in the `XML Script`.
14. Open the `Python Steppables` file. In this file, you find a Python class with methods objects. These methods will be called when the simulation is initialized (`__init__`), before the first Monte Carlo step (`start`), at every Monte Carlo step (`step`), after the last Monte Carlo step (`finish`) or when the user stops the simulation (`on_stop`).
15. You should use the `start` method to describe the initial configuration of our simulation. In the code snippet below you create three cells, one of each cell type. Please make sure that you assign the correct `TypeId` to the correct cell. The `TypeId` of a cell type is defined in the `CellType` plugin in the `XML Script`.

```
cell = self.potts.createCell()
cell.type = 1
adhesive = self.potts.createCell()
adhesive.type = 2
nonadhesive = self.potts.createCell()
nonadhesive.type = 3
```

Then, we assign lattice voxels to the different cell types. We create here an initial configuration with a `NonAdhesive` surface of height 1 with an `Adhesive` “H” shaped micropattern. We place the cell as a hemisphere with radius 28 centered on the surface.

```
self.cell_field[0:99,0:99,0:1] = nonadhesive
```

```
self.cell_field[20:25,15:85,0:1] = adhesive
self.cell_field[20:80,48:52,0:1] = adhesive
self.cell_field[75:80,15:85,0:1] = adhesive
x_0=50
y_0=50
z_0=1
radius=28
for x in range(22,79):
    for y in range(22,79):
        for z in range(2,30):
          if (x-x_0)**2+(y-y_0)**2+(z-z_0)**2<
          radius**2:
              self.cell_field[x,y,z] = cell
```

16. To gain some insight on the simulation you can print the cell surface and volume to the terminal at every 10 Monte Carlo steps by adding the following lines to `step`:

```
for cell in self.cell_list:
    if cell.type==1 and mcs%10==0:
        print("area=", cell.surface, "volume=",
        cell.volume)
```

    Save the `Python Steppables` file.

17. You can start the simulation now: right-click on the `.cc3d` file on the left panel and select `Open In Player`. You should see the initial configuration of the simulation as in Fig. 3. In the `Player`, a graphics window appears where you can change between a 2D and a 3D view. If you do not see the simulation correctly, you can use the `Visualization` menu to zoom or reset your view. You can use the `Model Editor` to change parameter values during a simulation to see the effects directly. The volume and area of the cell is printed to the terminal. After less than 100 Monte Carlo steps, the final cell shape is reached, *see* Fig. 4, and the shape only fluctuates slightly due to the finite simulation temperature T.

4.2 Morpheus

The Morpheus software is provided by the group of Andreas Deutsch at Technical University Dresden, Germany. You can obtain the software from https://morpheus.gitlab.io/. After the installation, you can open the graphical user interface, *see* Fig. 6. In the `Documents` section on the upper left part of the user interface, you can see the structure of a Morpheus simulation. We will go through these points and discuss how to set up the simulation of a 3D cell on an H-shaped micropattern in the following.

1. Start by saving the empty simulation. Select the `Save` button and choose a name and a directory for the simulation.

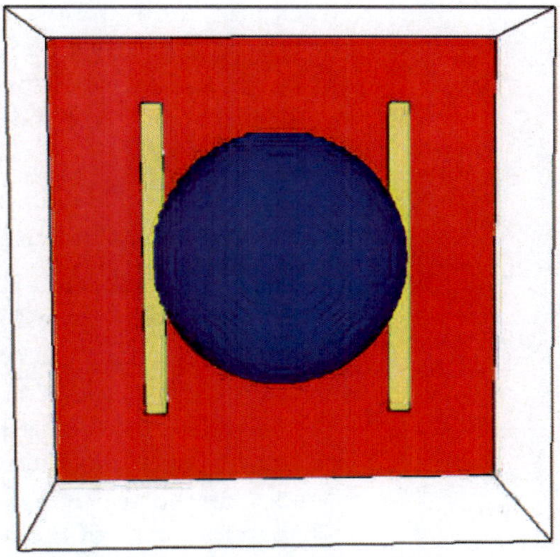

Fig. 3 After starting the simulation in CompuCell3D, the initial configuration of a single cell on an H-shaped micropattern can be seen in the graphics window

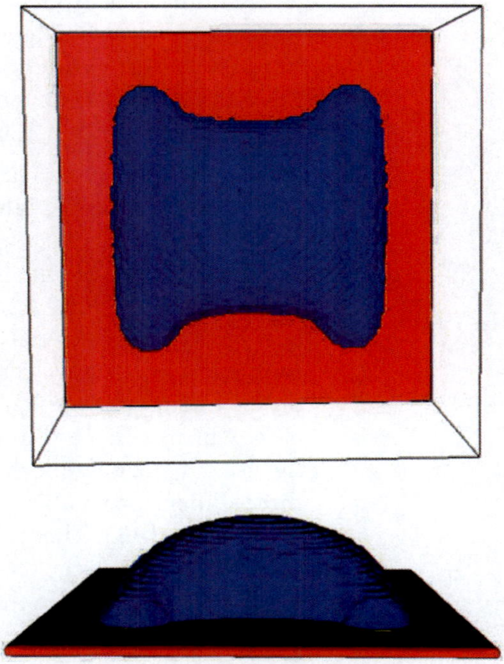

Fig. 4 Simulation result of a single cell on an H-shaped micropattern with CompuCell3D

Optionally, you can add a title and a description to the empty simulation.

2. Click on Space to move on to the next part. Select Lattice to change the lattice configuration of the simulation. In the Attributes window, change the lattice class from linear to cubic to change the space to three dimensions. Change the Neighborhood to 3. Change the size of the lattice to 100,100,50.
3. Move on to Time and change the StopTime to 1000.
4. Next, double-click on CellTypes. You need four different cell types, *see* Fig. 6. In the Morpheus environment, the order in which we introduce the cell types is important. Start with the medium: click on CellType, change the class from biological to medium and add the name Medium. Unselect CellType and add another element by clicking on the + symbol in the top left corner and then choose CellType. Change the name of the new cell type to Adhesive. This cell type is frozen in our simulation. Select the adhesive cell type and click again on +. Select FreezeMotion and change the Condition to 1.0 to freeze the adhesive cell type. Repeat the same process to add the NonAdhesive cell type and freeze it as well. Finally, add the Cell as the last cell type, and do not freeze it but instead add a VolumeConstraint and a SurfaceConstraint. Change the parameters of the VolumeConstraint to strength 2 and target 50000 and for the SurfaceConstraint change the mode to surface and choose the strength 1 and target 8000.
5. Double-click on CPM. Select Interaction and click on the + to add Contact. This defines the interaction energies *J*. Add a total of three Contact elements and define the interaction energies between cell medium, cell adhesive, and cell nonadhesive to 20.0, − 15000.0, and − 1.0, respectively. The interaction energy between cell and adhesive is much lower in this simulation because we cannot choose a higher neighborhood order. Change the Neighborhood Order of ShapeSurface to 7, which is much lower because Morpheus does not allow the use of large neighborhoods. For a small neighborhood on a three-dimensional cubic lattice the 7th level neighborhood is a good choice [42]. Then, click on + and add the MonteCarloSampler to define the temperature and the neighborhood size for choosing updates in the modified Metropolis algorithm. In MetropolisKinetics, change the temperature to 700.0.
6. Now you can move on to the CellPopulations, where you define the initial configuration for the simulation. Start with the Adhesive cell type. Change the type of the Population to Adhesive. Then add InitCellObjects and change the Arrangement from Sphere to Box. To initialize the H-pattern, set the origin of the Box to 20.0,15.0,0.0

and the size to 5.0,70.0,1.0. This is the first bar of the H-pattern. For the other bars, add two more Arrangements to InitCellObjects. The origin of the second Box is 20.0,48.0,0.0 with a size of 60.0,4.0,1.0 and for the third Box the origin is 75.0,15.0,0.0 with a size vector given by 5.0,70.0,1.0. To include the NonAdhesive cell type to the initial configuration, add another Population and change the type to NonAdhesive. Add again the InitCellObjects. Change the Sphere to a Box, and change the size of the Box to 100.0,100.0,1.0. This Box will not overwrite the previously defined Adhesive boxes, which is why the order of defining the cell types is important here. Finally, we can include the Cell in the initial configuration with a half sphere: Add the third Population and change the type to Cell. Add again InitCellObjects, but this time use the predefined Sphere. Change the center of the sphere to 50.0,50.0,1.0 and the radius to 28.0.

7. In the Analysis part, add the VtkPlotter and change the Channel to cell.type, which allows you to visualize the results, e.g., in ParaView [43].
8. Then, you can start the simulation. 3D simulations are not visualized in Morpheus, but because the VtkPlotter was added, the results can be visualized in other programs like ParaView, *see* Fig. 5.

4.3 Conclusions

In this chapter, we have described in detail how one can simulate 3D cell shape in structured environments using the CPM. We have chosen the H-pattern as a standard pattern for cell immobilization without polarization [40], but it is easy to extend this approach now to other patterns of interest. To learn more about the importance of the parameter choice, we recommend to use different values than the ones suggested above and to see what changes in the simulations. For example, it is instructive to change the volume and surface area rigidities $\lambda_V$and λ_A as well as the target volume V and target surface area A.

It is interesting to note that the two software packages gave different results: while the shape predicted by CompuCell3D as shown in Fig. 4 has the invaginated shape often seen in experiments, the shape predicted by Morpheus as shown in Fig. 5 is more rounded. Technically, this difference arises from the fact that Morpheus uses a smaller neighborhood area to evaluate the spin Hamiltonian [42]. In fact the same result as obtained by Morpheus can be obtained by decreasing neighborhood order in CompuCell3D. Due to its larger variability in this aspect, CompuCell3D might be better suited to simulate single cell shape, while Morpheus is well suited to simulate ensembles of cells for which individual shape is less important. For a detailed comparison with experiments, the

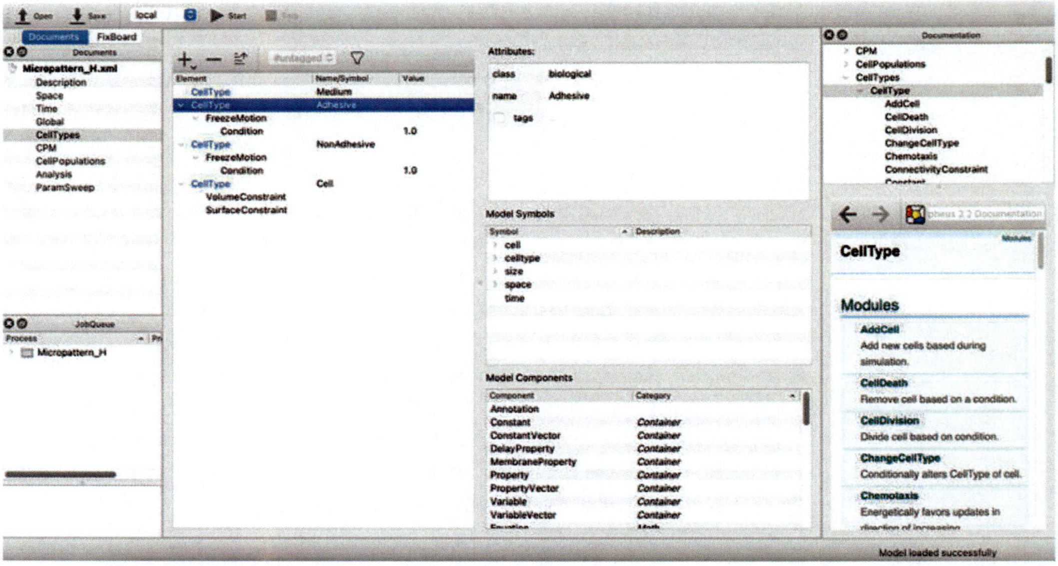

Fig. 5 Graphical user interface in Morpheus. The structure of the simulation can be seen in the `Documents` section. Parameters can be changed in the `Attributes` section

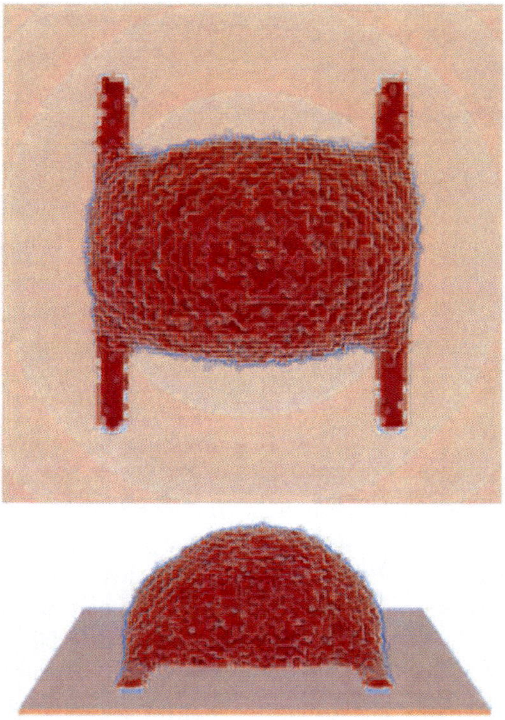

Fig. 6 Simulation result of a single cell on an H-shaped micropattern with Morpheus, visualized in ParaView

shapes obtained by the simulations can be compared with a variety of methods [6]. These include volume, surface area, and sphericity of the 3D cell shapes. In addition, one can compare the so-called landmark points, such as the height of the cell. More evolved methods to compare shapes would include Fourier descriptors and moment invariants, as implemented, for example, in the Konstanz information miner (KNIME) [44].

Acknowledgments

This research has been funded by the Deutsche Forschungsgemeinschaft (DFG, German Research Foundation) under Germany's Excellence Strategy via the Excellence Cluster 3D Matter Made to Order (EXC-2082/1 – 390761711). USS is a member of the Interdisciplinary Center for Scientific Computing (IWR) at Heidelberg.

References

1. Discher DE, Janmey P, Wang Y-L (2005) Tissue cells feel and respond to the stiffness of their substrate. Science 310:1139–1143. ISSN: 0036-8075, 1095–9203. http://www.sciencemag.org/content/310/5751/1139
2. Iskratsch T, Wolfenson H, Sheetz MP (2014) Appreciating force and shape—the rise of mechanotransduction in cell biology. Nat Rev Mol Cell Biol 15:825–833. ISSN: 1471–0072. http://www.nature.com/nrm/journal/v15/n12/abs/nrm3903.html
3. Hannezo E, Heisenberg C-P (2019) Mechanochemical feedback loops in development and disease. Cell 178:12–25. ISSN: 0092–8674. https://www.sciencedirect.com/science/article/pii/S0092867419306233
4. Ruprecht V et al. (2017) How cells respond to environmental cues – insights from bio-functionalized substrates. J Cell Sci 130: 51–61. ISSN: 0021-9533, 1477–9137. http://jcs.biologists.org/content/130/1/51
5. Schwarz US, Soine JRD (2015) Traction force microscopy on soft elastic substrates: a guide to recent computational advances. Biochim Biophys Acta Mol Cell Res Mechanobiol 1853: 3095–3104. ISSN: 0167-4889. http://www.sciencedirect.com/science/article/pii/S0167488915001822
6. Bodor DL, Pönisch W, Endres RG, Paluch EK (2020) Of cell shapes and motion: the physical basis of animal cell migration. Dev Cell 52: 550–562. ISSN: 1534-5807. https://doi.org/10.1016/j.devcel.2020.02.013
7. Albert PJ, Schwarz US (2016) Modeling cell shape and dynamics on micropatterns. Cell Adhes Migr 10:516–528. https://doi.org/10.1080/19336918.2016.1148864
8. Holmes WR, Edelstein-Keshet LA (2012) Comparison of computational models for eukaryotic cell shape and motility. PLoS Comput Biol 8:e1002793. ISSN: 1553-7358. https://doi.org/10.1371/journal.pcbi.1002793
9. Danuser G, Allard J, Mogilner A (2013) Mathematical modeling of eukaryotic cell migration: insights beyond experiments. Annu Rev Cell Dev Biol 29:501–528. https://doi.org/10.1146/annurev-cellbio-101512-122308
10. Liedekerke PV, Palm MM, Jagiella N, Drasdo D (2015) Simulating tissue mechanics with agent-based models: concepts, perspectives and some novel results. Comput Particle Mech 2:401–444. https://doi.org/10.1007/s40571-015-0082-3
11. Metzcar J, Wang Y, Heiland R, Macklin P (2019) A review of cell-based computational modeling in cancer biology. JCO Clin Cancer Inform 3:1–13. ISSN: 2473-4276. https://pubmed.ncbi.nlm.nih.gov/30715927/
12. Ziebert F, Swaminathan S, Aranson IS (2012) Model for self-polarization and motility of keratocyte fragments. J R Soc Interface 9:1084–1092. https://royalsocietypublishing.org/doi/abs/10.1098/rsif.2011.0433
13. Moure A, Gomez H (2021) Phase-field modeling of individual and collective cell migration. Arch Comput Methods Eng 28:311–

344. ISSN: 18861784. https://link.springer.com/article/10.1007/s11831-019-09377-1

14. Alt S, Ganguly P, Salbreux G (2017) Vertex models: from cell mechanics to tissue morphogenesis. Philos Trans R Soc B Biol Sci 372. ISSN: 14712970. https://doi.org/10.1098/rstb.2015.0520
15. Fletcher AG, Osterfield M, Baker RE, Shvartsman SY (2014) Vertex models of epithelial morphogenesis. Biophys J 106:2291–2304. ISSN: 0006-3495. https://doi.org/10.1016/j.bpj.2013.11.4498
16. Anderson A, Rejniak K (eds) (2007) Single cell based models in biology and medicine. ISBN: 978-3-7643-8101-1 and 3-7643-8101-9. Birkhäuser, Basel; Berlin [u.a.]
17. Scianna M, Preziosi L (2013) A Cellular Potts Model simulating cell migration on and in matrix environments. Math Biosci Eng 10: 235–261. https://doi.org/10.3934/mbe.2013.10.235
18. Graner F, Glazier JA (1992) Simulation of biological cell sorting using a two-dimensional extended Potts model. Phys Rev Lett 69:2013–2016. https://doi.org/10.1103/PhysRevLett.69.2013
19. Graner F, Glazier JA (1992) Simulation of the differential adhesion driven rearrangement of biological cells. Phys Rev E: Stat Nonlinear Soft Matter Phys 47:2128. https://doi.org/10.1103/PhysRevE.47.2128
20. Szabó A, Merks RMH (2013) Cellular Potts modeling of tumor growth, tumor invasion, and tumor evolution. Front Oncol 3: 87. ISSN: 2234-943X. http://www.ncbi.nlm.nih.gov/pubmed/23596570http://www.pubmedcentral.nih.gov/articlerender.fcgi?artid=PMC3627127
21. Marée AFM, Jilkine A, Dawes A, Grieneisen VA, Edelstein-Keshet L (2006) Polarization and movement of keratocytes: a multiscale modelling approach Bull Math Biol 5:1169–1211. ISBN: 1153800691. https://doi.org/10.1007/s11538-006-9131-7
22. Marée AFM, Grieneisen VA, Edelstein-Keshet L (2012) How cells integrate complex stimuli: the effect of feedback from phosphoinositides and cell shape on cell polarization and motility. PLoS Comput Biol 8:e1002402. ISSN: 1553-7358. https://doi.org/10.1371/journal.pcbi.1002402
23. Albert PJ, Schwarz US (2016) Dynamics of cell ensembles on adhesive micropatterns: bridging the gap between single cell spreading and collective cell migration. PLOS Comput Biol 12: e1004863. ISSN: 1553-7358. https://journals.plos.org/ploscompbiol/article?id=10.1371/journal.pcbi.1004863
24. Thüroff F, Goychuk A, Reiter M, Frey E (2019) Bridging the gap between single-cell migration and collective dynamics. eLife 8: e46842. ISSN: 2050084X. https://doi.org/10.7554/eLife.46842
25. Vianay B et al. (2009) Single cells spreading on a protein lattice adopt an energy minimizing shape. Phys Rev Lett 105:128101. https://doi.org/10.1103/PhysRevLett.105.128101
26. Albert PJ, Schwarz US (2014) Dynamics of cell shape and forces on micropatterned substrates predicted by a cellular Potts model. Biophys J 106:2340–2352. ISSN: 15420086. https://doi.org/10.1016/j.bpj.2014.04.036
27. Gradeci D et al. (2021) Cell-scale biophysical determinants of cell competition in epithelia. eLife 10. ISSN: 2050-084X. https://elifesciences.org/articles/61011
28. Swat MH et al. (2012) Multi-scale modeling of tissues using CompuCell3D. Methods Cell Biol 110:325–366. ISSN: 0091-679X. https://doi.org/10.1016/B978-0-12-388403-9.00013-8
29. Starruß J, De Back W, Brusch L, Deutsch A (2014) Morpheus: a user-friendly modeling environment for multiscale and multicellular systems biology. Bioinformatics 30:1331–1332. ISSN: 14602059. https://doi.org/10.1093/bioinformatics/btt772
30. Wortel IM, Textor J (2021) Artistoo, a library to build, share, and explore simulations of cells and tissues in the web browser. eLife 10. ISSN: 2050084X. https://doi.org/10.7554/eLife.61288
31. Daub JT, Merks RM (2015) Cell-based computational modeling of vascular morphogenesis using tissue simulation toolkit. Methods Mol Biol 1214:67–127. ISSN: 10643745. https://link.springer.com/protocol/10.1007/978-1-4939-1462-3_6
32. Cooper FR et al. (2020) Chaste: cancer, heart and soft tissue environment. J Open Source Softw 5:1848. ISSN: 2475-9066. https://joss.theoj.org/papers/10.21105/joss.01848
33. Meier-Schellersheim M, Mack G (1999) SIMMUNE, a tool for simulating and analyzing immune system behavior. arXiv: 9903017v1 [cs]. http://arxiv.org/abs/cs/9903017
34. Chen N, Glazier JA, Izaguirre JA, Alber MS (2007) A parallel implementation of the Cellular Potts Model for simulation of cell-based morphogenesis. Comput Phys Commun 176: 670–681. https://www.sciencedirect.com/science/article/pii/S0010465507002044